This book is to be returned on
or before the date stamped below

STP 1330

Composite Materials: Fatigue and Fracture, Seventh Volume

Ronald B. Bucinell, Editor

ASTM Stock #: STP 1330

ASTM
100 Barr Harbor Drive
West Conshohocken, PA 19428-2959
Printed in the U.S.A.

ISBN: 0-8031-2609-3

ISSN: 1040-3086

Photocopy Rights

Peer Review Policy

Each paper published in this volume was evaluated by two peer reviewers and the editor. The authors addressed all of the reviewers' comments to the satisfaction of both the technical editor(s) and the ASTM Committee on Publications.

The quality of the papers in this publication reflects not only the obvious efforts of the authors and the technical editor(s), but also the work of the peer reviewers. The ASTM Committee on Publications acknowledges with appreciation their dedication and their contribution of time and effort on behalf of ASTM.

Printed in Philadelphia, PA

September 1998

Foreword

This publication, *Composite Materials: Fatigue and Fracture, Seventh Volume,* contains papers presented at the Seventh Symposium on Composites: Fatigue and Fracture, held in St. Louis, Missouri, on 7–8 May 1997. The sponsor of the event was Committee D-30 on Composite Materials and Committee E-8 on Fatigue and Fracture.

The symposium chairman was Ronald B. Bucinell, Union College, Department of Mechanical Engineering, Schenectady, NY. He also served as editor of this publication.

Contents

Overview

The Seventh Symposium on Composites: Fatigue and Fracture was held on 7–8 May 1997 in St. Louis, Missouri. It was sponsored by ASTM Committee D-30 on Composite Materials and ASTM Committee E-8 on Fatigue and Fracture. The main purpose of the symposium was to provide a forum for presentations and discussions on the recent developments in fatigue and fracture of composites. Specifically called for were papers describing experimental and analytical research in the following areas of composite technology: failure mechanisms, nondestructive evaluation, environmental effects, prediction methodology, test method development, and impact. A total of 21 papers were presented in five sessions. The conference sessions were chaired by A. T. Nettles and M. K. Cvitkovich of NASA Langley Research Center, D. Cohen of Alient Tech Systems, J. E. Patterson of U.S. Army Missile Command, M. D. Lansing of University of Alabama at Huntsville, T. Chu of Southern Illinois University at Carbondale, and R. H. Martin of MERL. During the symposium T. K. O'Brien was awarded the Wayne Stinchcomb Memorial Award. As a result of the presentation evaluations taken during the symposium, M. K. Cvitkovich was given the award for the best presentation of a paper at the symposium.

Composite materials are found in many commercial, military, and aerospace structures. Most of these applications involve cyclic loads, foreign body impact, or thermomechanical loading. Optimizing the design of these structures requires full characterization of composite material response to the various load scenarios. Cost-effective characterization involves a combination of test methods that isolate specific phenomena of interest and models that can correlate the test method results with the actual structural behavior. The papers included in this volume address many of the important aspects of fatigue and fracture behavior of composite materials.

The papers included in this volume are classified into Fatigue and Fracture, Environmental Considerations, Impact, and Perspective sections. The papers include treatises on polymer, metal, and ceramic matrix composite materials. Included in the Fatigue and Fracture section are papers concerned with microstructural effects, damage, predictive tools, and test method development. The Environmental Considerations section focuses on the affects of temperature and other environmental factors on the long-term durability of composite structures. In the Impact section papers discuss impact response, damage formation, and the use of NDE techniques as a predictive tool. Finally, the Perspective section provides an artistic view of composite materials.

Fatigue and Fracture

O'Brien argues that the apparent G_{IIc} as typically measured is inconsistent with the original definition of shear fracture. He shows that the interlaminar shear failure actually consists of tension failures in the resin-rich layers between plies followed by the coalescence of ligaments created by these failures and not the sliding of two planes relative to one another that is assumed in the fracture mechanics theory. He presents several strain energy release rate solutions for delamination in composite laminates and structural components where failures have been experimentally documented. It is shown that failures typically occur at a location where the mode I component accounts for at least one half of the total G at failure. He argues that it is the Mode I and mixed-mode interlaminar fracture toughness data that will be most useful in predicting delamination failure in composite components in service.

Swanson presents biaxial tests to determine the stiffness and strength properties of carbon/epoxy material systems using tubular specimens. Loading includes biaxial tension, biaxial compression, mixed tension and compression, and compression under superposed pressure. The tests show a number of features that can be interpreted on both a macroscopic and a micromechanics level. He argues that relating the laminate failure values to the properties of the fiber and matrix requires a more detailed examination at the micromechanics level. He observes that ultimate fiber direction tensile strain values apparently depend on the details of the interaction of matrix cracking and fiber-matrix interphase strength, and thus in situ fiber strength in a laminate differs from that in a tow test.

Bucinell presents a stochastic model that predicts the growth of delamination in graphite/epoxy laminates subjected to cyclic loading. The advantage of this model is shown to be that both the mean and variance associated with the growth of delamination are predicted. He argues that understanding and predicting the variability associated with the delamination growth process is essential to the estimation of the reliability of composite structures. The empirical nature of the model has been minimized through the introduction of fracture mechanics parameters. The application of the model is demonstrated through an experimental evaluation that illustrates the ability of the model to predict both the mean and variance of the delamination growth process in composite laminates subjected to cyclic loading.

Ward and Hillberry present the development of an approach to fatigue crack propagation in titanium matrix composites that includes the effects on interfacial wear on the fiber-bridging behavior. They use a Coulomb friction-based fiber-bridging model, in which the effect of fiber surface roughness on the clamping stress between the fiber and matrix is included. They incorporate a previously developed wear model as a means to determine the reduction of the fiber surface roughness amplitude during fatigue cycling. They show that as the roughness decreases, its contribution to the clamping stress also decreases, resulting in a lower interfacial shear stress. The predictions of the developed model are shown to correlate well with experimental results for different loading conditions, especially those at the relatively high crack growth rates.

Joyce and Moon present their investigation of the effect of fiber waviness, which develops during the processing and manufacturing of fiber-reinforced composite structures, on compressive failure. They present data from a series of compression tests examining the effects of varying levels of in-plane fiber waviness. These tests use a novel combined shear/end loading compression test fixture to ameliorate problems typically associated with pure end-loading and pure shear loading. The fixture is shown to perform adequately when testing wavy specimens, but experienced repeated tab failures in the nonwavy specimens.

Cvitkovich, O'Brien, and Minguet present their investigation of the fatigue damage mechanisms and the influence of skin stacking sequence in carbon epoxy composite bonded skin/stringer constructions. A simple four-point-bending test fixture originally designed for previously performed monotonic tests was presented to evaluate the fatigue debonding mechanisms between the skin and the bonded frame. Microscopic investigations of the specimen edges were used to document the onset of matrix cracking and delamination, and subsequent fatigue delamination growth. The fatigue delamination growth experiments are presented and are found to include matrix cracking and delamination onset as a function of fatigue cycles as well as delamination length as a function of the number of cycles.

John, Jira, and Larsen present their results of an extensive characterization of the fatigue crack growth behavior of a model titanium alloy composite (TMC). The model TMC system used was $[0]_8$ SCS-6/Ti-6AL-4V. Presented are the results from tests conducted under tension fatigue loading at room temperature with a stress ratio of 0.1. The authors also discuss the ability of the shear lag model to predict the crack growth in these composites under a wide range of stress levels.

Peck presents her investigation of the transverse tension fatigue characteristics of IM6/3501 composite materials. To test the 90-degree laminae, she uses a three-point bend test. She argues that this potentially minimizes handling and gripping issues associated with tension tests. She presents the results of 50 specimens of nine different size configurations. She also presents the results of three-point flex fatigue testing on the smallest configuration for 59 specimens at various levels using an R ratio of 0.1 and a frequency of 20 Hz.

Environmental Considerations

Case, Iyengar, and Reifsnider present a life prediction method for ceramic matrix composites. Their model is based upon damage mechanics concepts included in the framework of the critical element model. One unique feature of the model is its ability to include general variations of temperature and applied loads as functions of time. They present a detailed description of the application of the model to elevated temperature fatigue. In addition, a validation example is presented that includes the combined effects of rupture and fatigue.

Johnston and Gates present their experimental investigation of the behavior of an open hole tension (OHT) graphite/bismaleimide composite specimen loaded in tension-tension fatigue under isothermal, fixed-frequency conditions. A range of stress levels and temperature levels were chosen to assess performance. The results of this work are shown to help explain the roles of aging and fatigue damage in the performance of OHT specimens of this material as well as providing insights to the individual and synergistic contributions of each of these processes.

Buchanan, John, and Goecke present their results of an experimental investigation of load-controlled isothermal low-cycle fatigue behavior of a titanium matrix (TMC). The TMC used in this investigation was composed of Ti-6Al-2Sn-4Zr-2Mo matrix (wire) reinforced with silicon-carbide (Trimarc-1™) fibers. The longitudinal fatigue data presented show good correlation with other TMC systems at both positive and negative stress ratios. The authors successfully use the Walker equation to correlate the longitudinal S-N data for stress ratios $R = -1.3$ and 0.1, and for predictions at $R = 0.5$ and 0.7. They show that the maximum fiber stress versus cycles to failure for several unidirectional TMC systems at similar test conditions consolidate to a narrow band, indicating that the life is fiber-dominated. The S-N behavior of the TMC, subjected to transverse fatigue loading, is successfully predicted using the matrix S-N data and a net-section model.

Liao, Schultheisz, Hunston, and Brinson study pultruded glass-fiber-reinforced vinyl ester composite coupons subjected to four-point-bend environmental fatigue to investigate long-term durability for infrastructure applications. Specimens were tested dry and while immersed in water and in solutions of water containing mass fractions of 5 and 10% NaCl salt. Some specimens were also preconditioned by soaking in the water or salt solution for 5 to 6 months without loading; the preconditioned are shown to fractionally decrease 5 to 13% in flexural strength compared to dry specimens. The authors find that for specimens cyclically loaded at or above 45% of the average flexural strength of dry coupons, no change in the fatigue life was observed for the specimens tested while immersed in the fluids as compared to specimens tested dry. The authors argue that the long-term environmental fatigue behavior is not controlled by the quantity of water absorbed; rather, it is governed by a combination of both load and fluid environment. However, they point out that a difference in fatigue life in the different fluid environments was not demonstrated.

Zaffaroni, Cappelletti, Rigamonti, Fambri, and Pegoretti discuss the accelerated hot-wet aging of glass-reinforced epoxy resin at 45 and 70°C at the same level of relative humidity

(RH = 84%). The authors compare the mechanical and physical properties of dry and differently saturated composites. The authors find that the higher the conditioning temperature the higher the equilibrium moisture content. The glass transition temperature is found to decrease for both the two moisture-saturated cases. The authors also found that the moisture absorption reduces the static properties while not modifying the endurance in fatigue tests.

Impact

Nettles presents the results of the low-velocity instrumented dropweight tests performed on carbon/epoxy laminates. The composite plates used in this study are 8-ply [+45/0/−45/90]$_s$ laminates supported in a clamped-clamped/free-free configuration with varying amounts of in-plane load applied. The author shows that for a given impact energy level, more damage is induced into the specimen as the external in-plane load is increased. The majority of damage observed is shown to consist of back face splitting of the matrix parallel to the fibers in that ply, associated with delaminations emanating from these splits. A simple free-edge delamination model is presented to explain the type and extent of the major delaminations caused by the preload/impact combinations.

Patterson presents a test program that was conducted to characterize the impact response of graphite/epoxy structures. The design of the test article utilized for this program was directed toward a generic thin-walled structure applicable for use as a rocket motorcase or launch tube. Low-energy impacts were imparted to empty cylinders and to cylinders whose casewalls were strengthened to simulate launch tube and rocket motorcase configurations. The author discussed the differences between the test configurations with regard to visual damage, impact load, absorbed energy, and casewall deflection.

Liu and Dang discuss their evaluation of the response of composite laminates under low-velocity impact using instrumented impact tests and computer simulations. The computational scheme developed by the authors included composite laminates with various thicknesses, fiber angles, and impact velocities. These results show that the peak contact force and maximum deflection are strongly affected by the thickness of composite laminates, while the fiber angles investigated seemed to play a less significant role.

Lansing, Walker, and Russell present the results of an experimental study in which two computer-sensing techniques are used to monitor filament-wound pressure vessels during pressurization. Acoustic emission was used to register the sound generated by microscopic damage propagation. Video image correlation is a noncontact computer vision technique that simultaneously measures full-field in-plane surface displacements and strains, both linear and angular, with subpixel accuracy. Neural networks were used to predict the burst pressures of impacted pressure vessels based upon data obtained at less than approximately one third of the expected burst pressure for an undamaged specimen.

Perspective

Reilly discusses the use of composite materials in sculpture and masonry murals. His discussion includes the effects of fracture by impact, thermal fatigue and fracture, multiaxial loading failure, new composite materials for art, and the monitoring of damage growth. He also shows how fracture of art can be caused by centuries of stress fatigue, pollution, seismic activity, and dynamic impact due to theft or bad custodial care.

Summary

In summary, the editor wishes to thank the authors, session chairmen, reviewers, and Dr. John Masters for working diligently to ensure that the papers included in the symposium and in this STP were of high quality. Also, thanks are extended to the ASTM staff for their efforts and perserverance in bringing the publication of this STP to fruition.

Ronald B. Bucinell

Union College
Department of Mechanical Engineering,
Schenectady, NY;
Symposium Chairman

Fatigue and Fracture

T. Kevin O'Brien[1]

Composite Interlaminar Shear Fracture Toughness, G_{IIc}: Shear Measurement or Sheer Myth?

REFERENCE: O'Brien, T. K., "**Composite Interlaminar Shear Fracture Toughness, G_{IIc}: Shear Measurement or Sheer Myth?**" *Composite Materials: Fatigue and Fracture, Seventh Volume, ASTM STP 1330,* R. B. Bucinell, Ed., American Society for Testing and Materials, 1998, pp. 3–18.

ABSTRACT: The concept of G_{IIc} as a measure of the interlaminar shear fracture toughness of a composite material is critically examined. In particular, it is argued that the apparent G_{IIc} as typically measured is inconsistent with the original definition of shear fracture. It is shown that interlaminar shear failure actually consists of tension failures in the resin-rich layers between plies followed by the coalescence of ligaments created by these failures and not the sliding of two planes relative to one another that is assumed in fracture mechanics theory. Several strain energy release rate solutions are reviewed for delamination in composite laminates and structural components where failures have been experimentally documented. Failures typically occur at a location where the Mode I component accounts for at least one half of the total G at failure. Hence, it is the Mode I and mixed-mode interlaminar fracture toughness data that will be most useful in predicting delamination failure in composite components in service. Although apparent G_{IIc} measurements may prove useful for completeness of generating mixed-mode criteria, the accuracy of these measurements may have very little influence on the prediction of mixed-mode failures in most structural components.

KEYWORDS: fractures toughness, shear, interlaminar fracture toughness, composites

One of the most common failure modes for composite structures is delamination. The remote loadings applied to composite components typically get resolved into interlaminar tension and shear stresses at discontinuities that create mixed-mode I and II delaminations. Over the past 10 to 20 years, it has become accepted practice to characterize the onset and growth of these mixed-mode delaminations using fracture mechanics. The strain energy release rate, G, and the Mode I component due to interlaminar tension, G_I, and Mode II component due to interlaminar shear, G_{II}, are calculated using the virtual crack closure technique [1]. In order to predict delamination onset or growth, these calculated G components are compared to interlaminar fracture toughness properties measured over a range from pure Mode I loading to pure Mode II loading.

Examples of mixed-mode delamination criteria for carbon fiber-reinforced composites with either a brittle epoxy (AS4/3501-6) or a toughened epoxy matrix (IM7/E7T1-2) are shown in Figs. 1*a* and 1*b*, respectively [2,3]. The critical G for delamination is plotted as a function of the Mode II percentage compared to the total G. Hence, at $G_{II}/G = 0$, the loading at the delamination front is a pure opening Mode I, whereas at $G_{II}/G = 1$, the loading at the

---

[1]U.S. Army Research Laboratory, Vehicle Technology Center, NASA Langley Research Center, Hampton, VA 23681.

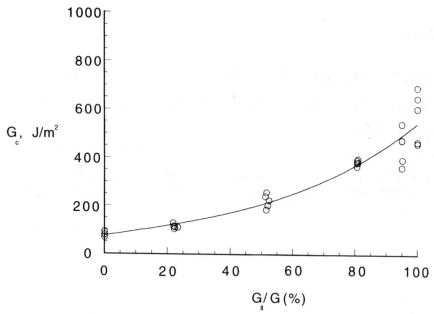

FIG. 1a—*Mixed-mode delamination criterion for AS4/3501-6.*

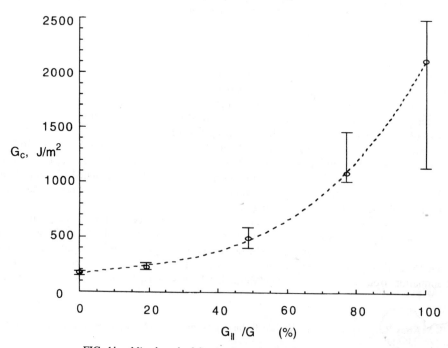

FIG. 1b—*Mixed-mode delamination criterion for IM7/E7T1-2.*

delamination front is a pure shear Mode II. As Fig. 1 indicates, the apparent Mode II toughness for a graphite epoxy material is typically much greater than the Mode I toughness and has significantly more variability or scatter. In this paper, some of the reasons for the differences in magnitude and scatter between G_{Ic} and G_{IIc} will be examined. The significance of these differences on the prediction of mixed-mode delamination in composite structural configurations will be discussed.

Background

In the 1950s, Irwin proposed a general theory of fracture [4,5], based on the method of Westergaard [6], that postulated the existence of three unique fracture modes that could occur at the tip of a crack (Fig. 2). These fracture modes included: (1) an opening Mode I, where the crack faces underwent opening displacements relative to one another as the crack grew, (2) an in-plane sliding shear Mode II, where the crack faces slid over one another in the direction of the crack growth, and (3) an out-of-plane scissoring (or tearing) Mode III where the crack faces slid relative to one another in a direction normal to the direction of crack growth. The elasticity solution for stress intensity factors associated with these three postulated fracture modes (K_I, K_{II}, K_{III}) were derived yielding a mathematically complete and consistent theory for fracture of materials and structures. Strain energy release rates may be related to these stress intensity factors squared through coefficients consisting solely of material properties. Solutions for cracked bodies with specific configurations and loadings were developed and applied to structural problems [7]. However, most of these problems consisted

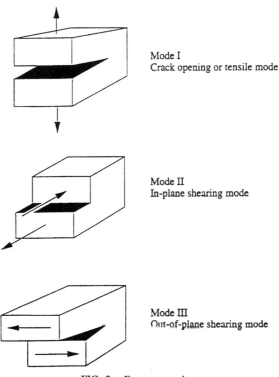

Mode I
Crack opening or tensile mode

Mode II
In-plane shearing mode

Mode III
Out-of-plane shearing mode

FIG. 2—*Fracture modes.*

of cracks in homogeneous materials (typically structural metallic materials) where cracks that may originally have all three fracture modes when loaded would typically turn immediately as the crack grew to assume a pure opening Mode I orientation. Hence, the resistance of these materials to fracture could be adequately described in terms of the opening mode fracture toughness, K_{Ic}, alone.

With the advent of adhesively bonded structures and laminated composite materials in the 1960s and 1970s, the problem of debonding of adhesive bonds, and delamination in composite materials, created a class of problems where cracks were constrained to grow in bond lines, or resin-rich regions between composite plies, such that macroscopically the cracks could not assume a pure opening Mode I orientation. Therefore, for this class of materials, the mixed-mode fracture problem that was resolved mathematically in the 1950s posed a challenge in terms of fracture toughness characterization.

The opening Mode I characterization proved relatively straight-forward with the advent of the double cantilever beam (DCB) test configuration, although complexities involving the influences of bondline and insert thicknesses, precracking techniques, and fiber bridging in composites delayed the standardization of this test method until the 1990s [8,9]. Development of test methods for characterizing the interlaminar shear fracture toughness, however, has proven to be a difficult task, both in terms of achieving an adequate configuration to yield a pure shear loading at the crack tip and in the interpretation of the test results [10]. It is the latter issue that will be the focus of this paper. As Shakespeare might have said, at issue is whether G_{IIc} is to be, or not to be, considered a generic property of the composite material. Since most of the data generated to date have been for the sliding shear Mode II fracture, the discussion will be limited to the measurement and interpretation of G_{IIc}.

Mode II Fracture Toughness Measurement Results

In the 1980s several test methods were proposed for measuring G_{IIc}. However, to date none of these have been standardized. The most popular methods are the end-notched flexure (ENF) and end-loaded split beam (ELS) shown in Fig. 3. The ENF test involves a simple three-point bend loading, but it results in an unstable delamination growth unless the initial crack is very long [11] or the test is controlled with a special shear displacement gage [12]. The ELS test involves a more complicated clamped boundary condition but results in a stable delamination growth [13]. Both of these test configurations have been analyzed and have been demonstrated to yield a pure sliding shear fracture Mode II at the delamination front [13,14].

Several interlaboratory "round robin" test programs have been conducted using both of these test methods [3,10]. However, interpretation of the test results has proven to be difficult to resolve. This difficulty may be illustrated by examining some typical results found in the literature [2,3,15–34]. For this review, papers where chosen that compared the influence of precracking versus testing from the embedded insert on G_{IIc} values measured on the same specimen. In addition, papers were chosen that reported the ratio of precracked, or insert, G_{IIc} values to G_{Ic} values measured only from the insert to avoid the complication due to fiber bridging. These data are summarized in Table 1.

Mode II Precracking Effects

One difficulty in measuring the interlaminar shear fracture toughness is the apparent inconsistency between G_{IIc} values measured by growing the crack from a thin midplane insert versus G_{IIc} values measured by growing the crack from an initial shear precrack. For G_{Ic} values measured using the DCB test, a single generic toughness value may be obtained as

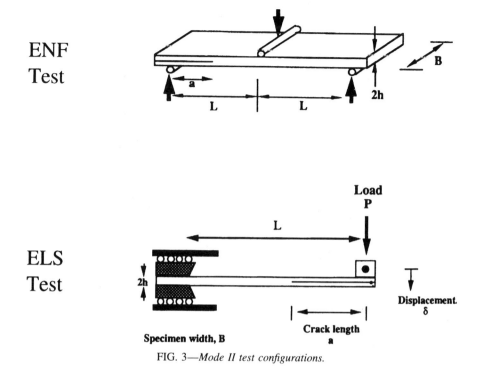

FIG. 3—*Mode II test configurations.*

the insert thickness is decreased [24], thereby achieving a composite material property where the insert successfully simulates an initial delamination crack without having to resort to precracking, which induces fiber bridging. However, G_{IIc} values decrease with insert thickness, but never reach a single value that may be considered a generic property of the composite material [24]. Furthermore, G_{IIc} values measured from the insert are sometimes greater, and sometimes less than, G_{IIc} values measured from a shear precrack. Figure 4 compares results from studies where G_{IIc} values were determined from both the insert and from precracking on the same specimen. In most cases the precrack value is lower than the insert value. However, for two materials (S2/SP250 glass epoxy and IM7/F3900 graphite epoxy) the reverse was true.

G_{IIc} Comparison to G_{Ic}

One result that is consistent in the literature for composite materials over a large range of toughnesses is that G_{IIc} *always* exceeds G_{Ic}. Figure 5 shows a plot of the ratio of G_{IIc} to G_{Ic} for composite materials with a large range of opening mode interlaminar fracture toughnesses. There is a general trend that the more brittle materials (lower G_{Ic} values) have G_{IIc} values that are much greater than the corresponding G_{Ic}, whereas the tougher matrix materials (higher G_{Ic} values) have G_{IIc} values that are close to, but still greater than, the corresponding G_{Ic}.

Some other interesting trends are noted when the total dataset in Fig. 5 is separated into two plots, one for relatively brittle thermoset (epoxy) matrix composites (Fig. 6) and the rest for relatively tough thermoplastic matrix composites, which for this literature search con-

TABLE 1—*Modes I and II interlaminar fracture toughness data.*

Reference	Material	G_{Ic}, J/m^2 (Insert)	G_{IIc}, J/m^2 (Insert)	G_{IIc}, J/m^2 (Precrack)
21	T300/914	185	795	598
19	IM6/6376	180	700	...
22	T300/BP907	400	1961	1519
17	AS4/3502	227	...	574
16	AS1-3501-6	132	460	...
16	HMS-3501-6	30	170	...
3	IM7/E7T1	146	1860	1220
3	IM7/F655	223	1204	660
3	IM7/F3900	571	2090	2571
23	IM6/5245C	...	979	758
23	IM7/8551-7	...	1185	1147
29	HTA/6376	360	1440	...
28	G30-500/5208	70	400	...
25	AS4-3501-6	111	...	814
26	IM7/8320	286	1243	...
26	IM7/5260	315	1692	...
27	T800H/3900-2	750	2090	1730
2	AS4-3501-6	80	572	...
2	IM7/977-2	283	1419	...
30	Toho/Q-1113	131	370	...
30	Toho/Q-C134	355	991	...
20	AS4/3502	227	...	574
15	T300/6376C	270	600	...
21	Eglass/DGEBA	264	3510	2510
24	S2/SP250	130	600	780
21	AS4/PEEK	1460	4250	2695
22	AS4/PEEK	1262	...	1586
23	AS4/PEEK	...	3186	3148
29	AS4/PEEK	1850	2220	...
27	AS4/PEEK	1320	2530	2040
2	AS4/PEEK	776	...	1201
30	AS4/PEEK	1260	1477	...
15	AS4/PEEK	1680	1740	...
18	AS1-3501-6	110	...	605
18	AS4/2220-3	160	...	750
18	AS4/PEEK	1330	...	1765
33	AS4-3501-6	137	1292	...
33	AS4/Dow-P6	165	1806	...
33	AS4/Dow-P7	340	1325	...
33	AS4/Dow-Q6	848	2836	...
32	AS4/PES	...	...	1250
31	AS4/PPS	196	...	933
34	AS1-3501-6	...	...	670

sisted solely of AS4/PEEK (Fig. 7). The G_{IIc}/G_{Ic} ratios for the epoxy matrix composite materials all exceeded 2 to 1, which gives some credence to the mechanistic fracture theory of Ref *35* (described in a later section) as an explanation of a lower bound for G_{IIc} values. However, for all but one of the results in the literature for the tough thermoplastic composite (AS4/PEEK), the G_{IIc}/G_{Ic} ratios range between 2 to 1 and 1 to 1. Hence, the micromechanisms at the delamination front may be quite different for these two classes of materials.

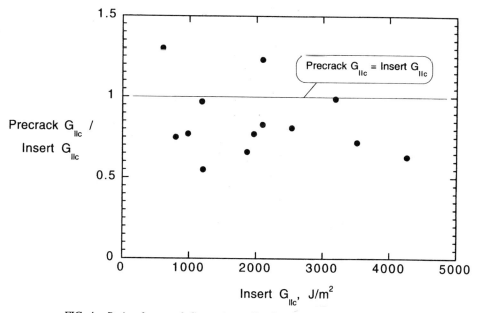

FIG. 4—*Ratio of precrack G_{IIc} to insert G_{IIc} for polymer matrix composites.*

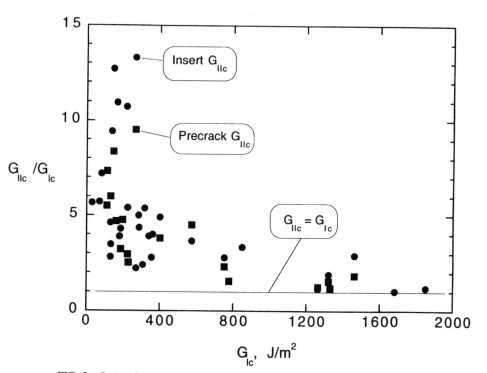

FIG. 5—*Ratio of Mode II to Mode I toughness for polymer matrix composites.*

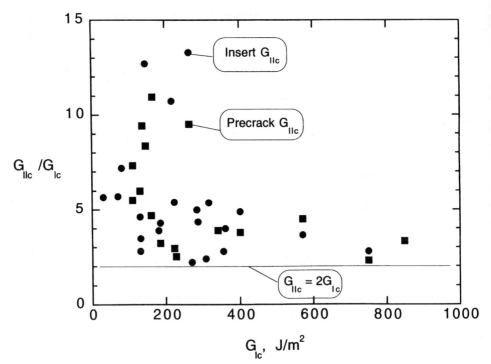

FIG. 6—*Ratio of Mode II to Mode I toughness for graphite and glass epoxy composites.*

Mode II Fracture Mechanisms

Figure 8 compares scanning electron microscope (SEM) photographs of Mode I and Mode II fracture surfaces for a brittle T300/5208 graphite epoxy composite [22]. For the brittle epoxy material, the Mode I fracture surface appears to have a fairly clean cleavage plane, whereas the Mode II fracture surface exhibits a very rough fracture plane with the characteristic "hackles" observed in Mode II delamination [16,20,22,33,36]. The sketch in Fig. 8 under the Mode II SEM image illustrates the principal tension stress at a 45° angle to the delamination plane that results from Mode II loading and is responsible for creating tension microcracks in the resin-rich region between plies. Once these cracks appear, the ligaments formed by them are forced to bend until they fracture and coalesce, creating the perceived extension of the original delamination via Mode II. Figure 9 shows an edge view generated by in situ testing in a scanning electron microscope [36] that illustrates the formation and coalescence of microcracks forming the final interlaminar shear fracture surface.

Figure 10 compares SEM photographs of Mode I and Mode II fracture surfaces for a tough AS4/PEEK graphite thermoplastic composite. In contrast to the brittle composite, for the tough thermoplastic matrix composite both Modes I and II fracture surfaces are similar, with extensive evidence of matrix yielding at the delamination front.

In both the brittle and tough matrix composite materials, the complex failure process under Mode II loading is far removed from the idealized sliding of two crack planes relative to one another as postulated in the fracture mechanics elasticity solutions. Hence, a mechanistic explanation of Mode II fracture is needed.

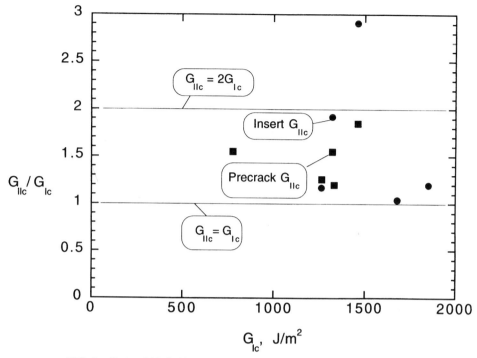

FIG. 7—*Ratio of Mode II to Mode I toughness for AS4/PEEK composites.*

Mode II Fracture Theory

In Ref *35*, a fracture-mechanics-based failure criteria was postulated based on the observation that for brittle homogeneous materials subjected to mixed-mode loading, "fracture occurs when the total Mode I component that is experiences is equal to a critical value, G_0." This idea was expressed mathematically as

$$G_I + \sin^2\omega G_{II} = G_0 \tag{1}$$

where ω corresponds to "the slope of the surface roughness." A worse case was postulated for omega of 45° corresponding to the "shear microcracks that form ahead of the main crack" in a composite interface that is delaminating. The authors noted that if $G_{II} = G_{IIc}$ when $G_I = 0$, then

$$\sin^2\omega = G_{Ic}/G_{IIc} \tag{2}$$

Equation 2 implies that G_{IIc} may be estimated based on a measured G_{Ic} and omega, which if assumed to be the worst case of 45° would yield a value of G_{IIc} equal to twice G_{Ic}. This is consistent with the initial damage mechanism postulated in the previous section, i.e., the formation of tension microcracks ahead of the delamination front. However, it does not account for the bending and fracture of the ligaments formed by the microcracks that releases much of the energy required to create the perceived macroscopic Mode II delamination

Mode I   Mode II

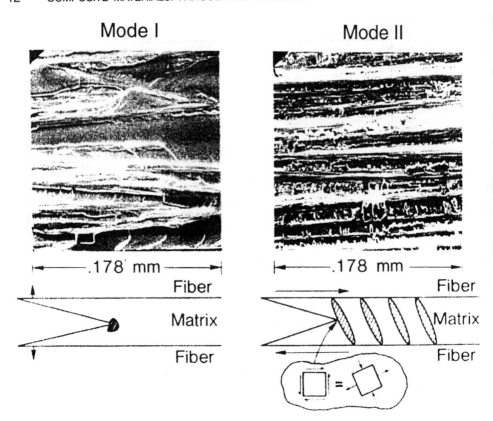

FIG. 8—*SEM photographs of delamination fracture surfaces in T300/5208 graphite-epoxy laminates.*

growth. Indeed, the authors cited an epoxy matrix composite example where $G_{IIc} = 2.22$ G_{Ic}. Further examples from the literature, as summarized in Fig. 5, also yield G_{IIc} values much greater than twice the corresponding G_{Ic}. Hence, the basic premise set forth in Ref *35* was sound, i.e., that the actual failure mechanism is due to tension, but a complete description of failure was missing. Therefore, Eq 2 at best represents only a lower bound on G_{IIc}.

The theory postulated in Ref *35* implies that the macroscopic shear fracture corresponds to the growth of the tension microcrack as an extension of the initial delamination due to the Mode I component of the mixed mode loading at the delamination front. For composite delamination under shear loading, however, a more realistic sequence of failure may be as follows: (1) a tension microcrack initiates ahead of the original delamination front wherever the weakest flaw exists in the resin-rich layer ahead of the crack and forms at a 45° angle to the original delamination; (2) further microcracks accumulate in front of the initial microcrack, with a spacing defined by classical shear lag considerations and the distribution of inherent flaws in the ply interface; and (3) these microcracks finally coalesce to form an extension of the delamination.

For the initial tension microcracking, a transverse tension strength criteria, using Weibull statistics to characterize the flaw sensitivity [37], may prove useful. This may also help to account for the increased scatter in Mode II toughness values compared to Mode I values as shown in Fig. 1. In fact, it may be argued that the transverse tension strength and Mode

Formation of tension microcracks

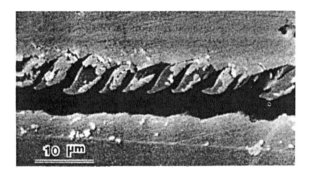

Coalescence of microcracks forming hackles

FIG. 9—*Formation and coalescence of tension microcracks* [36].

I interlaminar fracture toughness are the only true generic material properties that control the failure of a brittle material.

Beyond the initiation phase, however, a complex model of ligament bending and fracture for brittle materials, and ligament bending, yielding, and rupture for ductile materials, would be needed to adequately characterize the actual failure mechanism at the delamination front under Mode II loading. Such a model was recently proposed to account for this sequence of events [38]. The model was based on an adhesive bond containing a crack analogous to the composite delamination in the resin-rich layer between plies. A shear lag model was utilized to predict a regular spacing of microcracks ahead of the delamination front that depended on the opening mode fracture toughness of the resin, the shear modulus and yield strength of the resin, and the thickness of the resin-rich layers between the plies. A quantitative relationship was derived relating these constituent properties to the apparent composite G_{IIc} by utilizing a curve-fitting parameter, m, determined by plotting the effective shear modulus of the resin layer with multiple matrix cracks as a function of the crack spacing normalized by the resin layer thickness.

Although this model provides some useful insights into the relative influence of constituent properties on the apparent composite G_{IIc}, the predicted toughness is very sensitive to the

Mode I   Mode II

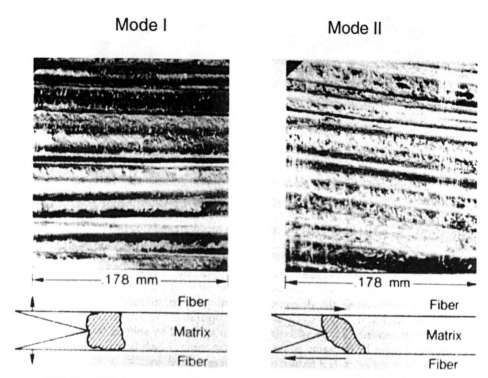

FIG. 10—*SEM photographs of delamination fracture surfaces in AS4/PEEK laminates.*

curve-fitting parameter chosen. Even if a model could be developed without introducing a new curve-fitting parameter, it would be difficult to demonstrate its value as a predictive tool because of the extreme variability of apparent G_{IIc} values relative to corresponding G_{Ic} values (Fig. 5). In addition, requiring a local characterization of the material, i.e., measuring transverse tension strength, resin fracture toughness, resin moduli, resin tension, and yield strengths, etc. needed for modeling crack-tip micromechanisms discreetly will be extremely labor intensive compared with simply measuring the apparent G_{IIc}. Furthermore, experience with mixed-mode delamination failures that occur in structural configurations under realistic loadings typically result in problems where the Mode I component is predominant, as illustrated in the next section.

Relevance to Mixed-Mode Delamination Onset Prediction

Since the mid-1970s, several strain energy release rate solutions have been developed for delaminations in flat uniform thickness composite laminates and in more realistic structural configurations (such as tapered, curved, and stringer reinforced laminates) where failures have been experimentally documented [39–47]. These studies all seem to indicate that it is the G_I component that controls the onset of delamination much more than the G_{II} component.

Initial studies on edge delamination of flat laminates with straight edges indicated that the onset of delamination depended strongly on the stacking sequence. Stacking sequences having large G_I components associated with delamination-prone interfaces yielded lower failure strains at delamination onset [39]. Failure occurred when the Mode I component reached G_{Ic}

for the composite regardless of the G_{II} component. Flat laminates with open holes, modeled as a series of laminates with varying layups using a rotated straight edge concept, yielded similar results where the G_I component appeared to control the onset of delamination around the hole boundary [40].

Recently, a number of studies have begun to appear in the literature for a variety of more typical structural configurations. These include studies on tapered laminates containing internal ply drops to vary thickness [41–43], curved laminates [44,45], and stringer reinforced laminates subjected to out-of-plane "pull-off" loading [46,47]. The material, configuration, and mixed-mode ratio (G_{II}/G) for these studies is summarized in Table 2. In the majority of these cases, the failure occurs at a location where the G_I component accounts for at least one half of the total G at failure, i.e., where G_{II}/G is less than 50%. Indeed, for the tapered and curved laminates studied, G_{II}/G never exceeded 10%.

In addition to these models of complex loadings on structural components, the problem of delamination resulting from low-velocity impact has been shown to consist of a sequence of events beginning with a matrix crack induced by transverse shear (principle tension stresses at 45° to the laminate axis) followed by mixed-mode I and II delamination [48]. Hence, the typical assumption that G_{IIc} is the most important parameter for characterizing low-velocity impact damage is questionable.

These structural configuration case studies indicate that the Mode I and mixed-mode interlaminar fracture toughness data will be most useful in predicting delamination failure in composite components in service. Even though the apparent G_{IIc} values will typically be measured for completeness of generating the mixed-mode criteria illustrated in Fig. 1, the accuracy of these measurements may have very little influence on the prediction of mixed-mode failures in most structural components.

Summary

The concept for G_{IIc} as a measure of the interlaminar shear fracture toughness of a composite material was critically examined. An extensive literature review was conducted to

TABLE 2—Delamination in composite structural configurations.

Ref. No.	Configuration	Material	Layup	G_{II}/G, %
41	tapered laminate	S2/SP250 glass-epoxy	0 and 45° plies	0–5
42	tapered laminate	S2/SP250 glass-epoxy	0 and 45° plies	0–5
42	tapered laminate	IM6/1827I graphite-epoxy	0 and 45° plies	0–5
43	tapered laminate	S2/SP250 glass-epoxy	0° plies	0–10
43	tapered laminate	S2/CE9000 glass-epoxy	0° plies	0–10
43	tapered laminate	IM6/18271 graphite-epoxy	0° plies	0–5
44	curved laminate	AS4/3501-6 graphite-epoxy	0° plies	5–10
45	curved laminate	AS4/3501-6 graphite-epoxy	0 and 90° plies	5–10
46	stringer reinforced laminate	IM6/3501-6 graphite-epoxy	0, 45, and 90° plies	13–40
47	stringer reinforced laminate	IM7/E7T1-2 graphite-epoxy	45° plies	28–55

identify studies where Mode I and Mode II fracture toughness measurements were compared. This review indicated that for composite materials over a large range of toughnesses, G_{IIc} always exceeds G_{Ic}. In addition, the more brittle materials had G_{IIc} values that were often much greater than the corresponding G_{Ic}, whereas the tougher matrix materials had G_{IIc} values that are close to, but still greater than, the corresponding G_{Ic}. However, it was noted that G_{IIc} values measured from the insert are sometimes greater, and sometimes less than, the G_{IIc} values measured from a shear precrack. Furthermore, examinations of the micromechanisms at the tip of the delamination front documented in the literature for a wide range of composite materials indicated that interlaminar shear failure actually consists of tension failures followed by the coalescence of ligaments created by these failures and not the sliding of two planes relative to one another that is assumed in fracture mechanics theory. Hence, the apparent G_{IIc} as typically measured is inconsistent with the original definition of shear fracture.

For the initial tension microcracking, a transverse tension strength criteria, using Weibull statistics to characterize the flaw sensitivity, may help to account for the increased scatter in Mode II toughness values compared to Mode I. Beyond the initiation phase, however, a complex model of ligament bending and fracture for brittle materials, and ligament bending, yielding, and rupture for ductile materials, would be needed to adequately characterize the actual failure mechanism at the delamination front under Mode II loading. Even if this could be achieved, it would be of questionable value because of the extreme variability of apparent G_{IIc} values relative to corresponding G_{Ic} values. In addition, requiring a local characterization of the material needed for modeling crack-tip micromechanisms discreetly will be extremely labor intensive compared with simply measuring the apparent G_{IIc}.

Several strain energy release rate solutions have been developed for both composite laminates and structural components where failures have been experimentally documented. In the majority of these cases, the failure occurs at a location where the Mode I component accounts for at least one half of the total G at failure. Hence, it is the Mode I and mixed-mode interlaminar fracture toughness data that will be most useful in predicting delamination failure in composite components in service. Although apparent G_{IIc} measurements may prove useful for completeness of generating mixed-mode criteria, the accuracy of these measurements may have very little influence on the prediction of mixed-mode failures in most structural components. Therefore, as Shakespeare might have said, the controversy over Mode II fracture toughness measurement may turn out to be much ado about nothing.

References

[1] Rybicki, E. F. and Kanninen, M. F., "A Finite Element Calculation of Stress Intensity Factors by a Modified Crack Closure Integral," *Engineering Fracture Mechanics*, Vol. 9, 1977, pp. 931–938.

[2] Reeder, J. R., "A Bilinear Failure Criterion for Mixed-Mode Delamination," *Composite Materials: Testing and Design, Eleventh Volume, ASTM STP 1206*, American Society for Testing and Materials, December 1993, pp. 303–322.

[3] Murri, G. B., Reeder, J. R., and Li, J., ASTM Subcommittee D30.06 Interlaboratory Round Robin Test Data, 1996.

[4] Irwin, G. R., "Analysis of Stresses and Strains Near the End of a Crack Traversing a Plate," *Transactions of the ASME Journal of Applied Mechanics*, 1957.

[5] Irwin, G. R., "Fracture Mechanics" in *Structural Mechanics*, Pergamon Press, New York, 1960.

[6] Westergaard, H. M., "Bearing Pressures and Cracks," *Transactions of the ASME Journal of Applied Mechanics*, 1939.

[7] Paris, P. C. and Sih, G. C., "Stress Analysis of Cracks," *Fracture Toughness Testing and its Applications, ASTM STP 381*, American Society for Testing and Materials, June 1964, pp. 30–83.

[8] O'Brien, T. K. and Martin, R. H., "Round Robin Testing for Mode I Interlaminar Fracture Toughness of Composite Materials," *Journal of Composites Technology and Research*, Vol. 15, No. 4, Winter 1993, pp. 269–281.

[9] Standard Test Method for Mode I Interlaminar Fracture Toughness of Unidirectional Fiber-Reinforced Polymer Matrix Composites, ASTM Standard D 5528-94a, *ASTM Annual Book of Standards*, Vol. 15.03, American Society for Testing and Materials, 1994, pp. 272–281.

[10] Davies, P., Ducept, F., Brunner, A. J., Blackman, B. A. K., and deMorais, A. B., "Development of a Standard Mode II Shear Fracture Test Procedure," *Proceedings,* Seventh European Conference on Composite Materials (ECCM-7), Vol. 2, Third Symposium on Composite Testing and Standardization (ECCM-CTS-3), London, May 1996, Woodhead Publishers, Ltd., Cambridge, U.K., pp. 9–15.

[11] Russell, A. J., "On Measurement of Mode II Interlaminar Fracture Energies," Defence Research Establishment Pacific (DREP) Victoria, British Columbia, Canada, Materials Report 82-0, December 1982.

[12] Kageyama, K., Kikuchi, M., and Yanagisawa, N., "Stabalized End Notched Flexure Test: Characterization of Mode II Interlaminar Crack Growth," *Composite Materials: Fatigue and Fracture, Third Volume, ASTM STP 1110,* American Society for Testing and Materials, September 1991, pp. 210–225.

[13] Davies, P., Moulin, C., Kausch, H. H., and Fischer, M., "Measurement of G_{Ic} and G_{IIc} in Carbon/Epoxy Composites," *Composite Science and Technology,* Vol. 39, 1990, pp. 193–205.

[14] Salpekar, S. A., Raju, I. S., and O'Brien, T. K., "Strain Energy Release Rate Analysis of the End-Notched Flexure Specimen Using the Finite Element Method," *Journal of Composites Technology and Research,* Vol. 10, No. 4, Winter 1988, pp. 133–139.

[15] Hashemi, S., Kinloch, A. J., and Williams, J. G. "The Analysis of Interlaminar Fracture in Uniaxial Fibre-polymer Composites," *Proceedings of the Royal Society of London,* Vol. A427, 1990, pp. 173–199.

[16] Russell, A. J. and Street, K. N., "Moisture and Temperature Effects on the Mixed-Mode Delamination Fracture of Unidirectional Graphite Epoxy," *Delamination and Debonding of Materials, ASTM STP 876,* American Society for Testing and Materials, October 1985, pp. 349–370.

[17] Bradley, W. L. and Cohen, R. N., "Matrix Deformation and Fracture in Graphite Reinforced Epoxies," *Delamination and Debonding of Materials, ASTM STP 876,* October 1985, pp. 389–410.

[18] Russell, A. J. and Street, K. N., "The Effect of Matrix Toughness on Delamination: Static and Fatigue Fracture under Mode II Shear Loading of Graphite Fiber Composites," *Toughened Composites, ASTM STP 937,* American Society for Testing and Materials, May 1987, pp. 275–294.

[19] Davies, P. and de Charentenay, F. X., "The Effect of Temperature on the Interlaminar Fracture of Tough Composites," *Proceedings,* Sixth International Conference on Composite Materials (ICCM VI), Vol. 3, July 1987, pp. 284–294.

[20] Corleto, C. R., and Bradley, W. L., "Mode II Delamination Fracture Toughness of Unidirectional Graphite/Epoxy Composites," *Composite Materials: Fatigue and Fracture, Second Volume, ASTM STP 1012,* American Society for Testing and Materials, April 1989, pp. 201–221.

[21] Prel, Y. J., Davies, P., Benzeggagh, M., and de Charentenay, F. X., "Mode I and Mode II Delamination of Thermosetting and Thermoplastic Composites," *Composite Materials: Fatigue and Fracture, Second Volume, ASTM STP 1012,* American Society for Testing and Materials, April 1989, pp. 251–269.

[22] O'Brien, T. K., Murri, G. B., and Salpekar, S. A., "Interlaminar Shear Fracture Toughness and Fatigue Thresholds for Composite Materials," *Composite Materials: Fatigue and Fracture, Second Volume, ASTM STP 1012,* American Society for Testing and Materials, April 1989, pp. 222–250.

[23] Russell, A. J., "Initiation and Growth of Mode II in Toughened Composites," *Composite Materials: Fatigue and Fracture, Third Volume, ASTM STP 1110,* American Society for Testing and Materials, September 1991, pp. 226–242.

[24] Murri, G. B. and Martin, R. H., "Effect of Initial Delamination on Mode I and Mode II Interlaminar Fracture Toughness and Fatigue Fracture Thresholds," *Composite Materials: Fatigue and Fracture, Fourth Volume, ASTM STP 1156,* American Society for Testing and Materials, June 1993, pp. 239–256.

[25] Hooper S. J. and Subramanian, R., "Effects of Water and Jet Fuel Absorption on Mode I and Mode II Delamination of Graphite/Epoxy," *Composite Materials: Fatigue and Fracture, Fourth Volume, ASTM STP 1156,* American Society for Testing and Materials, June 1993, pp. 318–340.

[26] Sriram, P., Khourchid, Y., Hooper, S. J., and Martin, R. H., "Experimental Development of a Mixed-Mode Fatigue Delamination Criterion," *Composite Materials: Fatigue and Fracture, Fifth Volume, ASTM STP 1230,* American Society for Testing and Materials, October 1995, pp. 3–18.

[27] Kageyama, K., Kimpara, I., Ohsawa, I., Hojo, M., and Kabashima, S., "Mode I and Mode II Delamination Growth of Interlayer Toughened Carbon/Epoxy Composite System," *Composite Materials: Fatigue and Fracture, Fifth Volume, ASTM STP 1230,* American Society for Testing and Materials, October 1995, pp. 19–37.

[28] Steinmetz, G. G. and Arendts, F. J., "Mixed-Mode Interlaminar Fracture Testing using the MMB-Fixture," *Proceedings* Tenth International Conference on Composite Materials (ICCM X), Vol. 1, August 1995, pp. 165–172.
[29] Kussmaul, K. and Alberti, M. V., "On the Delamination Behaviour of Carbon Fibre Reinforced Plastics under Mixed-Mode Fatigue Loading," *Proceedings,* Tenth International Conference on Composite Materials (ICCM X), Vol. I, August 1995, pp. 125–132.
[30] Tanaka, K., Kageyama, K., and Hojo, M., "Standardization of Modes I and II Interlaminar Fracture Toughness Tests for CFRPs in Japan," *Proceedings,* Second European Conference on Composites Testing and Standardisation (ECCM-CTS-2), Hamburg, Germany, September 1994, pp. 533–541.
[31] Davies, P., Benzeggagh, M. L., and de Charentenay, F. X., "The Delamination Behavior of Carbon Fiber Reinforced PPS," *Proceedings,* 32nd International SAMPE Symposium, Anaheim, CA, April 1987, pp. 134–146.
[32] Hashemi, S., Kinloch, A. J., and Williams, J. G., "Mechanics and Mechanisms of Delamination in a Poly(ether sulphone)-Fibre Composite," *Composites Science and Technology,* Vol. 37, No. 4, 1990, pp. 429–462.
[33] Hibbs, M. F. and Bradley, W. L., "Correlations Between Micromechanical Failure Processes and the Delamination Toughness of Graphite/Epoxy Systems," *Fractography of Modern Engineering Materials, ASTM STP 948,* American Society for Testing and Materials, 1987, pp. 68–97.
[34] Jurf, R. A. and Pipes, R. B., "Interlaminar Fracture of Composite Materials," *Journal of Composite Materials,* Vol. 16, September 1982, pp. 386–394.
[35] Charalambides, M., Kinloch, A. J., Wang, Y., and Williams, J. G. "On the Analysis of Mixed-mode Failure," *International Journal of Fracture,* Vol. 54, 1992, pp. 269–291.
[36] Corleto, C. R. and Bradley, W. L., "Correspondence Between Stress Fields and Damage Zones Ahead of Crack Tip of Composites under Mode I and Mode II Delamination," *Proceedings,* Sixth International Conference on Composite Materials (ICCM VI), Vol. 3, July 1987, pp. 378–387.
]37] O'Brien, T. K. and Salpekar, S. A., "Scale Effects on the Transverse Tensile Strength of Graphite Epoxy Composites," *Composite Materials: Testing and Design, Eleventh Volume, ASTM STP 1206,* American Society for Testing and Materials, December 1993, pp. 23–52.
[38] Lee, S. M., "Mode II Delamination Failure Mechanisms of Polymer Matrix Composites," to appear in *Journal of Materials Science,* Vol. 32, 1997, pp. 1287–1295.
[39] O'Brien, T. K., "Mixed-Mode Strain-Energy-Release Rate Effects on Edge Delamination of Composites," *Effects of Defects in Composite Materials, ASTM STP 836,* D. J. Wilkins, Ed., American Society for Testing and Materials, 1984, pp. 125–142.
[40] O'Brien, T. K. and Raju, I. S., "Strain-Energy-Release Rate Analysis of Delamination Around an Open Hole in a Composite Laminate," *Proceedings,* 25th AIAA/ASME/ASCE/AHS Structures, Structural Dynamics, and Materials Conference, Palm Springs, CA, May 1984, AIAA-84-0961.
[41] Salpekar, S. A., Raju, I. S., and O'Brien, T. K., "Strain Energy Release Rate Analysis of Delamination in a Tapered Laminate Subjected to Tension Load," *Proceedings,* American Society for Composites, Third Technical Conference, Seattle, WA, Sept. 1988, pp. 642–654.
[42] Murri, G. B., O'Brien, T. K., and Salpekar, S. A., "Tension Fatigue of Glass/Epoxy and Graphite/Epoxy Tapered Laminates," *Journal of the American Helicopter Society,* Vol. 38, No. 1, January 1993, pp. 29–37.
[43] Murri, G. B., Salpekar, S. A., and O'Brien, T. K., "Fatigue Delamination Onset Prediction in Unidirectional Tapered Laminates," *Composite Materials: Fatigue and Fracture, Third Volume, ASTM STP 1110,* American Society for Testing and Materials, September 1991, pp. 312–339.
[44] Martin, R. H., "Delamination Failure in a Unidirectional Curved Composite Laminate," *Composite Materials: Testing and Design, Tenth Volume, ASTM STP 1120,* American Society for Testing and Materials, January 1992, pp. 365–383.
[45] Martin, R. H. and Jackson, W. C., "Damage Prediction in Cross-Plied Curved Composite Laminates," *Composite Materials: Fatigue and Fracture, Fourth Volume, ASTM STP 1156,* American Society for Testing and Materials, June 1993, pp. 105–126.
[46] Minguet, P. J. and O'Brien, T. K., "Analysis of Skin/Stringer Bond Failure Using a Strain Energy Release Rate Approach," *Proceedings,* Tenth International Conference on Composite Materials (ICCM X), Vancouver, British Columbia, Canada, August 1995, Vol. 1, pp. 245–252.
[47] Li, J., O'Brien, T. K., and Rousseau, C. Q., "Test and Analysis of Composite Hat Stringer Pull-off Test Specimens," *Journal of the American Helicopter Society,* Vol. 42, No. 4, October 1997, pp. 350–357.
[48] Salpekar, S. A., "Analysis of Delamination in Cross-Ply Laminates Initiating from Impact Induced Matrix Cracking," *Journal of Composites Technology and Research,* Vol. 15, No. 2, Summer 1993, pp. 88–94.

Stephen R. Swanson[1]

Overview of Biaxial Test Results for Carbon Fiber Composites

REFERENCE: Swanson, S. R., **"Overview of Biaxial Test Results for Carbon Fiber Composites,"** *Composite Materials: Fatigue and Fracture, Seventh Volume, ASTM STP 1330,* R. B. Bucinell, Ed., American Society for Testing and Materials, 1998, pp. 19–33.

ABSTRACT: Biaxial tests to determine stiffness and strength properties have been carried out on carbon/epoxy material systems using tubular specimens. The tests have included AS4, IM7, and T800 carbon fibers in a variety of layups that include fibers in [0/ ± 60] layups, [0/ ±45/ 90] in quasi-isotropic layups, and layups with extra fibers in the loading or off-axis directions, as well as layups in which the loading directions were at an angle with respect to the fiber directions, and specimens with 2-D triaxial braids. Loadings included biaxial tension, biaxial compression, mixed tension and compression, and compression under superposed pressure. The tests show a number of features that can be interpreted on both a macroscopic and a micromechanics level. The results for specific layups and materials show that ultimate laminate failure can be accurately correlated by means of a critical fiber direction strain. However, relating the laminate failure values to the properties of the fiber and matrix requires a more detailed examination at the micromechanics level. Ultimate fiber direction tensile strain values apparently depend on the details of the interaction of matrix cracking and fiber-matrix interphase strength, and thus in situ fiber strength in a laminate differs from that in a tow test.

KEYWORDS: carbon fiber composites, composites, biaxial tests

Advanced fiber composites have excellent strength and stiffness-to-weight properties and are often used in strength-critical applications. The prediction of failure in fiber composites involves a number of aspects, and a number of potential failure modes must be considered. Although at present a complete theoretical basis for the prediction of strength has not been developed, both micromechanics models of failure and experimental data offer complementary viewpoints. The Mechanics of Composites Laboratory at the University of Utah has been involved in experimental measurements of the failure properties of carbon/epoxy composites, using biaxial tests of tubular specimens. This paper gives a review of the work and presents a discussion of both the experimental data and theoretical correlations.

In the following, some of the experimental data on failure of carbon/epoxy laminates under static biaxial loads are reviewed. A discussion of the mechanisms involved and an assessment of the theoretical understanding of the failure process are then given.

Effect of Multiaxial Stresses

The effect of multiaxial stresses must be considered in assessing the strength of fiber composites. The various plies or layers of a fiber composite with fibers in more than one direction will be in multiaxial stress even for uniaxial load because of Poisson ratio effects.

---

[1] Professor, Department of Mechanical Engineering, University of Utah, Salt Lake City, UT 84112.

Because of the many possible laminates or layups of interest, it is clearly desirable to be able to relate the strength of the entire layup to the properties of the individual layers. This approach will be utilized in the following. However, in this approach, the particular mode of failure of the individual plies must be carefully considered. In general, matrix cracking may or may not lead directly to failure, while fiber failure usually corresponds to the ultimate failure of the laminate.

Ply Failure Theories

Before discussing the various possible failure theories that can be applied to the individual plies of a laminate, it is useful to consider the physical processes involved in failure. In typical polymer-based composites, the resin has sufficient elongation capability so that fiber failure occurs before resin failure in the usual unidirectional tension coupon test with the load parallel to the fibers. On a more detailed level, it is believed that the matrix plays a significant role in bridging around the individual fiber breaks that occur at weak points in the fibers, and that the ultimate fiber failure occurs when a sufficient amount of these individual fiber breaks are coupled together. From a macroscopic, ply level viewpoint fiber failure is characterized by either the tensile stress or tensile strain at failure in the unidirectional specimen. The situation is more complicated in compression, as both fiber and matrix play a role in determining the strength.

When tested in a direction transverse to the fibers, the composite typically fails in the matrix at a transverse strain level often significantly less than the failure strain of neat resin, and also much lower than the fiber failure strain under axial loading. As a consequence of this lower transverse strain to failure, a laminate may exhibit matrix failure in the transverse plies (relative to the major loading axis) well before failure of the fibers that are in the loading direction. This matrix cracking is the first manifestation of laminate failure and has been studied extensively [1–5]. As the loading is increased, further matrix cracking will occur, forming a more or less regular spacing [2]. While transverse strain (or stress) is often used to characterize the propensity to produce matrix cracks, energetic approaches have also been applied [3] that indicate that the "in situ" matrix strength will actually depend on the thickness of adjacent ply groups, which has been reported in experiments [1–5].

It is important to note that matrix cracking may or may not lead directly to ultimate laminate failure. If the laminate loading is carried primarily by the matrix, then matrix cracking and subsequent softening can lead directly to failure. However, in most practical situations the laminate is designed so that the load is carried by fibers to take advantage of the strength of the fibers relative to the weak matrix. In these cases the ultimate strength of the laminate may be several times that of the load corresponding to the initiation of ply cracking and is caused by the failure of fibers.

The conclusion to be drawn from the above is that matrix cracking and ultimate laminate failure are typically separate events and must be considered separately.

Several failure criteria have been suggested for direct application to composites. The stress polynomial due to Tsai and Wu [6] is given in the usual quadratic form as

$$F_1 \sigma_1 + F_{11} \sigma_1^2 + F_2 \sigma_2 + F_{22} \sigma_2^2 + 2 F_{12} \sigma_1 \sigma_2 + F_{66} \tau_{12}^2 = 1 \qquad (1)$$

where the F terms are material constants, and the stresses are the in-plane ply stresses. This polynomial can be used directly to predict first ply failure, which may correspond to ultimate laminate failure if the major load is in compression and will usually correspond to matrix cracking if the major load is tension. This criterion does not directly differentiate between matrix and fiber failure, but can be interpreted as doing so indirectly by assuming that first ply failure corresponds to matrix failure and that last ply failure corresponds to ultimate failure of the laminate [7].

Two additional criteria available for ply failure prediction are based on separating the above polynomial into two parts, one describing matrix failure and the other describing fiber failure. Hahn, Erikson, and Tsai [8] recommend the following

$$\text{Fiber failure: } F_1 \sigma_1 + F_{11} \sigma_1^2 = 1 \tag{2}$$

$$\text{Matrix failure: } F_2 \sigma_2 + F_{22} \sigma_2^2 + F_{66} \tau_{12}^2 = 1 \tag{3}$$

A similar proposal has been made by Hashin [9], who recommends the following

$$\text{Fiber failure: } F_1 \sigma_1 + F_{11} \sigma_1^2 + F_{66} \tau_{12}^2 = 1 \tag{4}$$

$$\text{Matrix failure: } F_2 \sigma_2 + F_{22} \sigma_2^2 + F_{66} \tau_{12}^2 = 1 \tag{5}$$

In either of the above forms, there is no ambiguity about what type of failure is being predicted.

A failure criterion that is widely used for predicting fiber failure on a ply basis is that of maximum fiber direction strain. Because composites are often notably weaker in compression than in tension, two material property values are needed and the criterion becomes

$$\varepsilon_{1c} < \varepsilon_1 < \varepsilon_{1t} \tag{6}$$

where ε_{1c} and ε_{1t} are fiber direction failure strains in compression and tension, respectively.

Comparison with Experimental Data: Ultimate Laminate Failure

Experimental measurements of laminate failure have been carried out using tubular specimens, loaded by combinations of pressure and axial load. An earlier version of the specimen is shown in Fig. 1 [10]. A later version of this specimen is shown in Fig. 2, where the end grips have been modified to permit higher axial loads [11]. Both of these specimens are loaded by combinations of internal pressure and axial tension or compression. A smaller specimen, shown in Fig. 3, is loaded in biaxial compression by combinations of external pressure and axial compression [12].

Experimental evidence on ultimate laminate failure is given in Fig. 4 for AS4/3501-6 carbon/epoxy and IM7/8551-7 carbon/epoxy [13]. The laminates for both materials are quasi-isotropic [0/±45/90]$_s$ and were tested in the form of 96-mm (3.8-in.) inside diameter cylinders with the applied loading being internal pressure and axial tension or compression. Additional data of this type have been obtained on other laminates and materials, including [0/±60], and quasi-isotropic laminates loaded at an angle to the fibers for biaxial tension loading [14–17]. Additional failure data under biaxial loading are given in Fig. 5 for T800 /3900-2 carbon/epoxy in three different laminate layups, taken from Ref 18. Data on the failure of fiberglass/epoxy cross-ply laminates under biaxial compression are shown in Fig. 6, taken from Wang and Socie [19].

The data given in Figs. 4–6 indicate that the failure of laminates can be correlated within the accuracy of the experimental data by the maximum fiber strain criterion. This is a fun-

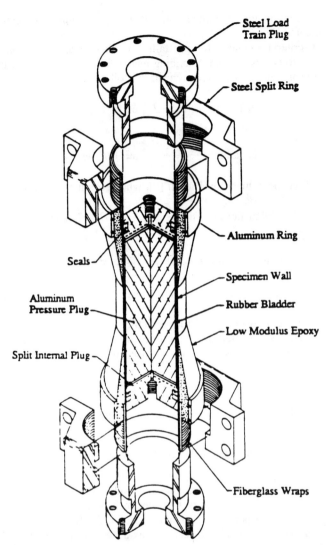

FIG. 1—*Specimen used for biaxial tests of fiber composite laminates. Specimen laoded by internal pressure and axial tension or compression.*

damental conclusion of the experimental studies reported in Refs *13–18*. Fiber strain has long been used as a laminate failure criterion in practical applications, so that the results shown in Figs. 4–6 and in Refs *13–18* are not particularly surprising. However, this carefully controlled experimental work does add credibility to a criterion that is sometimes regarded as being "too simple." The laminate ultimate stress predicted on the basis of the maximum fiber direction strain criterion is shown compared with the data for AS4/3501-6 in Fig. 4. Figure 5 shows the comparison of the maximum fiber strain failure criterion with the experimental data, where the same values for allowable fiber strain have been used for all three different laminate layups of T800/3900-2 carbon/epoxy.

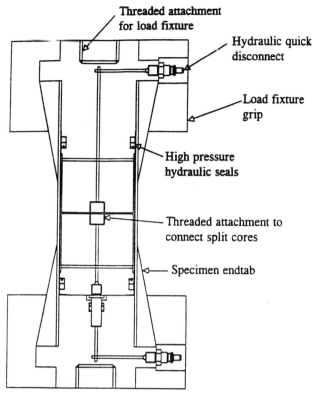

FIG. 2—*Biaxial specimen modified for higher axial loads. Specimen loaded by internal pressure and axial tension or compression.*

The value of the fiber failure strain in the laminates composed of AS4-3501-6 appears to be nearly the same as that measured in unidirectional tensile coupons. However, the laminate fiber strain value for IM7/8551-7 and for T800-3900-2 appears to be about 20 to 30% lower than the values measured in tensile coupons. Thus it may be necessary in general to establish allowables from laminate tests, rather than simply from using tensile coupon values. This creates a fundamental difficulty in that it is difficult to test laminates without introducing free-edge effects. This loss of fiber strain capability from tensile coupon to laminate is believed to be related to the high-toughness resin systems used for both of these two material systems [20].

Stress values can also be used as a laminate failure criterion. The use of a maximum fiber direction ply stress criterion can give accuracy equal to that of the strain criterion. While it is true that the value of stress in the fiber direction depends not only on the fiber direction strain but also on the transverse strain, in fact this dependence on transverse strain is quite small due to the very low value of the minor in-plane Poisson's ratio characteristic of fiber composites. Over the range of variables shown in the experimental data of Figs. 4 and 5, the difference between a maximum fiber stress and strain criterion is only a few percent and is within the scatter of the data.

The fiber failure criterion of Hahn, Erikson, and Tsai given in Eq 2 is equivalent to the use of two separate values for tensile and compressive fiber direction stress and is thus

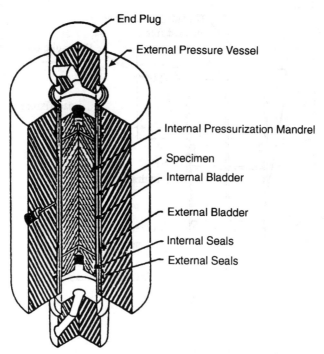

FIG. 3—*Specimen for biaxial compression tests of fiber composite laminates. Specimen loaded by external pressure and axial compression.*

identical to the fiber direction stress criterion discussed above. Thus if proper stress allowables are used, this criterion gives excellent agreement with the experimental data, essentially equivalent to the use of the strain criterion.

The Hashin fiber failure criterion of Eq 4 has an additional shear term, in addition to the fiber normal stress terms. For conditions where this shear stress is not large, this criterion is equivalent to the maximum fiber stress criterion and thus gives acceptable agreement with the data. However, for other conditions, the shear term is apparently overly conservative and agreement with the data is less good [*17*].

As mentioned above, the Tsai-Wu criterion of Eq 1 can be used as a fiber failure criterion by making special assumptions. In particular, the first ply to fail is assumed to be transverse matrix-dominated failure, while the last ply failure is considered to coincide with ultimate laminate failure. The Tsai-Wu approach for ultimate laminate failure is conservative by large factors under conditions of multiaxial laminate tensile stress and nonconservative for multiaxial laminate compressive stress. The inherent problem is that the matrix and fiber failure modes are not clearly differentiated. The transverse stress terms are overly weighted with respect to fiber failure by being based on matrix failure. Fiber failure may indeed be influenced by the transverse stresses, but the magnitude of these stresses is essentially limited by the ability of the matrix to transmit these stresses into the fiber. In general, a transverse stress that is large with respect to matrix allowables can still be small with respect to fiber allowables. Thus it appears that Eq 1 assigns too much weight to the transverse matrix stresses in the prediction of fiber failure.

Because the Tsai-Wu criterion is sensitive to the transverse stresses and overly conservative under conditions of multiaxial tensile stress, any reduction in the calculated transverse

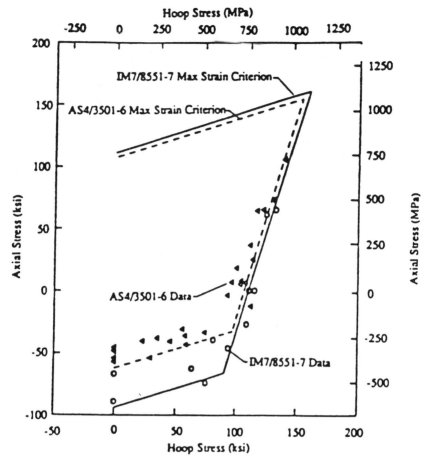

FIG. 4—*Laminate failure stresses for AS4/3501-6 and IM7/8551-7 carbon/epoxy laminates in quasi-isotropic* $[0/\pm45/90]_s$ *configuration under biaxial loading.*

stresses may serve to improve the accuracy of the criterion when applied to tensile stress states. For example, using the presumed reduction of effective transverse properties with matrix cracking and nonlinear shear response of Ref 21 gives a small improvement in the comparison of the Tsai-Wu criterion with experiment. An alternative procedure is suggested in Ref 7 in which the transverse properties E_{22}, G_{12}, and ν_{12} are multiplied by a degradation factor (DF), usually taken as 0.3, and the criterion of Eq 1 is used as a "first ply" failure criterion. This empirical approach does improve the predictive capability. This procedure is quite empirical, as the degradation factor essentially becomes a free constant. Also, the failure is predicted to be controlled by transverse plies rather than the plies in the loading directions, contrary to the usual interpretation of the experimental evidence.

Textile Form Materials

Fiber composites are often used in the form of textile products such as woven or braided fibers combined with various resin impregnation techniques, such as resin transfer molding

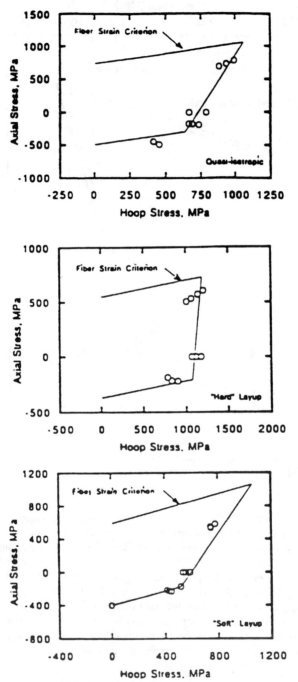

FIG. 5—*Comparison of data with failure envelopes predicted using the maximum fiber strain criterion for T800/3900-2 carbon/epoxy under biaxial stress loading. Layups are* $[0/\pm45/90]_{ns}$ *(quasi-isotropic),* $[0_3/\pm45/90]_s$ *(hard), and* $[0/(\pm45)_2/90]_s$ *(soft). The same critical strain values were used for all three layups.*

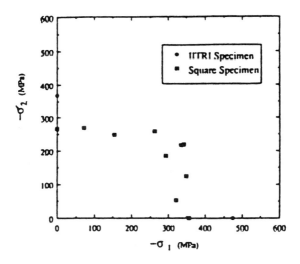

a) failure in stress space

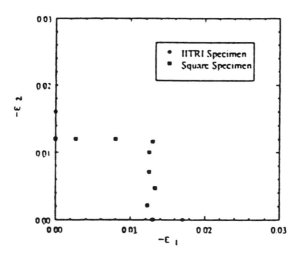

b) failure strains

FIG. 6—*Failure of cross-ply glass/epoxy laminates under biaxial compression loading, from Wang and Socie [19].*

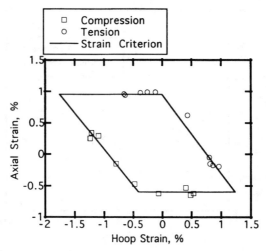

FIG. 7—*Biaxial failure strain envelope for 2-D triaxial braid, Architecture B. Line is prediction based on the maximum fiber direction strain failure criterion. Braid is $[0_{47\%}/\pm 45]$ AS4/1895 carbon/epoxy.*

(RTM). The results of a set of biaxial tests on a particular 2-D triaxial braid of AS4/1895 carbon/epoxy are shown in Fig. 7, taken from Smith and Swanson [22]. The triaxial braid is composed of a number of braided layers, each with straight axial fibers enclosed within the ± angle braids. A simplified method of analyzing the failure properties of the triaxial braid is to consider it to be analogous to a [0/±θ] laminate. A calculation of this type is illustrated in Fig. 7, where the measured failure strains are compared with a critical maximum strain criterion. The failure strain values are taken from uniaxial loadings of the braid material. It can be seen that this procedure provides an excellent correlation to the failure data. An additional comparison is shown in Fig. 8, where the failure stresses for this braid are

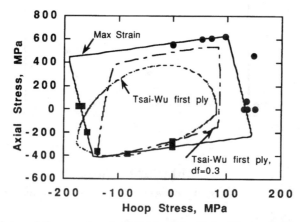

FIG. 8—*Biaxial stress failure envelope for 2-D triaxial braid, Architecture B. Lines show comparison of the maximum fiber direction strain criterion, the Tsai-Wu first ply criterion, and the Tsai-Wu first ply criterion with reduced transverse stiffness properties (DF = 0.3). Braid is $[0_{47\%}/\pm 45]$ AS4/1895 carbon/epoxy.*

compared with the predictions of the maximum fiber direction strain failure criterion, the Tsai-Wu first ply failure criterion, and the Tsai-Wu failure criterion with reduced transverse stiffness properties, using $DF = 0.3$. It can be seen that the fiber direction strain criterion gives a good prediction of the failure stresses.

Discussion

The major result of the evidence presented above is that laminate ultimate failure can be represented on a ply level analysis, which is thus applicable to all laminates. The most accurate failure criterion appears to be either maximum fiber direction stress or strain. These criteria have been used extensively in practical applications, and thus it is reassuring that the laboratory data also verify the validity of these criteria. To a more or less extent, criteria that include transverse normal or shear stress effects on fiber failure appear to be less applicable. While in general it would be expected that the transverse stresses do have an effect on fiber strength, the difficulty is in establishing how to accurately calculate these stresses. Clearly the maximum transverse stresses are limited by the strength of the matrix of matrix-fiber interphase. This stress level may be quite low with respect to the fiber strength of carbon fibers. Thus failure theories that include the effect of these stresses appear to be conservative, perhaps by as much as a factor of four as shown above. It may be that the transverse stresses are more important in the failure of aramid fiber laminates because of the weaker transverse strength of that fiber relative to carbon or glass fibers.

The use of a ply level criterion is of course desirable in that the criterion is presumed to apply to all (fiber-dominated) laminates of interest. While the number of laminates for which valid failure properties are available is not large, the data that exist appear to support this contention.

The question of how fiber strength or strain capability, say as measured in a unidirectional coupon test, translates into a "delivered" value in a laminate is not settled. While with the AS4/3501-6 carbon/epoxy system the fibers in the laminate appeared to have nearly the same strength properties as in a coupon, there was an apparent loss of strength of approximately 20 to 30% with the toughened resins of the IM7/8551-7 and T800/3900-2 carbon/epoxy systems. Thus it appears that at least two laminate tests are necessary to determine tension and compression "in situ" delivered fiber strengths.

The difference in delivered fiber strengths may be related to the fiber-matrix interphase properties. It has been conjectured that ply cracks in adjacent off-axis plies have a stress-concentration effect and thus reduce the strength of the load-carrying fibers. A mechanism for relieving this stress concentration has been proposed by Cook and Gordon [23], in which a microdelamination around the fiber at the tip of the crack relieves the stress. It is possible that higher matrix strength and toughness may tend to suppress this delamination and thus reduce the delivered fiber strength. Calculations to support this mechanism have been given in Ref 20.

The strength of laminates loaded in tension is largely governed by the strength of the fibers, and as shown above, can be correlated by critical values of strain or stress in the fibers. However, it is generally believed that the matrix plays a vital role in determining tensile strength, as indicated by the experimental observation that the strength of matrix-impregnated fiber bundles can be on the order of a factor of two higher than the measured tensile strength of dry fiber bundles, without matrix impregnation. The key to this apparently contradictory evidence lies in a synergistic effect between fiber and matrix.

In a dry fiber bundle, when a fiber breaks, it loses all load-carrying ability over its entire length and this load is shifted to the remaining fibers. When enough of the weaker fibers have failed, the strength of the remaining fibers is exceeded, and the bundle fails. In matrix-impregnated fiber bundles, it is believed that the matrix acts to bridge around individual fiber breaks, so that the fiber quickly picks up load-carrying ability. Thus the adjacent fibers only have to carry an increased load over a small axial distance. The statistical distribution of fiber defects makes it unlikely that the weakest location of each fiber would have the same axial location, so that failure must occur at a higher load value after enough fibers have failed in adjacent locations.

Rosen [24] assumed that the distribution of fiber strength was characterized by a Weibull statistical function, but did not consider any stress concentration effect on adjacent fibers. Zweben [25] and Zweben and Rosen [26] added detail to this model, including a proposed stress concentration effect. Harlow and Phoenix [27] carried out computer simulations that were limited to a small number of fibers because of numerical complexity. The essential features of the above models have been described by Batdorf [28]. Individual fibers are assumed to have a statistical population of weak regions characterized by a Weibull distribution function. As the composite is stressed, these weak points fail in an isolated manner. Although the fibers adjacent to a broken fiber are subject to increased stress, it is only over a small region as described above. Thus, it is statistically unlikely that weak regions will interact.

A challenge in the further understanding of failure in composites is to integrate the micromechanics approach to tensile failure with the experimental laminate failure data.

The compressive strength of fiber composites is often less than the tensile strength and thus can be a limiting factor in strength-critical applications. Several investigators have examined the inherent compressive strength of the fibers by supporting the fibers so that microbuckling is suppressed. Deteresa et al. [29] have shown that the relatively low compressive strength of aramid fibers may indeed be limited by the fibers themselves, as they tend to come apart laterally. Prandy and Hahn [30] have shown that carbon fibers made from a pitch precursor also have an inherently low strain to failure and fail by shear through the fiber. On the other hand, Deteresa [31] has measured failure strains for PAN precursor fibers such as AS4 that are in excess of −3%, far exceeding values seen in laminate tests. Thus, for typical PAN precursor carbon fibers, it appears that the fiber-matrix system is the limiting factor and not the inherent properties of the fibers.

One of the first attempts to understand the mechanics of compression failure of fiber composites was the work of Rosen [24], who developed a model for microbuckling of the individual fibers. A number of investigators have attempted to add detail to the basic Rosen model. Often the approach has been to incorporate a nonlinear stress-strain law for the matrix or initial defects in the fibers [32–40]. As would be expected, a lower and thus more realistic compressive strength is achieved. Another general approach has been to incorporate defects of various types, such as partial bonding of the fibers [41]. A review of this work has been presented by Guynn, Ochoa, and Bradley [42].

Many investigators have considered fibers that have initial waviness. Examinations of micrographs typically show that carbon fibers are not perfectly straight, although it is often difficult to quantify the description of just how wavy they are. The models with initially wavy fibers address an important weakness of the Rosen model, which is that as shown above it predicts that the compressive strength (for materials with sufficiently strong fibers) is affected only by the matrix shear modulus. However, experimental evidence by Madhukar and Drzal [43] and Swanson and Colvin [44] indicates that the fiber-matrix adhesion is also important. Models with initial fiber waviness provide a natural avenue for incorporating additional parameters into the model. With initially wavy fibers, increased loading produces additional deformation in the fiber. A limit is reached when either the fiber fails in bending or the fiber-matrix bond is exceeded. The failure mode is typically shown to be associated with the fiber-matrix adhesion for carbon fiber composites, in agreement with the experimental data mentioned above.

It is widely noted that compression failure values vary with the type of test performed. Some of this may be associated with experimental difficulties, as Berg and Adams [45] have demonstrated that careful attention to detail can result in higher measured failure strengths. However, there also seem to be systematic differences between tests. For example, the axial sandwich column tests reported by Shuart [46] and by Whitney et al. [47] give very high apparent strengths, with a compressive strain to failure of -2% for AS4 carbon/epoxy. Another example is in the bending tests reported by Jackson et al. [48] and by Wisnom [49], where the bend specimens failed in tension rather than in compression, implying that the compressive strengths are as high as or higher than the tensile strengths. It has also been shown that the strain to failure of critical plies can depend on the overall layup in a laminate [50,51]. Some models have explained these results by incorporating through-the-thickness effects in supporting the fiber [44,52,53]. The conclusion is that the compressive strength is expected to vary with the support conditions characteristic of each type of test.

The failure of fiber composites under compression is clearly complicated. The available experimental data as well as the theoretical models suggest that failure values can depend on a number of variables. As a practical matter, it would seem that the best procedure is to have experimental data for compression strength under conditions as close as possible to those seen in the actual structure.

Summary and Conclusions

Recent experimental evidence has shown that the ultimate failure of carbon/epoxy and glass/epoxy laminates can be approached on a ply level basis, so that a ply criterion can presumably be used for all laminates. If the laminate and loading is "fiber dominated," fiber failure is required to cause ultimate laminate failure. The ply level failure criterion that best represents the experimental data on the failure of a number of carbon/epoxy laminates is either fiber direction strain or stress. Allowable values for compression are often lower than for tension and are not well understood. Criteria that include interactions with transverse stresses tend to be conservative, in some cases by large factors.

References

[1] Piggott, M. R., *Developments in Reinforced Plastics,* Vol. 4, Elsevier Applied Science Publications, London, 1984, p. 131.
[2] Wang, A. S. D., "Fracture Mechanics of Sublaminate Cracks in Composite Materials," *Composites Technology Review,* Vol. 6, 1984, pp. 45–62.
[3] Master, J. E. and Reifsnider, K. L., "An Investigation of Cumulative Damage Development in Quasi-Isotropic Graphite/Epoxy Laminates," *Damage in Composite Materials, ASTM STP 775,* 1982, pp. 40–62.
[4] Flaggs, D. L. and Kural, M. H., "Experimental Determination of the In Situ Transverse Lamina Strength in Graphite/Epoxy Laminates," *Journal of Composite Materials,* Vol. 16, 1982, pp. 103–116.
[5] Parvizi, A., Garrett, K. W., and Bailey, J. E., "Constrained Cracking in Glass Fibre-Reinforced Epoxy Cross-ply Laminates," *Journal of Materials Science,* Vol. 13, 1978, pp. 195–201.
[6] Tsai, S. W. and Wu, E. M., "A General Theory of Strength of Anisotropic Materials," *Journal of Composite Materials,* Vol. 5, 1971, pp. 58–80.
[7] Tsai, S. W., *Composites Design,* 3rd ed., Think Composites Publications, Dayton, OH, 1987.
[8] Hahn, H. T., Erikson, J. B., and Tsai, S. W., "Characterization of Matrix/Interface-Controlled Strength of Unidirectional Composites," *Fracture of Composite Materials,* G. Sih and V. P. Tamuzs, Eds., Martinus Nijhoff Publications, 1982, pp. 197–214.

[9] Hashin, Z., "Failure Criteria for Unidirectional Fiber Composites," *Journal of Applied Mechanics,* Vol. 102, 1980, pp. 329–334.

[10] Toombes, G. R., Swanson, S. R., and Cairns, D. S, "Biaxial Testing of Composite Tubes," *Experimental Mechanics,* Vol. 25, 1985, pp. 186–192.

[11] Smith, L. V. and Swanson, S. R., "Design of a Cylindrical Specimen for Biaxial Testing of Composite Materials," *Journal of Reinforced Plastics and Composites,* Vol. 16, 1997, pp. 550–565.

[12] Smith, L. V. and Swanson, S. R., "Failure of Braided Carbon/Epoxy Composites Under Biaxial Compression," *Journal of Composite Materials,* Vol. 28, 1994, pp. 1158–1178.

[13] Colvin, G. E. and Swanson, S. R., "Characterization of the Failure Properties of IM7/8551-7 Carbon/Epoxy Under Multiaxial Stress," ASME, *Journal of Engineering Materials Technology,* Vol. 112, 1990, pp. 61–67.

[14] Swanson, S. R. and Nelson, M., "Failure Properties of Carbon/Epoxy Laminates Under Tension-Compression Biaxial Stress," *Proceedings,* Third Japan—U.S. Conference on Composite Materials, Tokyo, 1986.

[15] Swanson, S. R. and Christoforou, A. P., "Response of Quasi-Isotropic Carbon/Epoxy Laminates to Biaxial Stress," *Journal of Composite Materials,* Vol. 20, 1986, pp. 457–471.

[16] Swanson, S. R. and Trask, B., "Biaxial Tests of Off-Axis Quasi-Isotropic Laminates," *Proceedings,* American Society for Composites, 2nd Technical Conference, 1987, pp. 225–234.

[17] Swanson, S. R. and Trask, B. C., "An Examination of Failure Strength in [0/ ±60] Laminates Under Biaxial Stress," *Composites,* Vol. 19, 1988, pp. 400–406.

[18] Swanson, S. R. and Qian, Y., "Multiaxial Characterization of T800/3900-2 Carbon/Epoxy," *Composites Science and Technology,* Vol. 43, 1992, pp. 197–203.

[19] Wang, J. Z. and Socie, D. F., "Failure Strength and Damage Mechanisms of E-Glass/Epoxy Laminates under In-Plane Biaxial Compressive Deformation," *Journal of Composite Materials,* Vol. 27, 1993, pp. 40–58.

[20] Swanson, S. R., "Biaxial Failure Criteria for Toughened Resin Carbon/Epoxy Laminates," *Proceedings,* American Society for Composites, 7th Technical Conference on Composite Materials, Pennsylvania State University, 13–15 October 1992, pp. 1075–1083.

[21] Laws, N., Dvorak, G. J., and Hejazi, M., "Stiffness Changes in Unidirectional Composites Caused by Crack Systems," *Mechanics of Materials,* Vol. 2, 1983, pp. 123–137.

[22] Smith, L. V. and Swanson, S. R., "Strength Design with 2-D Triaxial Braid Textile Composites," *Composites Science and Technology,* Vol. 56 1996, pp. 359–365.

[23] Cook, J. and Gordon, J. E, "A Mechanism for the Control of Crack Propagation in All-Brittle Systems," *Proceedings of the Royal Society of London* A, Vol. 282, 1964, pp. 508–520.

[24] Rosen, B. W., "Mechanics of Fiber Strengthening," *Fiber Composite Materials,* American Society of Metals, Metals Park, OH, 1965, pp. 37–75.

[25] Zweben, C., "Tensile Failure of Fiber Composites," *AIAA Journal,* Vol. 6, 1968, pp. 2325–2331.

[26] Zweben, C. and Rosen, B. W., "A Statistical Theory of Material Strength with Application to Composite Materials," *Journal of the Mechanics and Physics of Solids,* Vol. 18, 1970, pp. 189–206.

[27] Harlow, D. G. and Phoenix, S. L. J., "The Chain-of-Bundles Probability Model for the Strength of Fibrous Materials II: A Numerical Study of Convergence," *Journal of Composite Materials,* Vol. 12, 1978, pp. 314–334.

[28] Batdorf, S. B., "Statistical Fracture Theories," *International Encyclopedia of Composites,* S. M. Lee, Ed., Vol. 6, 1991, pp. 395–404.

[29] DeTeresa, S. J., Allen, S. R., Farris, R. J., and Porter, R. S., "Compressive and Torsional Behavior of Kevlar 49 Fibre," *Journal of Materials Science,* Vol. 19, 1984, pp. 57–72.

[30] Prandy, J. M. and Hahn, H. T., "Compressive Strength of Carbon Fibers," *SAMPE Quarterly,* Vol. 22, 1991, pp. 47–52.

[31] DeTeresa, S. J., "Piezoresistivity and Failure of Carbon Filaments in Axial Compression," *Carbon,* Vol. 29, 1991, pp. 397–409.

[32] Chaplin, C. R., "Compressive Fracture in Unidirectional Glass-Reinforced Plastics," *Journal of Materials Science,* Vol. 12, 1977, pp. 347–357.

[33] Piggott, M. R., "A Theoretical Framework for the Compressive Properties of Aligned Fibre Composites," *Journal of Materials Science,* Vol. 16, 1981, pp. 2837–2845.

[34] DeFerran, E. M. and Harris, B., "Compression Strength of Polyester Resin Reinforced with Steel Wires," *Journal of Materials Science,* Vol. 4, 1970, pp. 62–72.

[35] Wang, A. S. D., "A Non-Linear Microbuckling Model Predicting the Compressive Strength of Unidirectional Composites," ASME Paper 78-WA/Aero-1, American Society of Mechanical Engineers, New York, 1978.

[36] Lanir, Y. and Fung, Y. C. B., "Fiber Composite Columns Under Compression," *Journal of Composite Materials,* Vol. 6, 1972, pp. 387–401.

[37] Herrman, L. R., Mason, W. E., and Chan, T. K., "Response of Reinforcing Wires to Compressive States of Stress," *Journal of Composite Materials,* Vol. 1, 1967, pp. 212–226.

[38] Hahn, H. T. and Williams, J. G., "Compression Failure Mechanisms in Unidirectional Composites," *Composite Materials: Testing and Design (7th Conference), ASTM STP 893,* American Society for Testing and Materials, West Conshohocken, PA, 1986, pp. 115–139.

[39] Hayashi, T., "Compressive Strength of Unidirectionally Fibre Reinforced Composite Materials," *Proceedings,* 7th International Reinforced Plastics Conference, British Plastics Federation, Oct. 1970, pp. 11/1–11/3.

[40] Chaudhuri, R. A., "Prediction of the Compressive Strength of Thick-Section Advanced Composite Laminates," *Journal of Composite Materials,* Vol. 25, 1991, pp. 1244–1276.

[41] Barber, J. and Triantafyllidis, N., "Effect of Debonding on the Stability of Fiber-Reinforced Composites," *Journal of Applied Mechanics,* Vol. 52, 1985, pp. 235–237.

[42] Guynn, E. G., Ochoa, O. O., and Bradley, W. L., "A Parametric Study of Variables that Affect Fiber Microbuckling Initiation in Composite Laminates: Part 1—Analyses," *Journal of Composite Materials,* Vol. 26, 1992, pp. 1594–1643.

[43] Madhukar, M. S. and Drzal, L. T., "Fiber-Matrix Adhesion and Its Effect on Composite Mechanical Properties. III. Longitudinal (0°) Compressive Properties of Graphite/Epoxy Composites," *Journal of Composite Materials,* Vol. 26, 1992, pp. 310–333.

[44] Swanson, S. R. and Colvin, G. E., "Compression Failure in Reduced Adhesion Fiber Laminates," ASME, *Journal of Engineering Materials and Technology,* Vol. 115, 1993, pp. 187–192.

[45] Berg, J. S. and Adams, D. F., "An Evaluation of Composite Material Compression Test Methods," *Journal of Composites Technology and Research,* Vol. 11, No. 2, 1989, pp. 41–46.

[46] Shuart, M. J., "An Evaluation of the Sandwich Beam Compression Test Method for Composites," *Test Methods and Design Allowables for Fibrous Composites, ASTM STP 734,* C. C. Chamis, Ed., American Society for Testing and Materials, West Conshohocken, PA, 1981, pp. 152–165.

[47] Whitney, J. M., Crasto, A. S., and Kim, R. Y., "Failure Criteria for Laminated Composites Subjected to Compression Loading," *Proceedings,* American Society for Composites, 7th Technical Conference Penn State University, Technomic, Westport, CN, 1992, pp. 604–612.

[48] Jackson, K. E., "Scaling Effects in the Flexural Response and Failure of Composite Beams," *AIAA Journal,* Vol. 30, 1992, pp. 2099–2105.

[49] Wisnom, M. R. "The Effect of Specimen Size on the Bending Strength of Unidirectional Carbon Fibre-Epoxy," *Composite Structures,* Vol. 18, 1991, pp. 47–63.

[50] Sohi, M. M., Hahn, H. T., and Williams, J. G., "The Effect of Resin Toughness and Modulus on Compressive Failure Modes of Quasi-Isotropic Graphite/Epoxy Laminates," *Toughened Composites, ASTM STP 937,* American Society for Testing and Materials, West Conshohocken, PA, 1987, pp. 37–60.

[51] Colvin, G. E. and Swanson, S. R., "In-Situ Compressive Strength of Carbon/Epoxy AS4/3501-6 Laminates," ASME, *Journal of Engineering Materials Technology,* Vol. 115, 1993, pp. 122–128.

[52] Swanson, S. R., "A Micro-Mechanics Model for In-Situ Compression Strength of Fiber Composite Laminates," ASME *Journal of Engineering Materials Technology,* Vol. 114, 1992, pp. 8–12.

[53] Swanson, S. R., "Constraint Effects in Compression Failure of Fiber Composites," *Proceedings,* Tenth International Conference on Composite Materials, ICCM 10, Whistler, B.C., Canada, Aug. 1995, pp. 14–18.

Ronald B. Bucinell[1]

Application of a Stochastic Model for Fatigue-Induced Delamination Growth in Graphite/Epoxy Laminates

REFERENCE: Bucinell, R. B., "Application of a Stochastic Model for Fatigue-Induced Delamination Growth in Graphite/Epoxy Laminates," *Composite Materials: Fatigue and Fracture, Seventh Volume, ASTM STP 1330*, R. B. Bucinell, Ed., American Society for Testing and Materials, 1998, pp. 34–54.

ABSTRACT: A stochastic model is used to predict the growth of delamination in graphite/epoxy laminates subjected to cyclic loading. The advantage of this model is that both the mean and variance associated with the growth of delamination is predicted. Understanding and predicting the variability associated with the delamination growth process is essential to the estimation of the reliability of composite structures. The empirical nature of the model has been minimized through the introduction of fracture mechanics parameters into the stochastic model. This has been accomplished by assuming that the mean of the stochastic model is represented by a fracture mechanics power law based on strain energy release rate. The applicability of the model is demonstrated through an experimental program that utilizes three laminate geometries. The results of the experimental evaluation illustrate the ability of the model to predict both the mean and variance of the delamination growth process in composite laminates subjected to cyclic loading.

KEYWORDS: composite materials, fatigue, delamination, experimental, stochastic model

Researchers have been investigating advanced composite materials for over 25 years—a relatively short period of time when compared with that for more traditional materials like metals. The attention this class of materials is getting is the result of their high specific properties and the ability to tailor these properties to a specific application. These features have proven to be extremely beneficial in many military and aerospace applications. The success of composite materials in these applications has led to their adoption in sporting goods, automotive, infrastructure, energy, and other applications. However, the complex anisotropic and heterogeneous nature of these materials has hindered the broad acceptance of their use. One of the most profound examples of this complexity is the accumulation and effect of the damage on the integrity of composite materials subjected to cyclic loading. In traditional homogeneous materials, failure results from the propagation of a single crack. In composite materials cracks form throughout the material during its life. These cracks interact and eventually coalesce and propagate to the final failure. This process is dependent on fiber architecture, materials used, and structural geometry. A thorough understanding of damage accumulation is critical to structural assessment.

The approaches that have been traditionally taken to address the structural assessment of composite materials subjected to cyclic loading include empirical theories [1–3], residual

---

[1] Union College, Department of Mechanical Engineering, Schenectady, NY 12308.

strength degradation theories [4–8], stiffness degradation theories [9–13], and damage mechanics theories [14–17]. Empirical, residual strength, and stiffness degradation approaches do not account for the underlying damage accumulation phenomenon that are inherent to the degradation of composite structures subjected to cyclic loading. As a result, their predictions are typically only applicable to a specific material and laminate combination. However, the underlying damage phenomena are intrinsic to the damage mechanics approach. The damage mechanics approach has the potential of predicting damage accumulation and residual properties from basic material properties without the need for specific laminate testing, although the fundamental understanding needed to make this approach successful has yet to be fully developed.

Cyclic loading of a composite structure will result in fiber, interlaminar, and intralaminar damage accumulation. Investigators have explored the use of the fracture mechanics principle of strain energy release rate [18] in modeling interlaminar [19–22] and intralaminar [23–25] cracking. However, few have addressed the high degree of variability associated with the damage growth processes. An understanding and accurate assessment of the variability is critical to reliability evaluations.

The first attempts to address the probabilistic nature of the composite damage accumulation process were based on the Weibull extreme value type distribution [26]. These attempts included fiber failure [27–29] and fatigue life [6,30,31] models. The Weibull approach is limited because the probability distribution has not been related to the damage process; it is not possible to separate and account for the major sources of variability in material response, the correlation characteristics of test data are not accounted for, and the nature of the physical process is not reflected. Further, there are only two limiting states of the material in Weibull distribution-based fatigue models: satisfactory (no damage) and failure. A rigorous stochastic approach can take into account all sources of variability in material response and lead to a reliable prediction of damage growth. Attempts at stochastic models for intralaminar [32] and interlaminar [33] damage growth are preliminary in nature.

In the remainder of this paper the fracture-mechanics-based stochastic delamination growth model developed in Ref 33 is further developed and applied to several laminate geometries. First the experimental program designed to generate data for the model is presented. Then the stochastic model development is summarized. This is followed by a discussion of the application of the model to the growth of delamination under constant-amplitude fatigue loading for three laminates. Following this, the parameters of the model are scrutinized and the arguments are made in favor of categorizing some of the parameters as material properties.

Experimental Investigation

In this section, the experimental study conducted to characterize the growth of delamination in laminated composite materials is discussed. The first objective of the experimental study is to determine the effects of the basic ply material properties, the laminate stacking sequence, and the maximum fatigue load amplitude on the mean growth of delamination. The second objective is to quantitatively describe the scatter in the growth of delamination. The results of the experimental investigation are used in the development and verification of the cumulative damage model presented later in this paper. Both static and cyclic fatigue tests were conducted to provide basic laminate property data for the model and to provide data for damage accumulation comparisons.

Material Description

The basic material system used to fabricate the laminates was AS4/3501-6 unidirectional graphite/epoxy prepreg tape. Several panels were fabricated in $[\pm 25/90]_s$, $[\pm 25/90_2]_s$, and $[\pm 45/90/0]_s$ laminate geometries. These panels were then autoclaved cured to the manufacturer's specifications. After curing, the panels were end-tabbed using either fiber-glass/epoxy or aluminum end-tabs. E-bond-600 adhesive was used to cement the end-tabs to the panels. The panels were then cut into straight tensile coupons 300 mm (12 in.) long by 19 mm (0.75 in.) wide.

Experimental Procedures

To achieve the experimental objective, three sets of tests were conducted. The first set of tests determined the elastic properties and the ultimate tensile strengths of the various laminates. The second set of tests established the delamination threshold, the load-delamination growth relationship, and located the delamination interface under a static loading. These data were used to characterize static delamination growth and to define the load levels for the constant-amplitude fatigue tests. The third and final set of tests established the load-time-delamination growth relationship and the location of the delamination interface under constant amplitude fatigue loading.

Uniaxial tensile static and fatigue tests were performed on a closed-loop servo-hydraulic Instron machine. For the static tests, the Instron was placed in displacement control and the crosshead speed was set to 0.25 mm (0.1 in.) per minute. For the fatigue tests, the Instron was placed in load control, and a sinusoidal wave form at the frequency of 6 Hz was used. The minimum to maximum load amplitude ratio was set to $R = 0.1$. Matrix damage in the test coupons was monitored nondestructively using an X-radiographic system (Hewlett Packard FAXTRON 800). A di-iodobutane solution was applied to the edge surfaces of the test coupon in order to enhance the X-radiographic image. Polaroid Type-55 positive/negative film was used in all X-radiographic photographs.

In the first set of tests, biaxial strain gage rosettes were mounted on each coupon for the purpose of characterizing the laminate properties. The longitudinal laminate modulus, E_x, Poisson ratio, v_{xy}, and the ultimate laminate strength were experimentally determined. The mean properties from these tests are summarized in Table 1. These results indicate that the thickness of the 90° layer has an adverse effect on the mean ultimate strength of the laminates. Law [34] observed that as the thickness of the 90° layer increased, the load level at which transverse cracking started decreased. The early presence of transverse cracks in a laminate enhances the susceptibility of the laminates to the modes of damage that lead to catastrophic failure.

The second set of tests, the incremental static loading, was conducted on all three laminate geometries. Each coupon in this set of tests was loaded to predetermined load increments

TABLE 1—*Mean laminate properties determined in laminate property characterization testing phase of study.*

Laminate Geometry	Replicates	E_x, GPa (Msi)	v_{xy}	σ_{ult}, MPa (ksi)
$[\pm 25/90]_s$	2	64.4 (9.34)	0.290	406 (59.0)
$[\pm 25/90_2]_s$	2	52.1 (7.55)	0.162	315 (45.8)
$[\pm 45/90/0]_s$	2	57.0 (8.27)	0.317	519 (75.3)

45
-45
90

0₂

90
-45
45

FIG. 1—*Typical edge-photomicrograph of* [±45/90/0]ₛ *laminate subjected to a qausi-static load.*

and then the load was relieved. The specimens were removed from the test fixtures and inspected for free edge delamination using X-radiography and then reloaded to a higher load level. The unloading and reloading of the specimens to higher predetermined levels was continued until the specimen failed. The free-edge interface where the delamination occurred was determined optically using a ×10 magnifying lens during the incremental loading. Several specimens were selected to be photomicrographed in order to record the location of the delamination interface. A photomicrograph of the edge of a [±45/90/0]ₛ laminate that was subjected to quasistatic loading is shown in Fig. 1. In this micrograph the delamination interface is seen to be the 90/0 interface. The experimentally observed location of the delamination interface for each of the laminates is summarized in Table 2 along with the static delamination onset stress and the breaking strength of the laminates. The ultimate strength of the laminates during the incremental static load testing (Table 2) compares well with the ultimate strength determined during the continuous static load testing (Table 1). This implies that the repeated loading and unloading of the incremental tests and the X-radiography enhancement solution did not effect the damage growth characteristics of these laminates under static loading.

TABLE 2—*Results of incremental static loading phase of experimental program.*

Laminate Geometry	Replicates	Delamination Onset Stress, MPa (ksi)	σ_{ult}, MPa (ksi)	Delamination Interface
[±25/90]ₛ	2	338 (49.0)	400 (58.0)	−25/90
[±25/90₂]ₛ	3	276 (40.0)	317 (46.0)	−25/90
[±45/90/0]ₛ	2	461 (66.0)	545 (79.0)	90/0

The third set of tests, constant amplitude fatigue, was conducted at a minimum of three fatigue load levels for each laminate. One load amplitude was chosen near the static delamination onset load, another was chosen so that the coupon would fail around one million cycles, and a third was chosen between the other two. During fatigue testing the cycling was halted at a series of predetermined cycles and the specimen was removed from the test fixture. Using the same X-radiography inspection procedure followed during the static testing, the growth of delamination was monitored. The specimens were then reloaded to a higher predetermined cyclic increment. This procedure was repeated until the specimen failed. The edge of the specimens were also examined for the location of the delamination interface. Figure 2 is a photomicrograph of the edge of a $[\pm 45/90/0]_s$ laminate that was subjected to fatigue loading. This photomicrograph shows the delamination forming in the $-45/90$ interface. A summary of the load levels at which the constant-amplitude fatigue tests were conducted along with the location of the delamination interface for all the laminates is found in Table 3. The fatigue delamination interface (Table 3) for the $[\pm 45/90/0]_s$ laminate does not correspond to the static delamination interface (Table 2). This difference, which implies that the mechanisms that lead to delamination growth may be different under static and fatigue loading, is discussed later.

Stochastic Delamination Growth Model

A stochastic process is a mathematical model of any dynamic process whose evolution with time is governed by some probabilistic law. The advantage of modeling fatigue delamination growth as a stochastic process is that the inherent variability in the growth of delamination can be predicted. The stochastic model in Ref *33* was developed to predict the growth of delamination in composite laminates. This model assumes that the number of cycles

FIG. 2—*Typical edge-photomicrograph of* $[\pm 45/90/0]_s$ *laminate subjected to a fatigue load.*

TABLE 3—*Summary of load levels at which specimens were tested during cyclic loading phase of experimental program.*

Laminate Geometry	Replicates	Maximum Fatigue Load, MPa (ksi)	Delamination Interface
$[\pm 25/90]_s$	3	200 (29.0)	−25/90
	3	228 (33.0)	−25/90
	3	276 (40.0)	−25/90
$[\pm 25/90_2]_s$	3	152 (22.0)	−25/90
	3	186 (27.0)	−25/90
	3	214 (31.0)	−25/90
$[\pm 45/90/0]_s$	3	269 (39.0)	−45/90
	5	317 (46.0)	−45/90
	6	365 (53.0)	−45/90
	2	414 (60.0)	−45/90

needed for a delamination to grow to a given length is best described by a random variable. The parameters of the probability distribution used to describe this random variable are assumed to vary with the applied fatigue load level. By describing these parameters in terms of the applied fatigue load level, the amount of testing required to correlate the model can be significantly reduced.

In Ref *33* the description of the stochastic model parameters in terms of the applied fatigue load level is based on the assumption that a deterministic fracture mechanics model can be equated to the mean expression of the stochastic model. Hence, the parameters in the stochastic model can be related directly to the applied fatigue load. This development is facilitated by discretizing both time and delamination size. A discretized time increment is referred to as a cycle. A cycle is defined as a repetitive period of cycling during which delamination can occur. In this development, each fatigue load reversal is considered a cycle. A discretized delamination size increment is referred to as a delamination state.

The following restrictions are imposed on the definitions of a cycle and a delamination state in the development of this model:

1. The increment of delamination damage at the end of a cycle depends only on the delamination state present at the start of that cycle.
2. Delamination damage can only increase during a cycle.
3. No initial delamination or manufacturing flaws exist in the virgin laminate that are on the order of a delamination state. Thus, all delamination damage starts in the undamaged state.
4. Between cycles, only the fatigue load level can change. The load frequency, R-ratio, environmental conditions, etc., are all assumed to remain constant.

Proceeding with these assumptions in mind, let the probability of the delamination advancing from the existing delamination state to the next during any cycle be assigned the value p. The probability of the delamination not advancing to the next delamination state is assigned the value q. Since it is assumed that the delamination can only increase by one delamination state during any cycle or remain in the same delamination state during that cycle, the following equality can be written

$$p + q = 1 \tag{1}$$

The values of p and q remain constant as long as the fatigue load level remains constant. Changes in the fatigue load level will result in changes in p and q.

The development in Ref 33 then went on to show that the constant-amplitude fatigue delamination growth process is described by the negative binomial distribution [35]. Where the mean ($E[\bullet]$) and variance (var[$\bullet$]) are given by:

$$E[T_n] = n/p \qquad (2)$$

and

$$\text{var}[T_n] = n(1 - p)/p^2 = n(p^{-2} - p^{-1}) \qquad (3)$$

In these expressions T_n is the random variable, defined as the number of cycles needed for a delamination to progress from the initial delamination state, (0)th, to the (n)th delamination state.

At this point the values of parameters p and n can be estimated from constant-amplitude fatigue delamination growth data using regression analysis. Since these parameters change with the fatigue load amplitude, they will have to be estimated for each load level under consideration unless a relationship between these parameters and the fatigue load level can be developed.

Experimental evidence [20,36,37] suggests that the fracture mechanics power law shown in Eq 4, written in terms of the strain energy release rate G and the critical strain energy release rate G_c, describes the mean growth behavior of delamination

$$\frac{da}{dN} = \alpha[G(\sigma_m, a)/G_c]^p \qquad (4)$$

Wang and Crossman [38] showed that G can be expressed explicitly in terms of the applied load σ_m as

$$G(\alpha_m, a) = C_e(a) \cdot t \cdot \sigma_m^2/E^2 \qquad (5)$$

where E is the elastic modulus of the laminate in the direction of the load, $C_e(a)$ is a coefficient function that depends only on the delamination size, a, and t is the thickness of a ply.

By assuming C_e is a constant [33], Eq 4 can be solved easily for the mean number of fatigue cycles required for the delamination to reach a given length, a

$$N = (a/\alpha)[(C_e \cdot t \cdot \sigma_m^2)/(G_c \cdot E^2)]^{-p} \qquad (6)$$

Equation 6 is a continuous representation of the mean delamination growth rate. Both Eqs 2 and 6 describe the same event. Discretizing Eq 6 enables the parameters of the probabilistic model (Eq 2) to be written in terms of fracture mechanics parameters.

The discretization of Eq 6 starts by letting a_{ds} be the size of a delamination state and n be the number of delamination states to a fixed crack length $a = a_f$

$$a_f = a_{ds} \cdot n \qquad (7)$$

It follows that

$$E[T_n] = \frac{n}{p} = N = (a_{ds} \cdot n/\alpha)[(C_e \cdot t \cdot \sigma_m^2)/(G_c \cdot E^2)]^{-p}$$

$$= \left(\frac{a_f}{\alpha}\right) \cdot [(C_e \cdot t \cdot \sigma_m^2)/(G_c \cdot E^2)]^{-p} \tag{8}$$

With a substitution of Eqs 6 and 7 into Eq 2, the probability of a delamination state advancing to the next state on a given cycle, in terms of fracture mechanics parameters, is given by

$$p = (\alpha/a_{ds})[(C_e \cdot t \cdot \sigma^2)/(G_c \cdot E^2)]^p \tag{9}$$

A similar procedure is used to write the variance in terms of fracture mechanics parameters. The development starts with the variance of the constant-amplitude fatigue delamination growth process, Eq 3. This development is based on the assumption that the values of p are on the order of 10^{-3} and smaller. With this assumption Eq 3 can be simplified to

$$\text{var}[T_n] = T_{sd}^2 = n/p^2 \tag{10}$$

where T_{sd} represents the standard deviation of the random variable T_n. Substituting Eqs 7 and 9 into Eq 10 and placing stress-related terms on the left-hand side of the equation yields

$$a_{ds} \cdot \sigma^{-4p} = T_{sd}^2 \left(\frac{\alpha^2}{a_{ds}}\right)\left[\frac{C_e \cdot t}{G_c \cdot E^2}\right]^{2p} \tag{11}$$

Load level is known to have a direct effect on mean fatigue delamination growth behavior. It is anticipated that load level will also have an effect on the variability associated with fatigue delamination growth behavior. Since the delamination state size a_{ds} is influenced by this variability, it can be concluded that a_{ds} is a function of load level. This conclusion is supported by data in Ref 39. The manner in which a_{ds} varies with the load level is assumed to be of the general form

$$a_{ds} = A\sigma^B \tag{12}$$

Substituting Eqs 7 and 12 into Eq 11 yields

$$\sigma^{B-4p} = T_{sd}^2 \cdot \left(\frac{\alpha^2}{A \cdot a_f}\right) \cdot \left(\frac{C_e \cdot t}{G_c \cdot E^2}\right)^{-2p} \tag{13}$$

The implication of this equation is that at a fixed delamination size, a_f, there is a relationship between the load level and the variance in the number of cycles to reach that delamination size, T_{sd}^2.

Taking the logarithm of both sides of Eq 13 and performing some simplification results in the following expression

$$\log(\sigma) = B_{v1} \cdot \log(T_{sd}) + B_{v0} \tag{14}$$

where

$$B_{v1} = \left(\frac{2}{B - 4p}\right) \tag{15}$$

and

$$B_{v0} = \left(\frac{1}{B - 4\rho}\right) \cdot \log\left\{\left(\frac{\alpha^2}{A \cdot a_f}\right)\left[\frac{C_e \cdot t}{G_c \cdot E^2}\right]^{2\rho}\right\}$$
(16)

Thus the variance of T_n can be written

$$\text{var}[T_n] = T_{sd}^2 = 10^{-2 \cdot B_{v0}/B_{v1}} \cdot \sigma^{2/B_{v1}}$$
(17)

At this point the model parameters p and n can be estimated directly from fracture mechanics parameters. Solving Eqs 2 and 9 simultaneously yields

$$p = \left[1 + \left(\frac{\text{var}[T_n]}{E[T_n]}\right)\right]^{-1}$$
(18)

$$n = E[T_n]\left[1 + \left(\frac{\text{var}[T_n]}{E[T_n]}\right)\right]^{-1}$$
(19)

where the mean, $E[T_n]$, and variance, $\text{var}[T_n]$, are defined in terms of fracture mechanics parameters in Eqs 8 and 17.

The stochastic delamination growth model has been developed completely for the case of constant-amplitude fatigue loading. Experimental data at several fatigue load amplitudes are required to estimate the parameters of the model, and then the model can be used to estimate the mean and variance in the growth of delamination at several other fatigue load levels. The following section discusses the correlation of the model with actual fatigue data.

Application of Model to Experimental Data

The stochastic model developed above can now be compared with experimental data. This process begins by using experimental data to estimate the parameters of the model (α, ρ, B_{v0}, and B_{v1}). This requires delamination growth versus fatigue cycle data at a minimum of three fatigue load levels for each of the specimen geometries (the load levels are summarized in Table 3). Additionally, a finite-element model simulating the site of delamination growth is required to estimate the energy released as the crack opens.

The parameters α and ρ are estimated using the delamination length versus number of cycles data for various fatigue load levels. A first-order linear regression equation is formed by taking the logarithm of both sides of Eq 6 and arranging the resulting equation in the form

$$Y = B_{m1} \cdot X + B_{m0}$$
(20)

where

$Y = \log(a_i/N_i)$,
$X = \log \sigma$,
$B_{m1} = 2 \cdot \rho$, and
$B_{m0} = \log \alpha \cdot \left[\dfrac{C_e \cdot t}{E^2 \cdot G_c}\right]^{\rho}$

TABLE 4—*Laminate-specific constants used in calculation of $\hat{\alpha}$ and $\hat{\rho}$.*

Laminate	E_s, GPa (Msi)	C_e, GN/m² (Mlb/in.²)	Interface	G_c, J/m² (lb/in.)	t, mm (in.)
$[\pm 25/90]_s$	64.4 (9.34)	20.1 (3.14)	$-25/90$	227 (1.30)	0.13 (0.0052)
$[\pm 25/90_2]_s$	52.1 (7.55)	21.0 (3.29)	$-25/90$	227 (1.30)	0.13 (0.0052)
$[\pm 45/90/0]_s$	57.0 (8.27)	11.8 (1.85)	$-45/90$	227 (1.30)	0.13 (0.0052)

The parameters α and ρ are estimated directly from the regression parameters B_{m1} and B_{m0}

$$\hat{\rho} = \hat{B}_{m1}/2$$

$$\hat{\alpha} = 10^{\hat{B}_{m0}} \cdot \left[\frac{C_e \cdot t}{E^2 \cdot G_c}\right]^{-\hat{B}_{m1}/2} \tag{21}$$

where the "$\hat{\ }$" indicates an estimated parameter. The laminate specific constants in B_{m0}, for each of the three laminates under consideration in this investigation, are summarized in Table 4. The parameter C_e is calculated using a quasi-three-dimensional finite-element model that employs the crack closure method, and the moduli were estimated from the previously discussed static experimental evaluation. The resulting estimates of α and ρ and their 95% confidence limits for these laminates are summarized in Table 5.

The stochastic model developed in the preceding section requires the estimation of the parameters B_{v0} and B_{v1} in Eq 14. This estimation requires that data for the variance in the time it takes the delamination to reach a width a_f be plotted in log T_{sd} versus log σ space. Equation 14 suggests that a first-order linear regression is appropriate for the estimation of B_{v0} and B_{v1}; however, because of the limited amount of this type of data, a least-squares analysis of variance cannot be performed. Instead, the jackknifing technique [40] was used.

Once B_{v0} and B_{v1} are estimated, the Parameters A and B in Eq 12 can be calculated as follows

$$B = \frac{2}{\hat{B}_{v1}} + 4\hat{\rho}$$

$$A = \frac{\hat{\alpha}^2}{a_f} \cdot \left(\frac{C_e \cdot t}{G_c \cdot E^2}\right)^{2\cdot\hat{\rho}} 10^{-\hat{B}_{v0}\cdot(\hat{B}-4\cdot\hat{\rho})} \tag{22}$$

TABLE 5—*First-order linear regression estimates of fracture mechanics parameters α and ρ from experimental data using Eqs 34 and 35.*

Laminate Geometry	Mean ρ	95% Conf. Level	Mean α, mm (in.)	95% Conf. Level, mm (in.)
$[\pm 25/90]_s$	6.115	5.481–6.749	4.81 (0.1892)	1.51–15.3 (0.0593–0.6042)
$[\pm 25/90_2]_s$	5.753	4.803–6.703	1.74 (0.0684)	0.31–9.67 (0.0123–0.3808)
$[\pm 45/90/0]_s$	5.300	4.706–5.894	0.09 (0.0030)	0.04–0.20 (0.0016–0.0077)

TABLE 6—*Summary of parameters* B_{v0}, B_{v1}, A *and* B *for laminate under consideration.*

Laminate Geometry	B_{v0}		B_{v1}		A	B
	Mean	Std. Dev.	Mean	Std. Dev.		
$[\pm25/90]_s$	1.820	0.047	−0.0720	0.0114	1.93×10^3	−3.338
$[\pm25/90_2]_s$	1.785	0.093	−0.0833	0.0212	1.86×10^{-2}	−1.010
$[\pm45/90/0]_s$	2.037	0.041	−0.0833	0.0117	1.86×10^3	−2.926

A summary of the regression Parameters B_{v0} and B_{v1} and the parameters A and B calculated from Eq 22 for the three laminates under consideration in this study are found in Table 6. The calculations of A and B are based on the mean values of B_{v0} and B_{v1} found in Table 6.

Now that all of the parameters of the model have been estimated, the model can be used to predict the behavior of the growth of delamination for the load levels under consideration. A summary of the model parameters $E[T_n]$, T_{sd}, p, and n, using the estimated parameters in Tables 5 and 6, for the load levels under consideration is found in Table 7. The model results are shown versus experimental data in Figs. 3 through 5. The model shows a good correlation to the experimental data. This correlation was also observed when the model was used to predict the fatigue delamination at the other load levels in Table 7.

Discussion

The intent of the stochastic delamination growth model is to predict the growth of delamination in laminated composite materials subjected to the cyclic loading. This model is currently in the process of being extended to include loading at multiple load levels (the cumulative damage case). Figures 3 and 5 show that the model is a good predictor of the mean and variance associated with the growth of delamination. There are issues related to this model that warrant further discussion. One relates to the prediction of the location of the delamination interface. As has been previously pointed out, the location of the delamination in the $[\pm45/90/0]_s$ laminate when subjected to static loading does not correspond to the location of the delamination when subjected to fatigue loading. Another set of issues involves the estimation of the model parameters and the application of the model; these are also discussed below.

TABLE 7—*Summary of parameters* T_{sd}, p, *and* n *for laminates under consideration in this study.*

Laminate Geometry	Load Level, MPa (ksi)	$E[T_n]$, Cycles	T_{sd}, Cycles	p	n
$[\pm25/90]_s$	200 (29.0)	1.59×10^5	9.28×10^4	1.85×10^{-5}	3
	228 (33.0)	3.28×10^4	1.54×10^4	1.38×10^{-4}	5
	276 (40.0)	3.13×10^3	1.07×10^3	2.73×10^{-3}	9
$[\pm25/90_2]_s$	152 (22.0)	3.25×10^5	3.41×10^4	2.80×10^{-4}	92
	186 (27.0)	3.09×10^4	2.92×10^3	3.62×10^{-3}	111
	214 (31.0)	6.31×10^3	5.56×10^2	2.04×10^{-2}	129
$[\pm45/90/0]_s$	269 (39.0)	4.05×10^5	2.60×10^5	5.99×10^{-6}	2
	317 (46.0)	7.05×10^4	3.55×10^4	5.59×10^{-5}	4
	365 (53.0)	1.57×10^4	6.43×10^3	3.80×10^{-4}	6
	414 (60.0)	4.22×10^3	1.44×10^3	2.04×10^{-3}	9

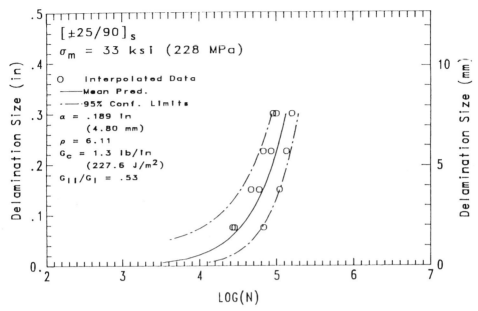

FIG. 3—*Delamination growth prediction for [+25/90]ₛ laminate at load level of 33.0 ksi (228 MPa).*

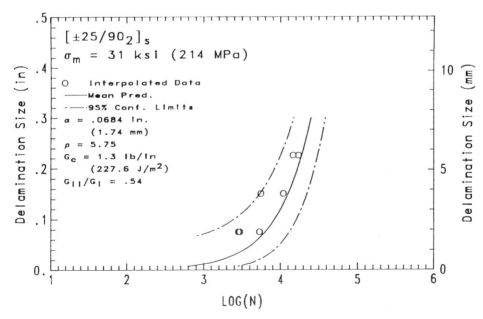

FIG. 4—*Delamination growth prediction for the [+25/90₂]ₛ laminate at load level of 31.0 ksi (214 MPa).*

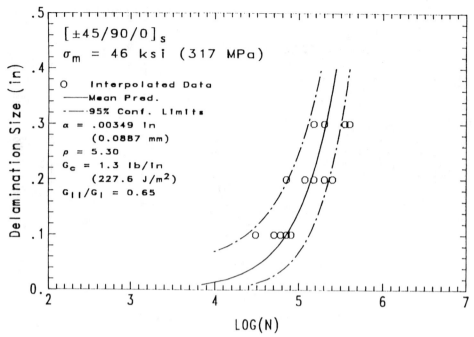

FIG. 5—*Delamination growth prediction for [+45/90/0]ₛ laminate at load level of 46.0 ksi (317 MPa).*

Delamination Interface

The specimens examined in the experimental program did not contain any type of foreign substance in the layer interface that forced the delamination growth. The site of delamination onset and growth was allowed to occur naturally. The site of delamination in both the $[\pm25/90]_s$ and the $[\pm25/90_2]_s$ laminates was observed in the $-25/90$ interface for both static and fatigue loading (see Tables 2 and 3). However, the site of the delamination in the $[\pm45/90/0]_s$ laminate was found to be in the 90/0 interface for static loading (Fig. 1) and the $-45/90$ interface for fatigue (Fig. 2) loading. O'Brien et al. [41,42] observed that delamination in laminates of the $[\pm0_2/\theta_2/\theta_2]_s$ family, where θ was 15, 20, 25, and 30°, formed in the $\theta/-\theta$ interface for both the static and cyclic loading conditions. This prompted an investigation to determine why the delamination interface in the $[\pm45/90/0]_s$ laminate was not the same under static and cyclic loading conditions. Since the stochastic model required a finite-element calculation of the strain energy release rate, both the stress distribution in the laminate and the modal components of the strain energy were readily available for examination.

Figures 6 and 7 show the distribution of the out-of-plane normal stress σ_{zz} and the shearing stress σ_{xz} along the free edge of the $[\pm45/90/0]_s$ laminate. The laminate coordinate system is arranged such that the x-axis is aligned with the load axis, the y-axis is in the plane of the laminate perpendicular to the x-axis, and the z-axis is perpendicular to the plane of the laminate. Figure 6 indicates that the normal opening stress is positive and singular at both the $-45/90$ and 90/0 interfaces. The normal opening stress at the edge of the $+45/-45$ interface is compressive and singular. Figure 7 shows that the shearing stress at the free edge is singular at the $+45/-45$ and the $-45/90$ interfaces. The positive singular normal stresses

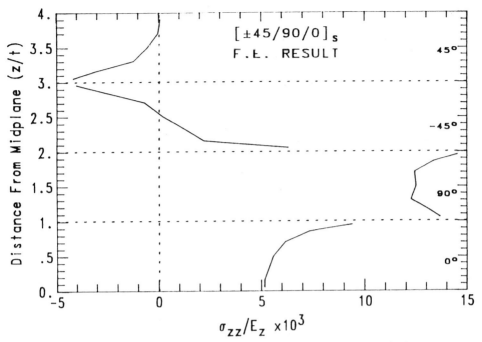

FIG. 6—*Finite-element calculated σ_{zz} stress along free edge of a $[\pm 45/90/0]_s$ laminate.*

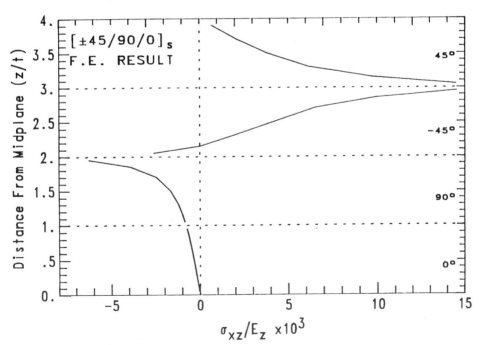

FIG. 7—*Finite-element calculated σ_{xz} stress along free edge of a $[\pm 45/90/0]_s$ laminate.*

in the $-45/90$ and $90/0$ interface do indicate that these interfaces are likely candidates for delamination growth. The singular compressive normal stresses in the $+45/-45$ interface explain why, even though the shearing stresses are singular, no delamination forms in this interface during static or fatigue loading. The stress states in the $-45/90$ and the $90/0$ interface provide some insight but no definitive explanation for the difference in the location of the delamination interface during static and fatigue loading.

The stochastic model developed here is based on the release of energy as the delamination crack grows. Figure 8 illustrates the finite-element calculation of the total energy coefficient function, which is proportional to the total energy released by crack in each of the $[\pm45/90/0]_s$ laminate interfaces. Figures 9 through 11 look at a breakdown of the modal components of the energy coefficient in each of the laminate interfaces. The $+45/-45$ interface is dominated by Mode III, the $-45/90$ interface has a mixture of Modes I and II, the $90/0$ interface is almost totally Mode I, and the midplane is totally Mode I. The laminate interface where the static delamination is observed to form, $90/0$, has the highest value of Mode I energy release in any of the interfaces other than the midplane. This trend is also observed in the $[\pm25/90]_s$ and the $[\pm25/90_2]_s$ laminates. The laminate interface where the fatigue delamination is observed to form, $-45/90$, has a lower Mode I component, a higher total energy release, and the highest Mode II component. In the $[\pm25/90]_s$ and the $[\pm25/90_2]_s$ laminates the fatigue delamination was located in the interface, $-25/90$, that also had a mixture of Modes I and II components, and where the total energy release was lower than the midplane and the $+25/-25$ interface that was nearly totally Mode I.

A few observations can be made about the location of delamination onset and propagation in composite laminates subjected to static and fatigue loading. First, in order for a delamination to form at an interface, the normal opening (σ_{zz}) edge stresses have to be very high and tensile. The finite-element analysis performed on all three of the laminates shows that the interfaces where the delamination did form had singular normal opening stresses at the free edge. If a high tensile state of stress exists at an interface and the loading is static, the

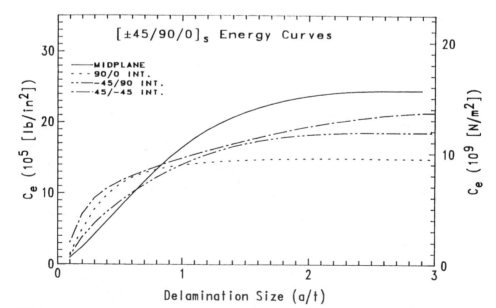

FIG. 8—*Total strain energy release rate coefficient curves for ply interfaces of* $[\pm45/90/0]_s$ *laminate.*

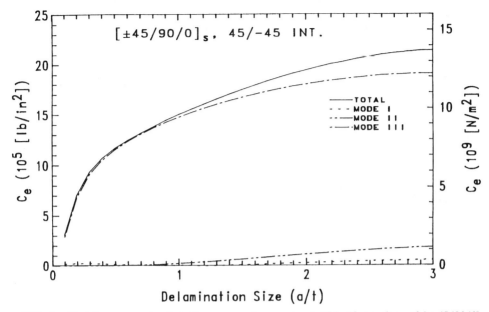

FIG. 9—*Modal components of strain energy release rate at 45/−45 interface of* [±45/90/0]$_s$ *laminate.*

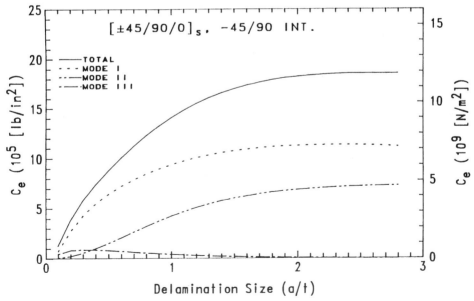

FIG. 10—*Modal components of strain energy release rate at* −45/90 *interface of* [±45/90/0]$_s$ *laminate.*

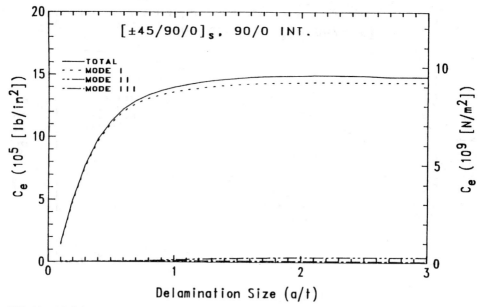

FIG. 11—*Modal components of strain energy release rate at 90/0 interface of* [±45/90/0]_s *laminate.*

growth of delamination will be dominated by Mode I fracture. In all the laminates considered here, the midplane was observed to be delamination free even though this interface has the highest Mode I component. This interface lacks the singular normal edge stresses. If a high interlaminar tensile state of stress exists at an interface and the loading is cyclic, it appears that the interface with the maximum Mode II component will control the growth of delamination. In the [±25/90]_s and [±25/90_2]_s laminates of both the above conditions are met by the same interface.

Application of the Stochastic Model

The stochastic model developed above provides a methodology for the prediction of the mean delamination growth rate in laminated composites subjected to constant-amplitude cyclic loading and an estimate of the variability associated with the growth of the delamination. The model is developed without any predisposition to a probability distribution. Additionally, fracture mechanics parameters have been introduced into the model in order to minimize the amount of data required to correlate the model and to maximize the predictive capability of the model.

The stochastic model developed here implies that the delamination grows uniformly along the length, toward the center of the specimen. The crack length being predicted is actually the width of the delamination measured from the edge of the specimen. Observations of the delamination growth for the laminates under consideration in this investigation show that the delamination does occur along the entire length of the laminate; however, it is not uniform. O'Brien et al. [42] also made a similar observation about the growth of edge delamination and went on to show that the error associated with assuming a uniform delamination front was negligible when compared with a detailed crack front model.

The stochastic model was developed around the assumption that a delamination may or may not advance during any fatigue cycle. These two events are assigned the probabilities p and q. The number of cycles to a given delamination size is found to be negative binomially distributed. This distribution captures the true discrete progressive nature of delamination growth during cycle loading. It is not surprising that this distribution is not the same as the distribution associated with the accumulation of damage in fibers and the strength of laminates. The damage associated with these phenomena occurs at multiple sites in the composite and then coalesces into a critical damage state. This type of process has been shown by Harlow and Phoenix [27–29] to follow a Weibull distribution.

The mean and variance predictions of the stochastic delamination growth model are compared with experimental data for each of the laminates in Figs. 3 through 5. All of the data obtained are found to lie within the 95% confidence limits predicted by the model. By changing the load parameter, σ, in Eqs 8 and 17, the model can be used to predict the mean growth of delamination and the variability associated with this growth at load levels other than those used to correlate the model. This prediction is expected to be most accurate within the range of the load levels used to correlate the model. There is always a danger in extrapolating models outside the range of correlation.

The forms of the equations used to estimate the mean and variance parameters for the delamination growth model from experimental data, Eqs 14 and 20, are both linear. Thus, in designing experiments to correlate the model for a given laminate, data should be generated at three load levels. Two of the load levels should correspond to the upper and lower bounds of the loading under consideration. The third load level should correspond to a load level midway between the upper and lower bound. The majority of the experimental data should be concentrated at the upper and lower bounds. This strategy will provide the most accurate estimates of the variability about the mean prediction of the model. This strategy was not employed during this experimental program. When this study was being carried out, the form of the estimation curves was not known; therefore data were arranged such that a relationship as high as a second-order polynomial could be evaluated.

The scant amount of data used in the estimation of Parameters B_{v0} and B_{v1} can be traced to the meaning of the data points. Each of the data points represents the variance in the time it takes a delamination to grow to a certain length at a given load level. Thus the generation of a single data point in this figure requires the use of all the data at a given load level. The number of data available for the estimation of B_{v0} and B_{v1}, at each load level, can be increased only if several batches of material are used. If this is done, several specimens from each batch of material can be tested at each load level. The variability of the delamination growth to a specified length, at a given load level, for each batch provides an independent data point that can be used in the estimation of B_{v0} and B_{v1}. In doing so, a more accurate estimate of the mean and variance of B_{v0} and B_{v1} will be estimated.

Equations 15 and 16 show B_{v0} and B_{v1} as functions of mostly fracture mechanics parameters. The nonfracture mechanics parameters in these equations are A and B. These parameters were introduced to mathematically model the effect of load level on the damage state size a_{ds}. The nature of A and B is summarized in Table 6. It does appear that A may be a material-related constant; however, the nature of B requires further investigation. From Eq 12, the negative sense of Parameter B indicates that as the load level increases, the size of a delamination state decreases. The implication of this is seen in Eq 12, where a decrease in delamination state corresponds to an increase in the probability of the delamination advancing during any fatigue cycle. If further investigation reveals the true nature of Parameters A and B, it could be possible to reduce the amount of data required to correlate the model. This could possibly lead to the use of data obtained with one material in a known laminate configuration predicting the mean and variance of the growth of delamination of the same material in a different laminate configuration. Further investigation into the nature of these parameters is required.

The experimental program in this paper concentrates on constant-amplitude loading. The development of the stochastic model presented here is not restricted in a similar manner. In the development of the model, the number of cycles for a delamination to grow to a specified size is developed assuming that the load levels are not the same [33]. Although this extension is not addressed in this paper, work is underway to extend the model to address cumulative damage scenarios and generate test data to validate the model.

Summary and Conclusions

A delamination growth model for composite laminates subjected to constant-amplitude fatigue loading has been developed. The parameters of the model have been defined in terms of fracture mechanics, material, and laminate geometry parameters. Experimental data verified that this model can be used to predict the mean and variance associated with the growth of delamination at several load levels. Further investigation is required to determine if the predictive capability of this model can be extended to the growth characteristics of delamination in laminates with geometries other than the geometries used to correlate the model.

The location of the delamination in laminates was found to be load-type specific. Under quasistatic loading the delamination formed in the interface with high singular opening edge stresses and a large Mode I energy release rate component. Under fatigue loading, the delamination formed in the interface with high singular opening edge stresses and a large Mode II energy release rate component.

This stochastic model should provide more accurate estimates of structural reliability than wearout [43] or damage tolerance [44] models since the nature of the delamination growth phenomenon is inherent to the model. This model will allow statistically significant damage thresholds to be established for given service loads. To truly be of assistance to designers of composite structures, the model must be extended to include variable-amplitude loading. This consideration is now being investigated.

References

[1] Mandell, J. F., Huang, D. D., and McGarry, F. J., "Tensile Fatigue Performance of Glass Fiber Dominated Composites," *Composites Technology and Review*, Vol. 3, No. 3, 1981, p. 96.

[2] Albrecht, C. O., "Statistical Evaluation of a Limited Number of Fatigue Test Specimens Including a Factor of Safety Approach," *Fatigue Test of Aircraft Structures, ASTM STP 338*, American Society for Testing and Materials, West Conshohocken, PA, 1962, p. 150.

[3] Sims, D. F. and Brogdon, V. H., "Fatigue Behavior of Composites Under Different Loading Modes," *Fatigue of Filamentary Composite Materials, ASTM STP 636*, American Society for Testing and Materials, West Conshohocken, PA, 1977, p. 185.

[4] Halpin, J. C., Johnson, T. A., and Waddoups, M. E., "Kinetic Fracture Models and Structural Reliability," *International Journal of Fracture Mechanics*, Vol. 8, 1972, p. 465.

[5] Wolff, R. V. and Lemon, G. H., "Reliability Prediction for Adhesive Bonds," Air Force Materials Laboratory Report, AFML-TR-72-121, March 1972.

[6] Hahn, H. T. and Kim, R. Y., "Proof Testing of Composite Materials," *Journal of Composite Materials*, Vol. 9, 1975, p. 297.

[7] Schaf, J. R. and Davidson, B. D., "Life Prediction Methodology for Composite Structures. Part I—Constant Amplitude and Two-Stress Level Fatigue," *Journal of Composite Materials*, Vol. 31, 1997, p. 128.

[8] Schaff, J. R. and Davidson, B. D., "Life Prediction Methodology for Composite Structures. Part II—Spectrum Fatigue," *Journal of Composite Materials*, Vol. 31, 1997, p. 158.

[9] Chou, P. C. and Croman, R., "Degradation and Sudden-Death Models of Fatigue of Graphite/ Epoxy Composites," *5th Conference on Composite Materials: Testing and Design, ASTM STP 674,* 1979, p. 431.
[10] Highsmith, A. L. and Reifsnider, K. L., "Internal Load Distribution Effects During Fatigue Loading of Composite Laminates," *Composite Materials: Fatigue and Fracture, ASTM STP 907,* American Society for Testing and Materials, West Conshohocken, PA, 1986, p. 233.
[11] Reifsnider, K. L., "The Critical Element Model: A Modeling Philosophy," *Engineering Fracture Mechanics,* Vol. 25, 1986, p. 739.
[12] Hwang, W. B. and Han, K. S., "Fatigue of Composite Materials-Damage Model and Life Prediction," *Composite Materials: Fatigue and Fracture 2nd Volume, ASTM STP 1012,* P. D. Lagace, Ed., American Society for Testing and Materials, West Conshohocken, PA, 1989, p. 87.
[13] Lee, L. J., Yang, J. N., and Sheu, D. Y., "Prediction of Fatigue Life for Matrix-Dominated Composite Laminates," *Composite Science and Technology,* Vol. 46, 1993, p. 21.
[14] Broutman, L. J. and Sahu, S., "A New Theory to Predict Cumulative Fatigue Damage in Fiberglass Reinforced Plastics," *Composite Materials: Testing and Design, ASTM STP 497,* American Society for Testing and Materials, West Conshohocken, PA, 1972, p. 170.
[15] Wang, A. S. D., Chou, P. C., Lei, C. S., and Bucinell, R. B., "Cumulative Damage Model for Advanced Composite Materials: Phase II Report," AFWAL-TR-84-4004, March 1984.
[16] Wang, A. S. D., Chou, P. C., and Bucinell, R. B., "Cumulative Damage Model for Advanced Composite Materials," AFWAL-TR-85-4104, October 1985.
[17] Timmer, J. S. and Hahn, H. T., "The Effect of Preloading on Fatigue Damage in Composites," *18th Annual Mechanics of Composites Review,* p. 144.
[18] Paris, P. C. and Sih, G. C., "Stress Analysis of Cracks," *Fracture Toughness Testing and Its Applications, ASTM STP 381,* American Society for Testing and Materials, West Conshohocken, PA, 1964, p. 30.
[19] O'Brien, T. K., "Characterization of Delamination Onset and Growth in Composite Laminates," *Damage in Composite Materials, ASTM STP 775,* American Society for Testing and Materials, West Conshohocken, PA, 1982, p. 140.
[20] Wang, A. S. D., Slomiana, M., and Bucinell, R. B., "Delamination Crack Growth in Composite Laminates," *Delamination and Debonding of Materials, ASTM STP 876,* American Society for Testing and Materials, West Conshohocken, PA, 1985, p. 135.
[21] O'Brien, T. K. and Hooper, S. J., "Local Delamination in Laminates with Angle Ply Matrix Cracks, Part I: Tension Tests and Stress Analysis," *Composite Materials: Fatigue and Fracture, Fourth Volume, ASTM STP 1156,* American Society for Testing and Materials, West Conshohocken, PA, 1993, p. 491.
[22] O'Brien, T. K., "Local Delamination in Laminates with Angle Ply Matrix Cracks, Part II: Delamination Fracture Analysis and Fatigue Characterization," *Composite Materials: Fatigue and Fracture, Fourth Volume, ASTM STP 1156,* American Society for Testing and Materials, West Conshohocken, PA, 1993, p. 507.
[23] Wang, A. S. D., "Fracture Mechanics of Sublaminate Cracks in Composite Materials," *Composites Technology and Review,* Vol. 6, 1984, p. 45.
[24] Suresh, N. and Wang, A. S. D., "Stress and Energy Based Fracture Condition for Fiber-Wise Splitting of Composite Materials," *Composite Materials: Fatigue and Fracture 5th Volume, ASTM STP 1230,* American Society for Testing and Materials, West Conshohocken, PA, 1993, p. 176.
[25] Sriram, P. and Armanios, E. A., "Shear Deformation Analysis of the Energy Release Rate of Transverse Cracking in Laminated Composites," *Composite Materials: Fatigue and Fracture 5th Volume, ASTM STP 1230,* American Society for Testing and Materials, West Conshohocken, PA, 1993, p. 215.
[26] Weibull, W., "A Statistical Distribution Function of Wide Applicability," *Transactions of the ASME: Journal of Applied Mechanics,* Vol. 18, 1951, p. 293.
[27] Harlow, D. G. and Phoenix, S. L., "The Chain-of-Bundles Probability for the Strength of Fibrous Materials I: Analysis and Conjecture," *Journal of Composite Materials,* Vol. 12, 1982, p. 195.
[28] Harlow, D. G. and Phoenix, S. L., "The Chain-of-Bundles Probability Model for the Strength of Fibrous Materials II: A Numerical Study of Convergence," *Journal of Composite Materials,* Vol. 12, 1978, p. 314.
[29] Harlow, D. G. and Phoenix, S. L., "Bounds on the Probability of Failure of Composite Materials," *International Journal of Fracture,* Vol. 15, 1979, p. 321.
[30] Ramani, S. V. and Williams, D. P., "Notched and Unnotched Fatigue Behavior of Angle-Ply Graphite/Epoxy Composites," *Fatigue of Filamentary Composite Materials, ASTM STP 636,* American Society for Testing and Materials, West Conshohocken, PA, p. 27.
[31] Talreja, R., *Fatigue of Composite Materials,* Technomic Publishing Company, Westport, CT, 1987.

[*32*] Wang, A. S. D., Chou, P. C., and Lei, S. C., "A Stochastic Model for the Growth of Matrix Cracks in Composite Laminates, *Journal of Composite Materials*, Vol. 18, 1984, p. 239.
[*33*] Bucinell, R. B., "Development of a Stochastic Free Edge Delamination Model for Laminated Composite Materials Subjected to Constant Amplitude Fatigue Loading," *Journal of Composite Materials*, in press.
[*34*] Law, G. E., "Fracture Analysis of [$\pm 25/90_n$]$_s$ Graphite-Epoxy Composite Laminates," Ph.D. thesis, Drexel University, Philadelphia, PA, 1981.
[*35*] Hastings, N. A. J. and Peacock, J. B., *Statistical Distributions*, Butterworths, London, 1975.
[*36*] Dahlen, C. and Springer, G. S., "Delamination Growth in Composites under Cyclic Loads," *Journal of Composite Materials*, Vol. 28, 1994, p. 732.
[*37*] Ye, L., "On Fatigue Damage Accumulation and Material Degradation in Composite Materials," *Composites Science and Technology*, Vol. 36, 1989, p. 339.
[*38*] Wang, A. S. D. and Crossman, F. W., "Initiation and Growth of Transverse Cracks and Edge Delaminations in Composite Laminates: Part 1—Energy Method," *Journal of Composite Materials*, Vol. 14, 1980, p. 71.
[*39*] Birnbaum, Z. W., Saunders, S. C., McCarty, R. C., and Elliott, R., "A Statistical Theory of Life-Length of Materials," Boeing Airplane Company, Document No. D2-1325, 1950.
[*40*] Miller, R. G., "The Jackknife—A Review," *Biometrika*, Vol. 61, 1970, p. 1.
[*41*] O'Brien, T. K., "Local Delamination in Laminates with Angle Ply Matrix Cracks, Part I: Tension Tests and Stress Analysis," *Composite Materials: Fatigue and Fracture (Fourth Volume), ASTM STP 1156*, American Society for Testing and Materials, West Conshohocken, PA, 1993, p. 491.
[*42*] O'Brien, T. K. and Hooper, S. J., "Local Delamination in Laminates with Angle Ply Matrix Cracks, Part II: Delamination Fracture Analysis and Fatigue," *Composite Materials: Fatigue and Fracture (Fourth Volume), ASTM STP 1156*, American Society for Testing and Materials, West Conshohocken, PA, 1993, p. 507.
[*43*] Kedward, K. T. and Beaumont, P. W. R., "The Treatment of Fatigue and Damage Accumulation in Composite Design," *International Journal of Fatigue*, Vol. 14, 1992, p. 283.
[*44*] O'Brien, T. K., "Towards a Damage Tolerance Philosophy for Composite Materials and Structures," *Composite Materials: Testing and Design (Ninth Volume), ASTM STP 1059*, American Society for Testing and Materials, West Conshohocken, PA, 1990, p. 7.

Geoffrey T. Ward[1] and Ben M. Hillberry[2]

An Approach to Include Interfacial Wear Effects in Modeling Fatigue Crack Growth of Titanium Matrix Composites

REFERENCE: Ward, G. T. and Hillberry, B. M., "An Approach to Include Interfacial Wear Effects in Modeling Fatigue Crack Growth of Titanium Matrix Composites," *Composite Materials: Fatigue and Fracture, Seventh Volume, ASTM STP 1330,* R. B. Bucinell, Ed., American Society for Testing and Materials, 1998, pp. 55–75.

ABSTRACT: An approach has been developed that includes the effects of interfacial wear on the fiber-bridging behavior of titanium matrix composites during fatigue crack propagation. This approach uses a Coulomb friction-based fiber-bridging model in which the effect of fiber surface roughness on the clamping stress between the fiber and matrix is included. A previously developed wear model has been incorporated into this bridging model as a means to determine the reduction of the fiber surface roughness amplitude during fatigue cycling. As the roughness decreases, its contribution to the clamping stress also decreases, resulting in a lower interfacial shear stress. In order to include this effect in model fatigue crack growth rates, the combined Coulomb friction and wear models have been applied to a discrete composite model formulation. Crack growth predictions were then performed using a single set of input parameters by allowing the fiber surface roughness to decrease due to wear over a discrete increment of fatigue crack geometry based on bridging conditions determined by the composite model. These predictions correlated very well with experimental results for different loading conditions, especially those at relatively high crack growth rates.

KEYWORDS: titanium matrix composites, silicon-carbide fibers, fiber bridging, wear, fiber/matrix interface, fatigue crack growth

Fatigue crack growth in titanium matrix composite (TMC) materials is a complex phenomenon that depends on the properties of both the fiber and the matrix, as well as those of the fiber/matrix interface. Considerable investigation has been performed on the mechanics of the interface and how it affects the crack growth behavior of the composite. The major emphasis of these investigations has focused around understanding the fiber-bridging process in which the fibers in the wake of a matrix crack remain intact as the crack continues to grow. These bridging fibers control the extent of crack opening displacement, thus reducing the overall crack growth rate as compared with a crack in an unreinforced matrix material.

The common approach to modeling the fiber-bridging process has been to assume that the relative displacement between the fiber and matrix is opposed by the shear stress at the interface. Several models, most notably the Marshall, Cox, and Evans (MCE) model [1], assume that this shear stress has a constant value over the entire range of relative motion. This shear stress value has typically been used as a curve-fitting parameter in order to

[1] Senior project engineer, Allison Engine Company, Indianapolis, IN 46206-0420; formerly research assistant, School of Mechanical Engineering, Purdue University, West Lafayette, IN 47907-1288.
[2] Professor, School of Mechanical Engineering, Purdue University, West Lafayette, IN 47907-1288.

correlate model predictions to experimental results. Ward and Hillberry [2,3] have modified the MCE model by including a Coulomb friction assumption to account for a distribution of shear stress and have combined this model with a wear model that modifies the shear stress during fatigue cycling as the interface undergoes mechanical wear. In addition, Herrmann and Hillberry [4] have developed a discrete composite model formulation that was used to model crack growth rates in TMCs. In this paper, an approach is developed to utilize the combined Coulomb friction and wear models within the discrete composite model, allowing wear to be calculated separately for each fiber/matrix unit cell. Crack growth predictions are then presented in which the fiber surface roughness was allowed to decrease due to wear over a discrete increment of fatigue crack growth based on bridging conditions determined from the composite model. These predictions were made using a single set of input parameters that were developed without requiring correlation to the experimental crack growth results.

Background of Fiber-Bridging Model

The Coulomb friction-based fiber-bridging model [2,3] considers a unit cell of fiber and matrix cylinders such that the fiber is subjected to an applied axial force, F_a, at the free end, and the composite is subjected to a balancing force at the other end (Fig. 1). The applied force on the fiber is assumed to cause relative motion over some slip length, l, with this motion resisted by a frictional shear stress, τ, at the fiber/matrix interface. The total amount of relative motion is designated u in Fig. 1. From stress equilibrium in the fiber direction, the stress in the fiber decreases with distance from the crack face due to load transfer to the matrix because of the frictional shear stress, τ, or

$$dF(y) = -2\pi r_f \tau(y) \, dy \tag{1}$$

where y is the distance along the fiber. The degree to which fiber bridging affects the composite behavior is based on the rate of load transfer between the constituents. Therefore, the interfacial shear stress plays a very important role in the fiber-bridging process. Kerans and Parthasarathy [5] assumed that the amplitude of surface roughness, A, causes an effective radial stress in addition to thermal residual stress, giving a total clamping stress of

$$\sigma_N(y) = \sigma_r^{res} - \frac{kE_f A(y)}{v_f r_f} \tag{2}$$

where

σ_r^{res} = radial residual stress at interface,
k = material constant

$$= \frac{E_m v_f}{E_f(1 + v_m) + E_m(1 - v_f)},$$

E_f, E_m = elastic moduli of fiber, matrix,
v_f, v_m = Poisson's ratios of fiber, matrix,
r_f = fiber radius, and
$A(y)$ = amplitude of fiber surface roughness (Fig. 2).

Because σ_r^{res} is compressive in nature, the surface roughness raises the magnitude of the clamping stress. Assuming Coulomb friction and including an additional term to the frictional

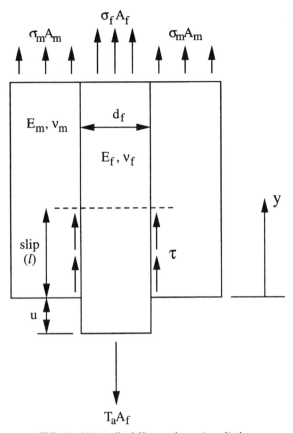

FIG. 1—*Unit cell of fiber and matrix cylinders.*

shear stress used by Kerans and Parthasarathy [5] to account for a constant axial thermal residual stress in the fiber, σ_f^r, of the resulting expression is

$$\tau(y) = -\mu \left[\sigma_N(y) + \left(k \frac{F(y)}{\pi r_f^2} + \sigma_f^r \right) \right] \qquad (3)$$

where μ = friction coefficient and σ_f^r = fiber axial residual stress. For initial loading, it is assumed that $A(y)$ has a constant value along the entire length of the interface, resulting in the following expressions for the relative displacement and total slip length

$$u = -\frac{T_a r_f}{2\mu E_f k} - \frac{r_f}{2\mu E_f k \Omega} \left[T_a(1 - \Omega) + \frac{\sigma_N}{k} - \frac{\sigma_R \Omega}{V_f} \right] \ln \left[\frac{T_a(1 - \Omega) + \frac{\sigma_N}{k} - \frac{\sigma_R \Omega}{V_f}}{T_a + \frac{\sigma_N}{k} - \frac{\sigma_R \Omega}{V_f}} \right] \qquad (4)$$

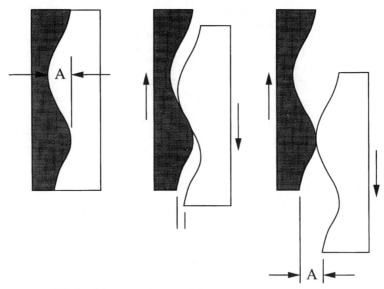

FIG. 2—*Schematic of assumed fiber surface roughness profile.*

$$l = - \frac{r_f \Omega (T_a + 2u\mu E_f k/r_f)}{2\mu k \left[T_a(1 - \Omega) + \dfrac{\sigma_N}{k} - \dfrac{\sigma_R \Omega}{V_f} \right]} \tag{5}$$

where

$$\frac{1}{\Omega} = 1 + \frac{E_f V_f}{E_m(1 - V_f)} = \frac{E_c}{E_m(1 - V_f)}$$

$$\sigma_R = \sigma_m^r \left[\frac{E_m(1 - V_f) + E_f V_f}{E_m} \right] = \sigma_m^r \frac{E_c}{E_m}$$

and

T_a = fiber-bridging traction due to applied force at free end of fiber,
V_f, V_m = fiber, matrix volume fractions,
E_c = elastic modulus of composite, and
σ_m^r = matrix axial residual stress.

Equations 4 and 5 were derived by following the same procedure as that used to develop the MCE model [1], with the exception of the use of Coulomb friction rather than a constant shear stress.

For fatigue conditions, similar expressions were developed with the extra condition that the amplitude of fiber surface roughness was not constant; rather, it had decreased due to mechanical wear of the interface. This wear was calculated using a general expression that was based on the assumption that wear is proportional to the clamping stress times the relative displacement between the two materials. In addition, it was assumed that the fiber surface roughness, $A(y)$, contributes to the fiber radius, and that only the roughness of the fiber is worn. Thus, only the contribution of surface roughness on the total clamping stress, σ_N, in Eq 2 is applied to the wear model. This allows the normal residual stress to exist without causing wear, thus providing the limiting condition that wear ceases when A goes to zero. If the material constants ν_f, E_f, and k are included with the proportionality constant, then

$$\frac{dA\ (y)}{dN} = -W\Delta u(y)A'(y)/r_f \qquad (6)$$

where

   W = dimensionless wear parameter,
   $A'(y)$ = previous roughness amplitude distribution, and
   $\Delta u(y)$ = change in relative displacement.

Note that the y-dependence of Δu is defined by the assumptions of the fiber-bridging model used, which typically result in a maximum value at the matrix crack face, or $y = 0$ [1]. Upon determining the wear following each reversal, the resulting $A(y)$ distribution was substituted into Eq 2 and resulting relationships were developed for the change in the relative displacement, Δu, as well as the reverse slip length, s, that occurs during unloading. These relationships may be found in Ref 3. As fatigue cycling continues, the slip lengths, l and s, change due to continuing wear, and these changes affect the rate of wear that occurs. Thus, it was determined that the resulting clamping stress distribution could be expressed as a series of piecewise quadratic functions, or

$$\sigma_N(y) = \begin{cases} B_1 + B_2y + B_3y^2; & 0 \leq y \leq s_0 \\ B_4 + B_5y + B_6y^2; & s_0 \leq y \leq l_{00} \\ B_7 + B_8y + B_9y^2; & l_{00} \leq y \leq l_0 \\ B_{10}; & y \geq l_0 \end{cases} \qquad (7)$$

where

   s_0 = reverse slip length from previous unloading,
   l_0 = total slip length from previous loading, and
   l_{00} = total slip length prior to l_0,

and the B_i coefficients are dependent on the wear parameter, W, and the roughness amplitude prior to the previous load reversal. The expressions for these coefficients are also reported in Ref 3, as are those used to determine W and the coefficient of friction, μ, which were found empirically to change during fatigue cycling.

Application of Fiber Bridging Model to a Discrete Composite Model

For metals such as titanium, the fatigue crack growth rate, da/dN, can be expressed in terms of the applied stress-intensity range, ΔK, using the following relation [6]

$$\frac{da}{dN} = f(\Delta K) \qquad (8)$$

where $f(\Delta K)$ is determined experimentally for each material. While the stress-intensity range may be calculated for various geometries of unreinforced metal using linear elastic fracture mechanics, the bridging fibers in composites affect the fatigue crack growth rate. Thus, this must also be considered when calculating the composite stress intensity range, ΔK_c.

The effect of fiber bridging may be incorporated by applying a traction to the fibers in the crack wake. In the case of no fiber bridging, the crack opens due to the applied load, σ_∞ (Fig. 3a). The bridging traction, $T(x)$, allows the fibers to remain intact and bridge the crack (Fig. 3b). Marshall, Cox, and Evans [1] first modeled these bridging tractions as an average pressure, $p(x)$, applied to the surface, where (Fig. 3c)

$$p(x) = V_f T(x) \qquad (9)$$

and V_f is the fiber volume fraction. This concept is currently used in virtually all fiber-

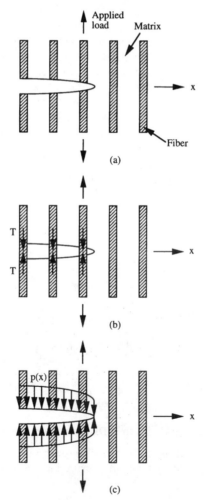

FIG. 3—(a) *Crack with no fiber bridging,* (b) *fiber bridging simulated with applied traction,* T, *and* (c) *pressure on crack surface to model fiber bridging.*

bridging models. The use of an average pressure enables the composite stress-intensity factor, K_c, to be calculated, assuming that the crack surface is subjected to a net pressure of $[\sigma_\infty - p(x)]$. Then for a straight, through-thickness crack, K_c can be found from [7]

$$K_c = 2 \int_0^a [\sigma_\infty - p(x)] \, G(a, x, W) \, dx \qquad (10)$$

where

   $2a$ = crack length,
   W = specimen width, and
   G = weight function to account for geometry.

Over the unbridged portion of the crack, this pressure, $p(x)$, is equal to zero. Several methods for determining $p(x)$ over the bridged region have been discussed in the literature, most of which use a continuum composite formulation, resulting in highly numerical solutions.

Recently, Herrmann and Hillberry [4] have introduced a discrete composite model that utilizes classical shear lag theory [8] while including the effect of matrix axial stiffness, neglecting, however, effects of shear deformation. Thus, this model requires that no longitudinal matrix splitting occurs, allowing longitudinal and transverse displacements to be decoupled. Including the effects of bridging pressure for the case when fiber bridging occurs, and also superposing axial residual stress as suggested by Cox and Marshall [9], the discrete model results in the following system of equations

$$[D]\{U\} = \{-1\} - \{\sigma_R/\sigma_\infty\} + \{\bar{p}\} \qquad (11)$$

where

   $[D]$ = constant coefficient matrix,
   $\{U\}$ = vector of nondimensional crack opening displacement,
   $\bar{p} = p/\sigma_\infty$, and
   σ_∞ = remote applied stress (Fig. 3a).

Note that $[D]$ has been derived in Ref 4 for a center-cracked composite panel with a total of $2N + 1$ broken cells (Fig. 4), resulting in a symmetric $(N + 1) \times (N + 1)$ matrix with constant elements. In addition, the $\{U\}$ vector is nondimensionalized such that $U = u/\psi$ ($U = u/2\psi$ for Cell 0), leading to [4]

$$\psi = \sigma_\infty \left[\frac{b^2}{G_m E_c} \right]^{1/2} \qquad (12)$$

where b = the width of the composite cell, and G_m = the shear modulus of the matrix.

Two distinct advantages result from using this discrete formulation. The first is that this model is capable of handling varying bridging geometries in which any combination of bridged and unbridged fibers in the cracked region may be considered. The second is that different shear stresses may be applied to each individual fiber/matrix interface. Thus, wear may be determined independently for each fiber.

To incorporate the fiber-bridging relationships discussed earlier, Eq 4 may be written in terms of p using Eq 9, and the resulting u-p relationship may be applied to Eq 11. However, it is very difficult to invert Eq 4 to solve for p (or T); thus, it is necessary to write the $\{U\}$ vector in terms of $\{\bar{p}\}$. Note that Eq 4 was derived for the condition of a fiber with a

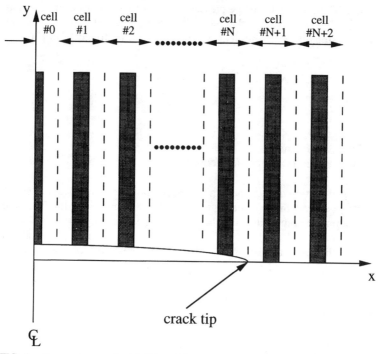

FIG. 4—*Composite panel with* 2N + 1 *broken cells (symmetric about centerline).*

constant surface roughness amplitude, thus does not include the effects of wear. The resulting equations for both the monotonic and fatigue conditions and their solution procedure is supplied in the next section.

The fiber-bridging relationships under worn conditions were developed for both load control and displacement control situations [3]. For the case of a fiber/matrix cell in a composite panel, the displacement control assumption provides a closer approximation due to the material around the unit cell constraining the crack-opening displacement. The prevailing consequence of the displacement control assumption is that for constant u_{max} and Δu the maximum bridging traction decreases every cycle due to wear, assuming that all global conditions, such as matrix crack length, remain constant. The result of this is that the increase in total slip length may be assumed to be very small, thus the increment of $l_{00} \leq y \leq l_0$ may be eliminated from Eq 7 to simplify the expression for the clamping stress distribution.

Taking the resulting fiber bridging relationship for the monotonic case and writing it in terms of $\{U\}$ and $\{\bar{p}\}$ gives

$$U_{max}^n = \alpha(\beta_5 V_f - \bar{p}_{max}^n) - \frac{\alpha}{\Omega}\left[\bar{p}_{max}^n(1 - \Omega) + \beta_4 V_f\right]\ln\left[\frac{\bar{p}_{max}^n(1 - \Omega) + \beta_4 V_f}{\bar{p}_{max}^n + \beta_6 V_f}\right] \quad (13)$$

where

$$\alpha = \frac{r_f \sigma_\infty}{2\mu E_f V_f k\psi}$$

and n refers to the individual cell number as shown in Fig. 4. In addition, the β_i^n terms are functions of the B_i coefficients of Eq 7 and of the wear parameter, W. Also note that for $n = 0$, the expression in Eq 13 is divided by 2.

If complete bridging occurs, then Eq 13 may be used directly in Eq 11, and the resulting series of equations may be solved for $\{\bar{p}\}$. This may not be done, however, if any broken fibers exist in the crack wake. In this case, the following function is defined

$$\xi_n = (1 - \gamma_n)U_n + \gamma_n \bar{p}_n \tag{14}$$

where $\gamma_n = 0$ for broken fibers and $\gamma_n = 1$ for unbroken fibers. Substituting ξ_n for $\bar{p}_n$ in Eq 13 allows for the final governing equation to be written as

$$[D]\{U\} = \{-1\} - [\Gamma]\{\sigma_R/\sigma_\infty\} + [\Gamma]\{\xi\} \tag{15}$$

where the $[\Gamma]$ matrix is a diagonal matrix with $\Gamma_{nn} = \gamma_n$, and $\{U\}$ is now defined as

$$\{U\} = [[I] - [\Gamma]]\{\xi\} + [\Gamma]\{U(\xi)\} \tag{16}$$

and $\{I\}$ is the identity matrix. Equation 15 is then solved for the $\{\xi\}$ vector using the Newton-Raphson iteration. If $\gamma_n = 0$, then $U_n = \xi_n$ and $\bar{p}_n = 0$, and, if $\gamma_n = 1$, then $\bar{p}_n = \xi_n$ and $U_n = U(\bar{p}_n)$ from Eq 13.

For the fatigue condition, the fiber-bridging relationships do not provide a closed-form solution for the reverse slip length, s, during unloading [3]. However, it is possible to solve for the change in bridging pressure for the given cell as a function of s. Thus, it becomes logical to use the resulting fiber bridging relationships to iterate on s, where the governing equation for fatigue using the discrete composite model is [4]

$$[D]\{\Delta U\} = \{-1\} + \{\Delta\bar{p}\} \tag{17}$$

Selecting the appropriate relationships for fatigue conditions under displacement control from Ref 3, and rewriting in nondimensional terms, results in the following expressions for $\{\Delta U\}$ and $\{\Delta\bar{p}\}$

$$\Delta U_n = \left[\frac{\alpha'}{e^{-2\mu ks_n/r_f} - 1 + \Omega}\right]\left[\sigma_1^n\left\{2 + \left[\Omega - 2 + (1 - \Omega)\left(\frac{2\mu ks_n}{r_f}\right)\right]e^{2\mu ks_n/r_f}\right\}\right.$$

$$+ \sigma_2^n\left\{\left[\Omega - (1 - \Omega)\left(\frac{2\mu ks_n}{r_f}\right)\right]e^{-2\mu ks_n/r_f}\right\}$$

$$+ V_f\left\{-\frac{2}{k}[B_4^n + B_5^n s_n + B_6^n s_n^2] - \frac{B_6^n r_f^2}{\mu_2 k_3} - \frac{2}{V_f}\sigma_3^n\right\}[e^{-2\mu ks_n/r_f} - 1 + \Omega]$$

$$\left. - \frac{r_f V_f}{\mu k^2}[B_5^n + 2B_6^n s_n]\left\{e^{-2\mu ks_n/r_f} - 1 + (1 - \Omega)\left(\frac{2\mu ks_n}{r_f}\right)\right\}\right] \tag{18a}$$

$$\Delta \bar{p} = \left\{ \frac{1}{\Delta \sigma_\infty [e^{-2\mu k s_n/r_f} - 1 + \Omega]} \right\} \left[\sigma_2^n e^{-2\mu k s_n/r_f} - \sigma_1^n e^{2\mu k s_n/r_f} + \frac{V_f r_f}{\mu k^2} (B_5^n + 2B_6^n s_n) \right] \quad (18b)$$

where

$$\alpha' = \frac{r_f}{2\mu E_f V_f k \Omega \psi} \quad (\psi \text{ is same as in Eq 12, but with } \Delta \sigma_\infty),$$

$$\sigma_1^n = \bar{p}_{max}^n \sigma_\infty + V_f(\beta_1^n - \beta_{11}^n e^{-2\mu k s_0^n/r_f}),$$

$$\sigma_2^n = \bar{p}_{max}^n \sigma_\infty + V_f(\beta_7^n - \beta_{12}^n e^{2\mu k s_0^n/r_f}),$$

$$\sigma_3^n = \bar{p}_{max}^n \sigma_\infty - \sigma_R \Omega$$

Once again, for $n = 0$, the expression for ΔU_0 must be divided by 2.

As with the monotonic case, the above procedure is not necessary for the cells that contain unbroken fibers. Therefore, a function similar to that of Eq 14 is defined as

$$\xi_n' = (1 - \gamma_n)U_n + \gamma_n s_n \quad (19)$$

This changes the governing equation to

$$[D]\{\Delta U\} = \{-1\} + [\Gamma]\{\Delta \bar{p}(s_n)\} \quad (20)$$

where

$$\{\Delta U\} = [[I] - [\Gamma]]\{\xi'\} + [\Gamma]\{\Delta U(\xi')\}$$

Equation 20 may now be solved using Newton-Raphson iteration as before.

Procedure for Predicting History-Dependent Crack Growth Behavior

The modeling and procedures developed to this point have been done for the purpose of generating history-dependent crack growth predictions. Prior research discussed earlier has provided successful correlation to experimental data at given crack lengths with the use of a constant shear stress as a curve-fitting parameter. These methods have fallen short, however, in their ability to make outright predictions of crack growth rates and bridging relationships given only the initial geometry and the loading profile. This section provides a simple approach to do this in which the wear behavior defines the history-dependent behavior.

Consider a center-crack composite panel, such as the one shown in Fig. 5, with fibers parallel to the remote load, which is in the y-direction. The panel contains an initial notch with the tip at the edge of Cell N, thus which covers a width of $2N + 1$ cells (Fig. 4), each of which contains a broken fiber that was cut when forming the notch. Next, assume that a precrack is formed that carries through Cell $N + 1$, which contains a bridging fiber that has not yet been worn. The procedure used to predict history-dependent behavior, then, is as follows. Given the remote stresses, σ_∞ and $\Delta \sigma_\infty$, determine the composite load response and bridging behavior from the discrete model governing Eqs 15 and 17. From the resulting closing pressure, $p(x)$, determined from the bridging relationships and the updated crack length, the net stress-intensity factor including bridging may be determined from Eq 10. In order to relate the resulting composite stress-intensity factor to the stress intensity seen by the matrix, the approach used by McMeeking and Evans [10] was assumed. This approach states that the composite and the matrix are subjected to the same stress intensity, or

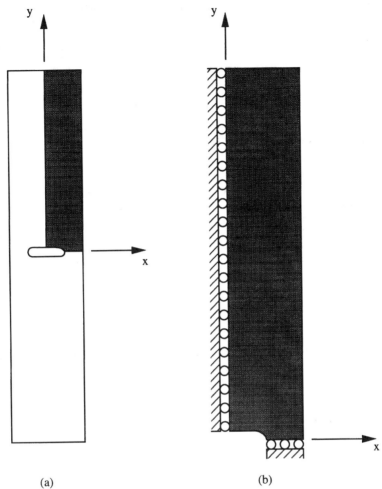

(a)   (b)

FIG. 5—(a) *Specimen geometry, and* (b) *portion modeled due to symmetry.*

$$\Delta K_m = \Delta K_c \tag{21}$$

Other approaches considered to relate the stress-intensity factors were the strain compatibility approach of Marshall et al. [1] and the energy balance approach of McCartney [11]. Because the stress-intensity factors were calculated from crack-opening displacements (CODs) obtained using the discrete composite model, the relation in Eq 21 was determined to be consistent with these calculations, while the other approaches are generally considered to be more accurate when using a continuum model. The crack growth rate, da/dN, for the composite may then be obtained from matrix crack growth rate data.

As the matrix crack advances through the fiber/matrix cell, the interface of that cell will begin to cycle when the crack reaches the point when the fiber is in the crack wake, which may be well before the crack tip reaches the other side of the entire cell, as shown in Fig. 4. Therefore, it was assumed that the calculated response is the same as if the crack tip was

TABLE 1—*Mechanical and thermal properties of fiber and matrix at room temperature.*

Property	Fiber (SCS-6)	Matrix (Timetal®21S)
Young's modulus, E (GPa)	393 [17]	112 [18]
Poisson's ratio, v	0.25 [17]	0.35 [19]
Yield stress, σ_Y (MPa)	N/A	1050 [19]
CTE* (mm/mm/°C)	3.794 × 10⁻⁶ [17]	9.476 × 10⁻⁶ [20]
Initial roughness, A_{in}		0.3 mm [21]

* CTE values are average values for ΔT from 621.1 to 21.1°C.

at the edge of the fiber. Once da/dN is determined from this response, the number of cycles required to grow the crack from the edge of the fiber to the edge of the cell may be found, assuming that da/dN remains constant throughout those cycles. Due to the discrete nature of the composite model, it is not possible to analyze wear on a cycle-by-cycle basis, thus it was assumed that the fiber remains in a displacement-control state as the crack advances. The wear is then determined either for this number of cycles or until a steady-state condition has been reached. At this point, the load response is determined from the discrete model, and the da/dN calculation will be made for the case where the fiber nearest the crack tip has been worn. This value of da/dN is taken as the prediction for the current stress intensity range.

For this value of da/dN, the number of cycles required to extend the matrix crack to the far edge of the next fiber is calculated, and the wear is determined for each fiber already bridging the crack. Once the global load response is calculated, the previous steps are repeated. Thus, this procedure requires that the wear cycling be performed twice for each increment of crack growth through a unit cell.

Comparison of Predictions to Experimental Data

As a test of the procedure described in the previous section, crack growth predictions were made for the SCS-6/Timetal®21S (Timet Corp., Henderson, NV) composite system for the conditions of the experiments performed by Herrmann and Hillberry [12]. The thermal and mechanical properties of the composite constituents used in the predictions are given in Table 1. Using these properties, the fiber and matrix thermal residual stresses at the interface were calculated using the concentric cylinder model as before. The resulting residual stresses used in the predictions are supplied in Table 2. The specimens used by Herrmann and Hillberry [12] were center-cracked panels of [0]₄ composite laminates with a fiber volume fraction of $V_f = 0.36$. The nominal geometry of these specimens was used to make the predictions (Table 3). The tests were run with a stress ratio of $R = 0.1$ at three different levels: $\Delta\sigma_\infty = $ 200, 300, and 392 MPa. Predictions were made for all three stress levels. The reference

TABLE 2—*Thermal residual stresses at fiber/matrix interface at room temperature.*

Residual Stress, σ_{res}	Fiber	Matrix
Radial	−204 MPa	−204 MPa
Axial	−579 MPa	326 MPa

TABLE 3—*Geometry of specimen, $V_f = 0.36$ [12].*

Dimension	Magnitude
Width, w (mm)	25.5
Thickness, t (mm)	0.216
Initial notch length, a_0 (mm)	2.57
Cell width, b (mm)	0.204
Initial cell index, N	12
Fiber diameter, d_f (mm)	0.142

crack growth data for the Timetal®21S matrix was obtained from experimental results of John and Jira [13].

Fully Bridged Case

In the experiments performed at the lowest stress level of $\Delta\sigma_\infty = 200$ MPa, all fibers remained intact that were not initially cut, resulting in full fiber bridging outside of the notch [12]. This caused a steady decrease in crack growth rate as the crack continued to advance around additional bridging fibers. The growth rate predictions correlated very well with the trend of the experimental results. Unfortunately, the extremely steep slope of the da/dN-ΔK curve greatly magnifies any small differences in predicted and experimental values (Fig. 6). This may be seen more clearly when comparing the prediction with the experimental results on an a versus N basis (Fig. 7).

Periodically throughout the experiments run by Herrmann and Hillberry [12], COD measurements were taken using an Elber gage [14]. The Elber gage is an effective point extensometer with a 1.5-mm gage length, and the two points are placed on opposite sides of the crack. Due to additional strain in the composite between the matrix crack face and the Elber gage points, Herrmann and Hillberry [12] used the discrete formulation of the shear lag model to include the additional displacement in order to compare the predictions with the measured displacements.

To predict Elber gage displacement when the matrix crack is at a certain length, the load response may be determined from the discrete composite model for a crack extending through the number of cells that roughly correlates to the desired crack length, and then the COD predictions may be modified for the additional displacement using the aforementioned procedure. Because the entire history was known, these predictions were made with the fibers having been worn up to that point. For the case of $\Delta\sigma_\infty = 200$ MPa, model predictions correlated extremely well with the measured Elber gage displacements. The "left" side of the crack was measured to be 3.85 mm, and the "right" side was measured to be 3.87 mm (Fig. 8). The prediction was made for the bridging geometry for a crack length of 3.86 mm.

For this simulation, the average shear stresses and the predicted slip lengths were investigated. Due to Poisson's effect, $\tau_2(y)$ for unloading was typically greater than $\tau_1(y)$ for loading. However, since the unloading occurs only over the reverse slip length, and τ_1 acts over a distance up to l, the average values of τ_2 were usually lower than those for τ_1. For this case, τ_1 and τ_2 ranged from 25 to 13.4 MPa and 23.8 to 12.5 MPa, respectively, depending on the fiber number for the longest crack length. The average shear stress value used by Herrmann and Hillberry [12] to curve fit the model to the data for this particular test was $\tau = 12$ MPa. With regard to the slip lengths, the predicted values ranged from approximately 0.9 to 1.9 mm. This compares well with the slip lengths observed experimentally, in which l typically varies between about 1 and 3 mm [15].

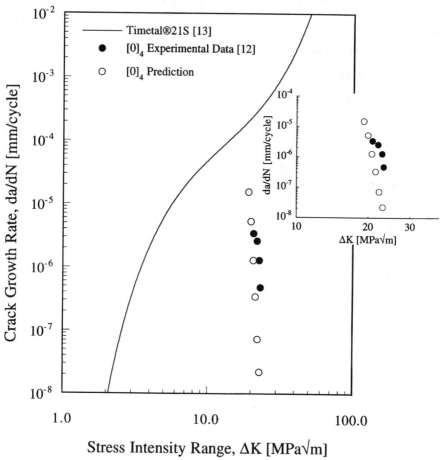

FIG. 6—*Comparison of predicted composite crack growth rate with experimental data for full bridging outside of notch: $\Delta\sigma_\infty$ = 200 MPa. Close-up view shown to emphasize slope of* da/dN *values.*

Partially Bridged Case

In the experiments performed at the two higher stress levels, the fiber stresses were sufficient to cause failure of some fibers in the crack wake, leading to a partially bridged region outside of initial notch [12]. These fiber failures occurred at somewhat random locations within the crack wake due to the variation in strengths of the SCS-6 fibers. Prior to fiber failure, the predictions correlated very well with the experimental data for completely bridged conditions at both stress levels (Figs. 9 and 10). Note in these figures that the point at which the experimental data changed from decreasing to increasing *da/dN* values represents the time when fiber failure began to occur.

Since the maximum fiber stresses are calculated in each prediction of load response, the fiber strength may be used to predict fiber failure. Gambone and Wawner [16] have found that the mean value of fiber strength for the SCS-6 fibers is approximately σ_{ult} = 3665 MPa. However, the predicted fiber stresses from the model never reached this level. Thus, to predict the effects of fiber failure, it was assumed that a weak fiber must break first in order to cause

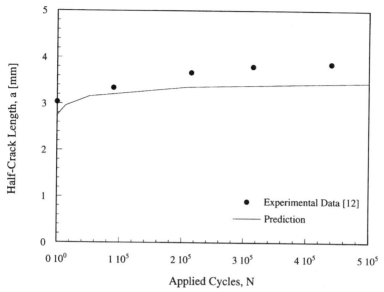

FIG. 7—*Comparison of predicted and experimental* a *versus* N *results for complete bridging outside of the notch:* $\Delta\sigma_\infty = 200$ *MPa.*

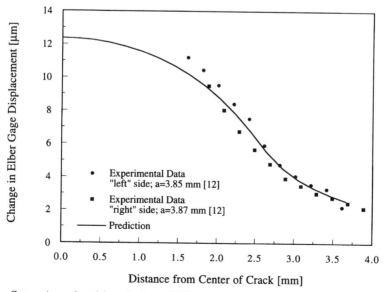

FIG. 8—*Comparison of model prediction and Elber gage displacements for complete bridging outside of the notch:* $\Delta\sigma_\infty = 200$ *MPa.*

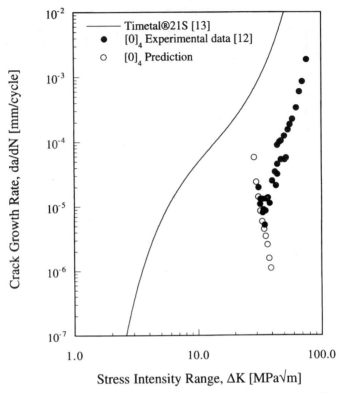

FIG. 9—*Comparison of predicted composite crack growth rates for complete fiber bridging with experimental data including fiber failure:* $\Delta\sigma_\infty = 300$ *MPa.*

the remaining fibers to carry the extra load and reach their failure stress. To do this, the fiber strength used in the model was reduced to a value that would cause fiber failure near the point when the crack growth rates began to increase. For the case with $\Delta\sigma_\infty = 300$ MPa, this change resulted in a versus N predictions that correlated very well with the experimental data (Fig. 11). It is seen that as fibers began to fail, the bridging effect decreased and the crack growth rate started to increase. To achieve this, the fiber strength was decreased to 825 MPa, which is much lower than the expected value. It should be noted that this was really only necessary to cause the first bridging fiber to fail. The predicted fiber stresses at the times when failure was defined were typically greater than 825 MPa, although they were still much lower than expected. For the highest stress range of $\Delta\sigma_\infty = 392$ MPa, a fiber strength of $\sigma_{ult} = 1320$ MPa resulted in a strong correlation between prediction and experiment (Fig. 12). Once again, failure of most fibers was predicted at stress levels much higher than the assumed strength.

When predicting the Elber gage displacements for partially bridged cases, one must be aware that the method used above will typically predict fiber failures in chronological order. In other words, the first bridging fiber closest to the initial notch will usually be the next fiber to fail. Since this is not necessarily the case in reality, knowledge of the actual bridging geometry compared to the predicted is important to understand the predictions. For the stress range of $\Delta\sigma_\infty = 300$ MPa, Elber gage data were available at a crack length of $a = 4.04$ mm. The second and third fibers from the notch were broken in the given specimen, with the first

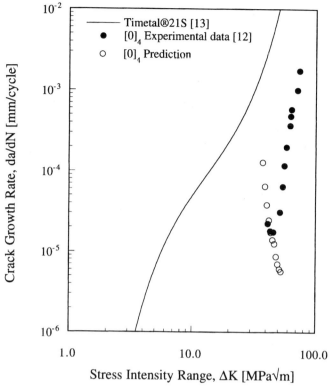

FIG. 10—*Comparison of predicted composite crack growth rates for complete fiber bridging with experimental data including fiber failure:* $\Delta\sigma_\infty = 392$ *MPa.*

bridging fiber still intact [*12*]. From the model with fiber failures considered, the first two fibers had been predicted to fail, suggesting that the predicted COD may be slightly high. It turned out that while the prediction was in fact slightly higher than measured, the difference was very small (Fig. 13). For the final case of $\Delta\sigma_\infty = 392$ MPa, the Elber gage displacements were measured at a crack length of $a = 5.31$ mm for a given specimen. While the bridging geometry was not available, the model predicted that all but the last two fibers had failed. For this case, the predicted displacements were much less than that of the measured values (Fig. 14). One possible explanation could be that one or both of those two fibers had also failed prior to the measurements.

Discussion of Results

The wear-based Coulomb friction fiber-bridging model can be a useful tool for making history-dependent fatigue crack growth predictions. In most cases, the model performed very well in predicting both crack growth rates and crack-opening displacements for various loading conditions with a single set of input parameters. It should be noted that the discrete composite model using a constant shear stress assumption also performed very well for this set of tests [*12*]. To do so, however, a separate shear stress value was required to fit the model to the experimental data. The distinct advantage with the current approach is that all material properties were determined or derived separately [*3*]; thus, the model predictions

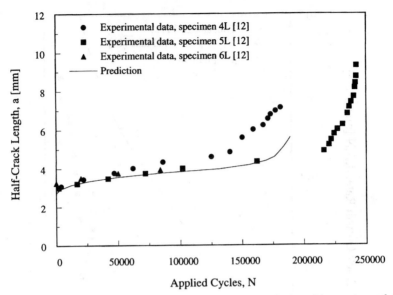

FIG. 11—*Comparison of Coulomb friction model a versus N prediction with experimental data, with fiber failure included in prediction:* $\Delta\sigma_\infty = 300$ *MPa.*

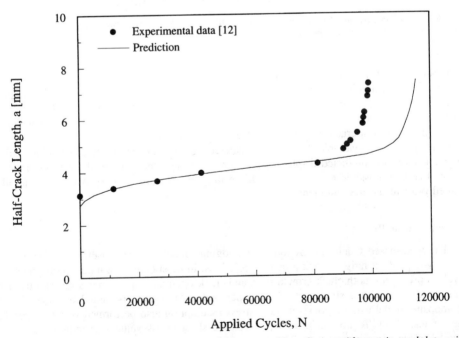

FIG. 12—*Comparison of Coulomb friction model a versus N prediction with experimental data, with fiber failure included in prediction:* $\Delta\sigma_\infty = 392$ *MPa.*

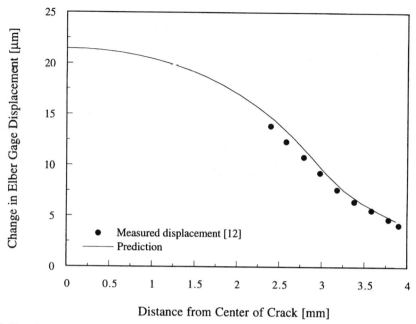

FIG. 13—*Comparison of model prediction and Elber gage displacements for partial bridging outside of the notch: a = 4.04 mm, $\Delta\sigma_\infty$ = 300 MPa.*

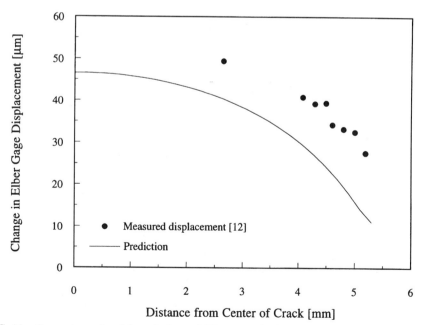

FIG. 14—*Comparison of model prediction and Elber gage displacements for partial bridging outside of the notch: a = 5.31 mm, $\Delta\sigma_\infty$ = 392 MPa.*

were not dependent of the results of the crack growth tests. It should also be noted that predictions were also made using the two alternate approaches to Eq 21 discussed earlier. It was found that the current approach provided the best correlation between the predictions and experimental results, thus justifying its use.

The main area of concern seems to be in accurately predicting crack growth rates at stress intensity ranges approaching the threshold value for the matrix material. The most likely reason for this problem is the use of discrete increments of crack extension. With the discrete formulation, it was assumed that da/dN was constant throughout crack advancement over an entire unit cell. At such low crack growth rates, this required a very large number of cycles to extend the crack the width of a cell. Due to wear of the interface, it is likely that as the number of cycles increases, the validity of the displacement control assumption decreases. Thus, changes in the global load-displacement response during the course of crack extension through a cell is probably affecting the crack growth rates, and this effect cannot be accounted for with the discrete approach. These changes would also affect the wear behavior due to modified peak displacements and tractions. At higher crack growth rates, this does not appear to be a problem in that the predictions seemed to improve as the applied stress increased.

Conclusions

An approach has been developed that includes the history-dependent effects of mechanical wear of the fiber/matrix interface on fatigue crack growth behavior in TMCs. It was found that this approach provides predictions that correlate very well with experimental data at higher crack growth rate levels. At lower levels, the model underpredicts the crack growth rates, which may be due to the fact that the discrete model formulation does not allow for the model to account for changes in the global composite behavior that may occur over a large number of applied cycles. In addition, this approach calculated fiber stress values much lower than expected to predict fiber failure during fatigue crack tests. While this may be due in part to the stress calculations from the model, these results also suggest that the strength of the reinforcing fibers in the material that was tested may decrease during fatigue cycling. Overall, it is felt that for some specific applications, especially involving relatively high crack growth rates, this approach can be a powerful tool in making predictions for different conditions based only on a single set of independently developed input values.

Acknowledgment

Support for this research was provided through NASA Langley Research Center. The contract monitor was Dr. W. S. Johnson.

References

[1] Marshall, D. B., Cox, B. N., and Evans, A. G., "The Mechanics of Matrix Cracking in Brittle-Matrix Fiber Composites," *Acta Metallurgica*, Vol. 33, No. 11, 1985, pp. 2013–2021.
[2] Ward, G. T. and Hillberry, B. M., "A Wear Model for Fiber/Matrix Interface Fatigue in Continuous Fiber Titanium Matrix Composites Under Fiber Bridging Conditions," *Fatigue 96 Proceedings*, G Lütjering and H. Nowack, Eds., Berlin, Germany, May 1996, pp. 1457–1462.
[3] Ward, G. T., "Fatigue Crack Growth in Unidirectional Titanium Matrix Composites: A Wear-Based Approach to Modeling," Ph.D. dissertation, Purdue University, West Lafayette, IN, May 1996.
[4] Herrmann, D. J. and Hillberry, B. M., "A New Approach to the Analysis of Unidirectional Titanium Matrix Composites with Bridged and Unbridged Cracks," *Journal of Engineering Fracture Mechanics*, Vol. 56, No. 5, March 1997.

[5] Kerans, R. J. and Parthasarathy, T. A., "Theoretical Analysis of the Fiber Pullout and Pushout Tests," *Journal of the American Ceramics Society*, Vol. 74, No. 7, 1991, pp. 1585–1596.

[6] Hertzberg, R. W., *Deformation and Fracture Mechanics of Engineering Materials*, John Wiley and Sons, New York, 1976.

[7] Tada, H., Paris, P. C., and Irwin, G. R., *The Stress Analysis of Cracks Handbook*, Del Research Corporation, St. Louis, MO, 1985.

[8] Hedgepath, J. M., "Stress Concentrations in Filamentary Structures," NASA TN D-882, National Aeronautics and Space Administration, Washington, DC, May 1961.

[9] Cox, B. N. and Marshall, D. B., "Crack Bridging in the Fatigue of Fibrous Composites," *Fatigue and Fracture of Engineering Materials and Structures*, Vol. 14, No. 8, 1991, pp. 847–861.

[10] McMeeking, R. M. and Evans, A. G., "Matrix Fatigue Cracking in Fiber Composites," *Mechanics of Materials*, Vol. 9, 1990, pp. 217–227.

[11] McCartney, L. N., "Mechanics for the Growth of Bridged Cracks in Composite Materials: Part I. Basic Principles," *Journal of Composites Technology & Research*, Vol. 14, No. 3, Fall 1992, pp. 133–154.

[12] Herrmann, D. J. and Hillberry, B. M., "Effects of Fiber Bridging and Fiber Fracture on Fatigue Cracking in a Titanium-Matrix Composite," *Composite Materials: Fatigue and Fracture (Sixth Volume), ASTM STP 1285*, E. A. Armanios, Ed., American Society for Testing and Materials, West Conshohocken, PA, 1997, pp. 103–125.

[13] John, R. and Jira, J. R., USAF Materials Laboratory, personal correspondence with D. J. Herrmann, March 1994.

[14] Elber, W., "Fatigue Crack Closure Under Cyclic Tension," *Engineering Fracture Mechanics*, Vol. 2, 1970, pp. 37–45.

[15] Bakuckas, J. G., Jr. and Johnson, W. S., "Application of Fiber Bridging Models to Fatigue Crack Growth in Unidirectional Titanium Matrix Composites," *Journal of Composites Technology & Research*, Vol. 15, No. 3, Fall 1993, pp. 242–255.

[16] Gambone, M. L. and Wawner, F. E., "The Effect of Elevated Temperature Exposure of Composites on the Strength Distribution of the Reinforcing Fibers," *Proceedings*, MRS symposium on Intermetallic Composites III, Spring, 1994.

[17] Mirdamadi, M. and Johnson, W. S., "Stress-Strain Analysis of a $[0/90]_{2S}$ Titanium Matrix Laminate Subjected to a Generic Hypersonic Flight Profile," NASA TM-107584, National Aeronautics and Space Administration, Washington, DC, March 1992.

[18] Ward, G. T., Herrmann, D. J., and Hillberry, B. M., "Stress-Life Behavior of Unnotched SCS-6/Ti-β21S Cross-Ply Metal Matrix Composites at Room Temperature," *Fatigue 93 Proceedings*, J.-P. Bailon and J. I. Dickson, Eds., Montreal, Canada, May 1993, pp. 1067–1072.

[19] Ashbaugh, N., personal correspondence, June 1991.

[20] Mirdamadi, M., personal correspondence, June 1992.

[21] Mackin, T. J., Warren, P. D., and Evans, A. G., "Effects of Fiber Roughness on Interface Sliding in Composites," *Acta Metallurgica et Materialia*, Vol. 40, No. 6, 1992, pp. 1251–1257.

Peter J. Joyce[1] and Tess J. Moon[2]

Compression Strength Reduction in Composites with In-Plane Fiber Waviness

REFERENCE: Joyce, P. J. and Moon, T. J., "Compression Strength Reduction in Composites with In-Plane Fiber Waviness," *Composite Materials: Fatigue and Fracture, Seventh Volume, ASTM STP 1330,* R. B. Bucinell, Ed., American Society for Testing and Materials, 1998, pp. 76–96.

ABSTRACT: The effect of fiber waviness, which develops during the processing and manufacture of fiber-reinforced composite structures, on compressive failure was investigated. Analytical and experimental evidence has shown that out-of-plane waviness (also referred to as layer or ply waviness) is a major contributing factor in compressive strength reduction; however, there is a paucity of data concerning the effects of in-plane waviness or wrinkling on the compressive response of composites. In this paper, we present data from a series of compression tests examining the effects of varying levels of in-plane fiber waviness. These tests used a novel combined shear/end loading compression test fixture (WTF combined loading compression test fixture) in order to ameliorate problems typically associated with pure end-loading (brooming and end-damage) and pure shear loading (tab debonding and high stress concentrations due to discontinuity stresses). The fixture performed adequately when testing wavy specimens, but we experienced repeated tab failures in the non-wavy specimens. The compression test results exhibit a distinct linear trend of decreasing compressive strength with increasing waviness severity as represented by the maximum off-axis angle of the wavy fibers. Optical microscopy revealed that kink bands, leading to catastrophic failure, initiate at the most severe fiber misorientation sites in the wavy regions.

KEYWORDS: composite materials, in-plane fiber waviness, layer waviness, ply waviness, finite element analysis, compression testing

Introduction

Outline and Motivation

A great deal of effort has gone into developing suitable manufacturing techniques for fiber-reinforced composites. While raw material costs continue to plague the advanced composites market, composites manufacturers have brought increasing automation and sophistication on-line in recent years, which has yielded higher quality parts with increased consistency. Still, a number of manufacturing issues cloud what promises to be a bright future for advanced composites, chiefly process-induced defects such as fiber waviness, fiber misalignment, porosity, and so forth [1,2].

The objective of this research is to highlight the effect of process-induced waviness on the compressive response of unidirectional fiber-reinforced composites; the effects of layer (or out-of-plane) waviness on the mechanical response of composites are well documented

[1] Graduate research assistant, Materials Science & Engineering, The University of Texas at Austin, Austin, TX 78712.
[2] Associate professor, Mechanical Engineering Department, The University of Texas at Austin, Austin, TX 78712.

elsewhere. Our work investigates the compression response of thin composite laminates with in-plane fiber waviness. Our discussions with others in the composites industry as well as a survey of the literature suggests this type of waviness is both rare and often overlooked; in many cases, it is assumed to be a secondary effect in the presence of larger-scale out-of-plane waviness. There are two principal reasons for this: (1) there have been no practical, reliable tools developed in the open literature for characterizing this type of fiber waviness and (2) the driving mechanisms that cause this type of waviness during composites processing have not been studied in great detail until recently [3,4]. In conjunction with our own in-house investigation of the development of process-induced waviness, we decided to take a closer look at the defect criticality of in-plane fiber waviness in composite laminates. The test results and fractography results are collected for comparison with data in the literature.

Definition of Fiber Waviness

Following the definition set forth in our earlier work [5], we distinguish between fiber misalignment and fiber waviness according to geometry and, to a lesser degree, the point of origin in composites processing. Fiber misalignment is the partial or localized misorientation of the reinforcing fibers with respect to the nominal fiber axis (Fig. 1). Fiber misalignment is often used more to describe the spatial orientation imperfections in the fibrous reinforcement, commonly attributed to the prepregging process or ply layup error, whereas the sinusoidal-like deformations commonly induced during processing are usually referred to as fiber waviness (Fig. 2). Fiber waviness is usually modeled as a sinusoidal wave in the form $y = A \sin(2\pi x/L)$, where the mean fiber position is parallel to the nominal fiber axis (x), and y represents the transverse deformation of the fibers.

Fiber waviness can result from a variety of manufacturing-induced phenomena; a thorough review of the literature on process-induced fiber waviness and waviness-inducing mechanisms is given by Bhalerao [3], as well as Kugler and Moon [4]. For the most part, the current state of the art in composites manufacturing precludes the manufacture of fiber-reinforced composites with perfect alignment and good control of fiber waviness, since the mechanisms which cause fiber waviness are—at best—poorly understood by most in the composites community.

FIG. 1—*Photomicrograph of fiber misalignment in a carbon-fiber-reinforced thermoplastic composite laminate (fiber diameter = 7 μm).*

FIG. 2—*Photomicrograph of fiber waviness in a carbon-fiber reinforced thermoplastic composite laminate (fiber diameter = 7 μm).*

Fiber waviness can be categorized as either in-plane or out-of-plane waviness. Out-of-plane waviness, also known as layer waviness or ply waviness, involves the cooperative motion of multiple plies through the thickness of the laminate. In-plane waviness involves the cooperative deformation of many fibers within the plane of the laminate (see Fig. 2).

Background

Effects of Fiber Waviness on Composite Mechanical Response

Compressive Failure—Compressive failure mechanisms in fiber-reinforced composite materials present unique challenges to the development of a comprehensive theory for predicting compression failure. The consensus in composite materials research is that compression failure is triggered by buckling instability, or microbuckling [6]. This instability is thought to initiate in regions of local fiber misalignment [7] and to propagate, causing compressive failure by kink banding or shear crippling [8–11]. For a more detailed and thorough review of the literature on microbuckling and kinking of composites, the reader is referred to the studies of Hahn and Williams [9] and Schultheisz and Waas [12,13].

Many researchers have observed decreased compressive strength with increasing fiber waviness [14–20]. Mrse and Piggott [14] observed different levels of process-induced in-plane fiber waviness in thermoplastic composites fabricated from different prepreg materials; they were further able to show that increasing fiber divagation degrades compressive strength. Camponeschi [15] used optical microscopy to quantify lamina or out-of-plane waviness in carbon/epoxy and glass/epoxy systems resulting from standard autoclave manufacturing methods. His compression test results indicate a linear relationship between compression strength and lamina waviness for initial misalignment angles on the order of 1° or an amplitude/wavelength of $A/L = 0.002$. Highsmith et al. [16] observed in-plane fiber waviness in carbon-fiber-reinforced thermoplastic composites; furthermore, they observed cooperative waviness in discrete "wrinkled regions" rather than the waviness of individual fibers. These observations led them to develop a micromechanics model, based on the kinematics of the wavy fibers, to study the influence of fiber waviness and matrix nonlinearity on the compressive behavior of unidirectional composites. Reasonable agreement between predictions and experimental data for compressive strain to failure was obtained.

Due to the difficulty involved in measuring natural fiber waviness, many researchers have resorted to creating artificial fiber/ply waviness to investigate the effect of waviness on compressive strength [17–22]. However, it is not a straightforward task to mimic the waviness of real composite structures in the laboratory. Most researchers studying the compression response of wavy composites have attempted to produce test coupons with known levels of fiber/ply waviness by transferring a prescribed waviness geometry from an exaggerated tool surface; seldom have verification measurements of the resulting fiber/ply waviness been presented in conjunction with compression strength testing. Naley [17] introduced varying levels of in-plane fiber waviness into carbon/polysulfone laminates by first consolidating the laminae into a macroscopically deformed plate and then flattening the plate, thereby translating the macroscopic deformations into localized, wavy imperfections in the composite's microstructure similar to the "wrinkled regions" reported by Highsmith et al. [16]. Although no quantitative waviness measurements were established, Naley found that even low levels of waviness affected composite performance. Further, he observed a significant drop in the compressive strain to failure from 1.25 to 0.36% in the most wavy specimens. Adams and Hyer [19] measured compressive strength in carbon/polysulfone laminates with artificially induced layer waviness. They observed greater reductions in compressive strength as their induced waviness levels became more severe, with the most severe waviness being that which has a large amplitude and a short wavelength. Failure modes were also found to change with the degree of ply waviness. The experimental results were correlated with finite element predictions to quantify the effects of ply waviness on both static compressive strength and compression fatigue life [20]. Although the results differ as to the level of waviness severity that significantly affects compressive strength, they agree that strength is reduced by ply waviness.

As part of an ongoing ONR-sponsored investigation, Hsiao et al. [21,22] have studied compressive failure in unidirectional and cross-ply specimens. They observed the development of layer waviness in their cross-ply specimens, which they attributed to lamination residual stresses. Specimens with uniform, controlled levels of out-of-plane waviness exhibited a 42% drop in axial stiffness when loaded in compression [21]. Unidirectional specimens were also fabricated with graded layer waviness using a three-step curing method described in detail by Hsiao [22]. Their compression test results demonstrated a 30 and a 6% reduction in strength and stiffness, respectively, compared with a normal aligned specimen.

Using live video to monitor the failure mechanisms, Hsiao [21] described the failure sequence in detail. Under an applied load, the fibers are observed to bend, inducing different levels of interlaminar stresses along the fiber length. These interlaminar stresses may cause local delamination, which in turn causes a reduction in the constraint on the fibers. The resulting delamination yields non-uniform stiffness across the specimen width and thus load eccentricity, which in turn may lead to delamination growth and further multiple delaminations. Eventually, the delaminated sublayers initiate global buckling, leading to final failure in the form of broad kink bands. While the results of such investigations of artificially induced fiber waviness are important in understanding the fundamental effects of fiber waviness and small-angular-fiber misalignments, it is important to note that real fiber waviness is very difficult to achieve in the laboratory.

Fatigue—While compressive strength is the property most strongly affected by wavy fibers, fatigue failure and stiffness are also affected. A theoretical model for the relationship between fiber waviness of single fibers and compression-compression fatigue failure was developed by Slaughter and Fleck [23] based on initial imperfections, such as fiber misalignment and matrix yielding. Additional research in this area has been conducted by Piggott and Lam [24], who observed both a reduction in stiffness (the flexural modulus decreased dramatically while Young's modulus exhibited a less dramatic reduction) and an increase in Poisson's ratio under tensile-tensile (1:10) fatigue loading. They hypothesize the existence of some initial fiber waviness that evolves under fatigue loading into more severe waviness; this damage sequence is used to explain the observed stiffness degradation and the apparent increase in Poisson's ratio in fatigue-loaded wavy composites.

Tensile Failure and Stiffness—Several researchers have investigated the effect of fiber waviness on tensile strength and Young's modulus. Work by Van Dreumel and Kamp [25] showed no influence of fiber misalignment on Young's modulus, a modest loss of tensile strength, and an increase of Poisson's ratio with increasing in-plane waviness. Wang et al. [26] observed a definite reduction in tensile strength, but they could not ascertain whether the source was misalignment or waviness. Finally, Rai et al. [27] found that the reduced stiffness of their airframe components could be attributed to a combination of both fiber misalignment and fiber waviness (both in-plane and out-of-plane).

Experimental Methodology

Material System

The specimens in this investigation were removed from a series of $[0]_{12}$ laminates fabricated by hand layup of aligned fiber prepreg tape, consisting of Thornel®[3] T-300 carbon fiber reinforcement and Udel®[3] P-1700 polysulfone resin. This material was selected as part of a related experimental program investigating the development of process-induced fiber waviness [3,4]. A considerable body of research regarding the effect of layer waviness on stiffness and strength reduction in laminates and cylinders fabricated from this material system has already been developed by investigators at Virginia Polytechnic Institute and State University [18–20].

All of the test panels were processed using a symmetrical bagging procedure with aluminum tooling on both faces and a modified autoclave cycle. The autoclave cycle was modified from the prepreg manufacturer's recommended cycle to facilitate processing in air instead of a nitrogen environment, thereby reducing project costs. The modified autoclave cycle consisted of a hold temperature of 290°C for 1 h, pressure of 0.7 MPa, and a vacuum of 0.1 MPa. The parts were allowed to cool to 150°C before releasing the pressure in the autoclave and the vacuum on the part. Since preliminary observations showed that our modifications to the autoclave cycle had no adverse affects on the laminate consolidation [4], this cycle has been adopted as the baseline for all of our subsequent work. Further details on the selection of the processing cycle are available in Kugler and Moon [4].

Test Fixture and Specimen Geometry

Compression test fixtures fit into three distinct categories according to how the load is introduced to the specimen: through shear, by end loading, or through bending. An excellent survey of the literature regarding compression test fixtures can be found in the review paper by Schultheisz and Waas [12]. A review of the available literature suggested that no one fixture has been universally accepted by the composites community. For this investigation,

---

[3] Amoco Performance Products Inc., Marietta, GA.

the geometry and characteristic dimensions of the fiber-wrinkled regions dictated that our fixture selection utilize relatively small specimens, with a gage section large enough to accommodate and isolate a single packet of fiber waviness. Moreover, since our research involves qualification of as-manufactured composites, it was important that the fixture be easily adaptable to a wide range of thicknesses.

The consensus seems to be that shear-loaded specimens provide the best measure of compressive strength since end loading too often leads to undesirable failure by splitting and brooming at the ends of the specimen [28–31]. However, shear-loaded specimens have other drawbacks: they are susceptible to tab debonding as well as high stress concentrations due to discontinuity stresses at the tab tip adjacent to the gage section. In consultation with Wyoming Test Fixtures (WTF) [32], we selected the WTF combined loading compression test fixture (see Fig. 3) for this investigation. Our objective in selecting the WTF combined loading test fixture was to experiment with combined shear and end loading as proposed by Haeberle and Matthews [33] and Hsiao et al. [34] to ameliorate some of the problems associated with pure end loading and pure shear loading [29–31]. This new fixture modifies the WTF end-loaded side supported fixture by incorporating roughened clamping surfaces for shear load introduction into the specimen. By insetting the end of the specimen by 2 mm with respect to the ends of the clamp blocks, the fixture can be used to first introduce shear loading; eventually the specimen slips in the grips until the ends of the specimen come in contact with the loading platens, whereupon end loading is introduced. The fixture can be used to achieve any combination of shear and end loading depending on the bolt torque applied to the clamping bolts. When our fixture was purchased, the fixture had not yet been demonstrated using tabbed specimens—or for use with unidirectional composite materials.

The post and linear bearings design used to align the top and bottom halves of the specimen (see Fig. 3) minimizes friction carried in the alignment pins and thereby eliminates redundant load paths in the fixture. The grip surfaces are flame sprayed to minimize damage to the specimen from clamping, which may negate the need to use tabbed specimens. The high resultant coefficient of friction in the grips optimizes the shear loading without building up disproportionate transverse normal stresses, since lower clamping loads are required than in the knurled jaws of the IITRI fixture.

Since no database existed to describe the optimal shear/end loading combinations for the WTF combined loading fixture, a two-dimensional finite element (FE) model incorporating

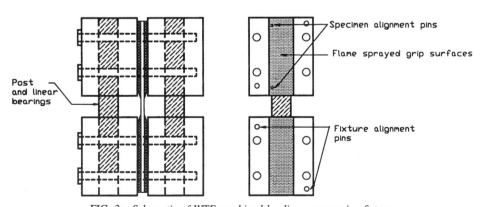

FIG. 3—*Schematic of WTF combined loading compression fixture.*

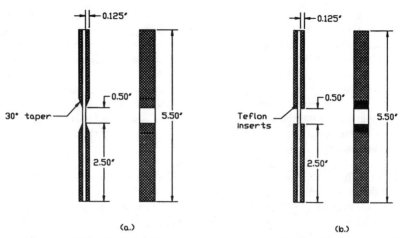

FIG. 4—*Schematic of candidate specimen geometries:* (a) *tapered tabs (30° taper) and* (b) *square tabs with Teflon inserts.*

the specifications of the WTF combined loading fixture to define the boundary conditions for 50% shear loading/50% end loading was developed using ABAQUS, V5.1[4] to evaluate the resulting stress concentrations [35]. The FE model included an explicit analysis of the specimen, adhesive layer, tab material, and clamp blocks. Model results were developed for two candidate specimen geometries, square tabs weakly bonded in the vicinity of the gage section, similar to the specimen designs proposed by Rolfes-Sendeckyj [36] and Haeberle and Matthews [33] and 30° tapered tabs as proposed by Adams and Odom [30] (Fig. 4). In both specimen designs, the specimen coupon is 140 mm long and 13 mm wide, with a 13-mm-long unsupported gage length.

The stresses thru the thickness near the tab tip and along the specimen/tab interface were reported to examine the effect of combined loading on the stress development in the specimen coupon. In order to compare our FE results with the FE analyses of Tan [37] and Ault and Waas [38], the stresses through the thickness near the tab tip were normalized with respect to the average value of the axial stress along the specimen centerline to examine the stress field in the gage section. The results of the FE model in Ref 35 demonstrated a marked advantage to using combined shear/end loading over the IITRI and Celanese fixtures and specimens [37], or the RAE specimen analyzed by Ault and Waas [38]. The stresses along the specimen/tab interface exhibited a distinct stress concentration at the tab tip due to shear loading with only a marginal reduction in the 30° tapered tab specimen design. On the other hand, the through-thickness stresses exhibit better uniformity in the specimen design with square tabs and Teflon inserts, since the stress concentration is under the tab material, well away from the tab tip. The FE model also predicted that the shear and transverse normal stress magnitudes are several orders of magnitude less than that of the induced normal stresses, indicating that both specimens should produce accurate compression strength results.

To verify the results of the FE model [35], an initial battery of six tests with each specimen geometry was performed on a series of non-wavy specimens to evaluate the suitability of the two candidate specimen geometries (Fig. 5). For this investigation, the specimens were

---

[4] Hibbitt, Karlsson, & Sorenson, Inc., Pawtucket, RI.

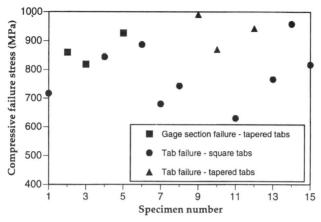

FIG. 5—*Initial test results on candidate specimens of T300/P1700 composite.*

cut from $[0]_{12}$ laminates approximately 2.5 mm thick. In this first set of tests, while the failure stresses were similar, none of the specimens with square tabs and Teflon inserts failed in the gage section. To further investigate the specimen design with square tabs, three more tests were performed with relatively compliant fiberglass tabs; these specimens also exhibited tab failures. These preliminary tests convinced us to proceed with the tapered tab design for the remainder of this investigation.

Wavy Test Specimens

Initial observations of the laminates produced as part of this investigation revealed a pattern of distributed fiber-wrinkled regions on the surface of all the laminates (Fig. 6). Using the methodology outlined in Ref 5, several waviness parameters—amplitude, wavelength,

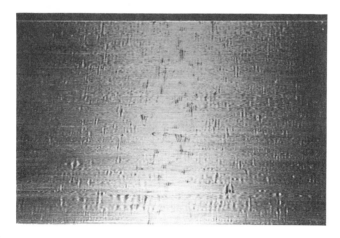

FIG. 6—*Photo illustrating the pattern of fiber-wrinkled regions on the laminate surface (15 by 22 mm) in T300/P1700.*

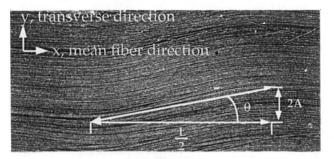

FIG. 7—*Definition of waviness parameters.*

length, and off-axis angle—were carefully measured and recorded in these fiber-wrinkled regions. Figure 7 shows how the waviness parameters were defined and measured using video microscopy and image analysis. Since compression failure is considered an instability problem, this investigation focused on the maximum off-axis angle, θ, or the rotation angle of the fibers [19].

Specimens were selected by reviewing the waviness measurements to identify wavy regions within each laminate. Then, using a template of the specimen geometry, the laminates were screened for prospective specimens in which a single wavy region of interest (ROI) could be isolated within the specimen gage section. Figure 8 shows a schematic of a wavy ROI in the specimen gage section, illustrating the characteristic dimensions. A wavy ROI typically encompassed between 200 to 1000 fibers, exhibiting cooperative in-plane waviness over a single wavelength. Based on prior observations [5], a typical wavy ROI ranged from 2 to 35 mm² on the specimen surface and propagated 2 to 6 plies through the thickness from a single face. Both surfaces of the laminate were thoroughly inspected to ensure that the opposite face of the wavy specimens is nominally straight in the vicinity of the gage section.

To remove wavy specimens, the panels were first sectioned lengthwise using a large horizontal milling machine with specially adapted diamond tooling. This equipment and procedure are outlined in Ref *17*. Once the wavy ROI was isolated to a single strip of composite,

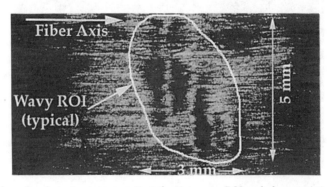

FIG. 8—*Schematic of specimen gage section showing wavy ROI and characteristic dimensions.*

the specimen coupon was measured and removed using an Isomet®[5] 2000 Precision Saw with a high-concentration diamond wafering blade, taking care to center the wavy ROI in the center of the specimen gage section. The use of diamond-grinding wheels from start to finish in sectioning the specimens resulted in minimal specimen damage and required little or no polishing prior to testing.

To prepare the surfaces for adhesive bonding, the specimen coupons and fiberglass tabbing strips were sanded with 120-grit paper and wiped clean of dust and grease with acetone. The tab material used for this investigation was a standard, bi-directional woven glass/ polyester material (G10). The tabs were cut approximately 16 mm wide and 70 mm long from a 3-mm-thick sheet. The tapered tabs were prepared individually by grinding a 30° taper angle onto the tabs. The tabs were then bonded to the specimen coupon using a two-part, room-temperature-curing adhesive, Epoxi-Patch®.[6]

In the case of the specimens with square tabs and Teflon inserts, the Teflon inserts were cut from a sample of stiff Teflon tape with a silicone adhesive on one face from Specialty Tapes.[7] The inserts were cut slightly oversize and applied to the tabs, sticky side down, approximately 6 mm from the centerline of the specimen, taking care that the inserts laid flat and that the edges were straight and perpendicular to the long axis of the tabs; the tabs were then adhesively bonded to the specimen coupons.

In both cases, when the adhesive had reached its final cure strength, the specimens were carefully ground using a belt grinder to remove excess adhesive and ensure parallelism of the specimen and tabs on at least one edge—the reference edge. Since surface finish has been demonstrated to affect compressive failure mode and load, care was taken to avoid grinding the specimen itself.

Testing Procedure

All of the tests in this investigation were performed on a 110-kN, servohydraulic load frame. The tests were performed in displacement control with a constant strain rate of 0.25 mm/min. Approximately half the specimens were instrumented front and back with single pattern strain gages to monitor the longitudinal strains as a method of detecting excessive bending in the unsupported gage section. The strain gages were centered in the unsupported gage section of the specimens and aligned with respect to the loading axis. The strain data were collected and conditioned using the testing interface constructed by Naley [*17*]. All of the data acquisition and much of the subsequent analysis was performed using in-house software developed using LabView®.[8]

Each specimen was carefully mounted in the fixture and aligned using the specimen alignment pins in the top pair of clamp blocks to ensure good alignment of the specimen with the load axis during the test. The specimens were mounted in the clamp blocks with the clamp blocks extending the full length of the tabs, making sure to leave a gap on both ends of the specimen coupon of approximately 2 mm. The transverse normal force in the clamp blocks was applied using a high-resolution torque wrench to tighten the four clamp bolts on either end. A series of preliminary tests suggested that a bolt torque of approximately 3.5 N · m provided relatively balanced shear/end loading when the grip surfaces were well cleaned after each test [*35*].

[5] Buehler, Ltd, Lake Bluff, IL.
[6] Dexter Aerospace Products, Seabrook, NH.
[7] Specialty Tapes Div. of RSW, Inc., Racine, WI.
[8] National Instruments, Austin, TX.

The fixture and specimen were loaded using specially machined compression platens, including a fixed platen on the bottom and a spherically seating platen on the top to remove any remaining load-axis eccentricity. Initially, the specimen is loaded in pure shear until the friction interface between the grip surfaces and fiberglass tabs can no longer transmit additional shear stress and the specimen slips, usually in one end and then the other, separated by approximately 10% applied load. Once both ends have slipped, end loading initiates through direct contact between the specimen and the loading platens until final failure. A detailed description of the loading sequence for the WTF combined loading compression test fixture is presented in Ref 35.

Results

Figure 9 summarizes the results of our compression tests. On first inspection, the debilitating affect of increasing fiber waviness severity on compression strength is quite clear. Gage section failures dominate the trend as the maximum off-axis angle in the wavy packets increased to relatively large angles. While tab-related failures dominate in the low-angle specimens, the data show a nearly linear reduction in compressive strength of 30 to 50% for moderate off-axis angles ($\theta \approx 5°$) and nearly 70% for more severe waviness ($10° < \theta < 15°$). Gage section failures are driven by an instability due to the initially wavy fibers, while those for lower misalignment angles are driven by the stress concentration effect of the tab.

Based on visual inspection, the specimens were observed to fail primarily by in-plane kink banding and, in some cases, by out-of-plane kink banding. Longitudinal splitting frequently coincided with excessive bending as indicated by the strain data record. To get a closer look at the failure mode in the specimens suspected of in-plane kink banding, the failure zone was removed from the specimen coupons and scrubbed with acetone to remove the residual strain gage adhesive. The specimens were then inspected using dark field microscopy, which yielded very clear images of the anticipated kink band failures (Figs. 10, 11, and 12).

Approximately 75% of the specimens were observed to fail by in-plane kink banding, which appeared to propagate from the original wavy fibers and spread across the width of the specimen until the remaining ligament failed by catastrophic fracture. In some cases, the failure mode appeared to change across the width of the specimen—as in the case of the

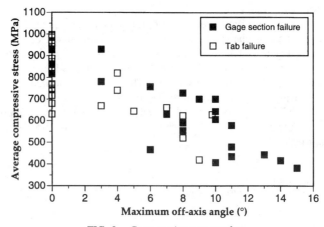

FIG. 9—*Compression test results.*

FIG. 10—*Photomicrograph of wavy fibers in the vicinity of kink band failure in T300/P1700 (fiber diameter = 7 μm).*

specimen in Fig. 10, in which the kink banding appears to have given way to longitudinal splitting, followed by fast fracture.

Almost 50% of the specimens exhibited complementary kink bands with mixed in-plane and out-of-plane character: In Figs. 11 and 12, both specimens exhibited complementary kink bands propagating tangent to the wavy fibers at the point of inflection. The different failure modes observed suggest a complicated three-dimensional stress state in the vicinity of the wavy fibers just prior to fracture. Inspection before and after compressive failure of the specimen in Fig. 12 showed that although the kink band broadens from the initial 1000 to a final 5000 wavy fibers across the width of the specimen, the maximum fiber angle remains unchanged.

Analysis and Discussion

Before examining the data further, several potential error sources are identifiable. First, inability to detect low levels of waviness and other errors associated with the fiber waviness measurement technique contribute to the scatter in Fig. 9. In addition, the compressive stress at failure may not be adequately represented by the average compressive stress (Fig. 9), which is computed from the applied load and specimen cross-sectional area, since a majority of the specimens tested in this investigation exhibited noticeable—albeit small—levels of bending.

The fiber waviness measurement errors may produce the largest contribution to scatter in Fig. 9. For one, the degree of scatter in nominally non-wavy specimens suggests a threshold

FIG. 11—*Photomicrograph of complementary kink bands propagating from a region of fiber waviness in T300/P1700 (fiber diameter = 7 μm).*

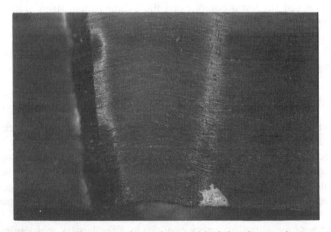

FIG. 12—*Photomicrograph illustrating the evolution of kink banding in the presence of an initial imperfection in T300/P1700 (fiber diameter = 7μm).*

of observability. In general, visual inspection using light microscopy cannot detect shallow waviness—characterized by long wavelengths, low amplitude, and small off-axis angles. While this may be due to the limitations of our inspection technique, it is accurate to say the non-wavy specimens can exhibit fiber misalignment on the order of 1 to 3°, originating from any or all of the following: the prepregging process, hand layup, specimen removal, the existence of low-angle waviness, and/or misalignment of the specimen in the fixture. Furthermore, the waviness measurements are also susceptible to a number of error sources, chiefly due to the perception of wavy fibers at the optical microscope. Based on thorough uncertainty analysis [3], the relative error in off-axis angle measurements is dependent on the unique wave geometry for each observation. The relative error $d\theta/\theta$ is seen to decrease for larger values of θ; for 2°, $d\theta/\theta \approx 0.87$, while for higher values of θ, e.g., 8°, $d\theta/\theta \approx 0.25$. Hence, the more severe the fiber waviness, the greater the accuracy in the waviness measurements.

The WTF combined loading compression fixture seems relatively insensitive to bending; specimen bending in the WTF combined loading compression fixture is believed to originate from the localized asymmetry of the specimen on the face containing wavy fibers; however, bending may also result from uneven friction in the grips, non-uniform torque in the clamp bolts, and end effects due to the specimen machining. Approximately half of the specimens were strain gaged, front and back, to monitor the amount of bending in each test; if the longitudinal strains (front and back) diverged, this was an indication the specimen was experiencing bending in the unsupported gage section. As the magnitude of bending increases, the incremental strain on the convex face becomes increasingly tensile while that on the concave face becomes increasingly compressive. In the extreme case, as the specimen approaches final failure due to buckling, the curved geometry may become apparent to the naked eye. Since noticeable levels of bending were observed in the latter half of most tests, it became relevant to consider not only the average compressive stress, but also the maximum compressive stress and the ratio of bending strain to compressive strains.

To quantify the effect of specimen bending, the ratio R of the bending stress to average axial stress in the gage section at failure was calculated using

$$R = \frac{|\varepsilon_f - \varepsilon_b|}{\varepsilon_f + \varepsilon_b}$$

where ε_f and ε_b are the compressive strain values immediately prior to failure on the front and back of the specimen, respectively. Here, R ranges between [0.02,0.27] or from 2% in the non-wavy specimens to 27% in the most wavy specimens (Fig. 13).

Figure 13 demonstrates a clear increasing trend in the bending deformation with increasing waviness severity as depicted using a best fit line. If we attribute bending in the non-wavy specimens to problems associated with the fixture and tabs, then $R(0) \cong 2\%$ suggesting that the stress state was nearly uniaxial and that the fixture performed well. The scatter in Fig. 13 may be due to a number of effects. For example, the wavy region geometry and the amount of fiber waviness may play a more significant role than originally suspected. Also, the specimen face containing the wavy ROI was not recorded consistently, thereby making it impossible to definitively correlate the concave side of the bending deformation with the coincidence of the wavy ROI. This might explain some of the scatter in Fig. 13, since the waviness lying on the convex side of the specimen may have a less crippling effect relative to a wavy ROI on the concave side of the specimen.

The maximum compressive stresses for all strain-gaged specimens were computed using the scheme presented by Naley [17]. His "correction" scheme uses classical beam theory

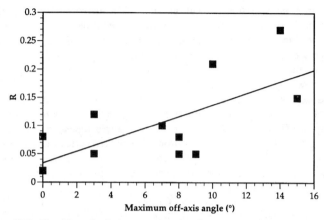

FIG. 13—*Plot of relative specimen bending versus off-axis angle.*

(plane sections remain plane and normal to the neutral axis) and estimates the maximum strain to be the sum of the bending and axial compressive strains. A non-linear stress-strain curve is fit to the average stress-strain response, as illustrated in Fig. 14; a third-order polynomial was generally adequate. The maximum compressive stress was then estimated from the curve fit using the appropriate maximum compressive strain value from the concave side of the specimen.

The complete compression test results for all gage-section failures are presented in Table 1. If we look first at the average compressive strength of the nominally non-wavy T300/P1700 specimens from Table 1, $\bar{\sigma}_c(0)$ = 868 MPa. This value is 10% lower than the average compressive strength for T300/P1700 composite laminates measured by Adams and Hyer [20] using the IITRI fixture. However, it is almost 27% higher than the baseline compressive strength (without correction) measured by Naley [17] for T300/P1700 laminates using a modified D-695 fixture, but nearly 50% lower than the compressive failure stress when Naley applied the bending correction. Our results suggest that applying the bending correction may tend to overestimate the failure stress in the instance of global buckling, where excessive

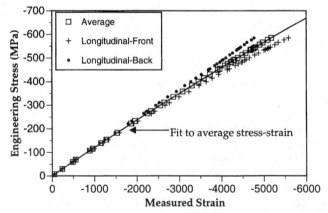

FIG. 14—*Stress-strain plot exhibiting typical bending and bend fit formulation.*

TABLE 1—*Compression test results, gage section failures only.*

Panel	Spec.	A, mm	L mm	θ_{max}, °	Average Compressive Strength, MPa	Maximum Compressive Stress, MPa	R
1	1	...^{a}	...^{a}	...^{a}	926	996	0.08
1	2	...^{a}	...^{a}	...^{a}	818	835	0.02
1	3	...^{a}	...^{a}	...^{a}	860	874	0.02
2	1	0.02	1.38	3	781	864	0.12
2	2	0.08	2.44	7	629	689	0.10
2	3	0.08	2.02	9	700	734	0.05
2	4	0.09	2.02	3	929	984	0.05
3	1	0.05	1.56	8	729	767	0.05
4	1	0.06	1.79	8	554	...^{b}	...^{b}
4	2	0.09	1.97	10	407	...^{b}	...^{b}
4	3	0.24	3.60	15	382	410	0.15
4	4	0.12	2.51	11	434	...^{b}	...^{b}
4	5	0.08	1.70	11	479	...^{b}	...^{b}
5	1	0.12	2.33	11	578	...^{b}	...^{b}
5	2	0.11	1.95	13	443	...^{b}	...^{b}
5	3	0.04	1.70	6	466	...^{b}	...^{b}
5	4	0.12	2.66	10	606	730	0.21
6	1	0.09	1.56	10	701	...^{b}	...^{b}
6	2	0.12	2.01	14	416	549	0.27
6	3	0.05	1.26	8	590	635	0.08
6	4	0.26	3.81	15	643	...^{b}	...^{b}
6	5	0.03	0.94	6	758	...^{b}	...^{b}

^{a} Nominally non-wavy specimen; no waviness measurement recorded.
^{b} Strain data not available; specimen was not strain gaged or gages failed.

bending occurs. This may explain the discrepancy between the relatively low measured average compressive stresses and the high-failure stresses calculated by Naley; therefore, for comparison we will consider only the average stresses reported in Ref *17*. These differences are typical of the variability from one compression test fixture to another observed in the literature. For example, in their evaluation of compression test fixtures, Daniels and Sandhu [*39*] reported the average compressive strength of AS4/3501-6 composite laminates between 772 MPa using the Boeing (open hole) fixture and 1398 MPa using the Rolfes fixture and specimen.

Figure 15 shows the maximum compressive stress applying Naley's bending correction scheme to all specimens for which front and back strain data were available. Applying a linear fit to the gaged specimen data, $\sigma_c(\theta) = -28.9\,\theta + 939$, which relates the compression strength, σ_c, to fiber waviness severity in terms of the off-axis angle, θ, in T300/P1700. Judging from the error inherent in the off-axis angle measurements, this appears to be a reasonable fit. Thus we observe a compression strength reduction of 25 to 30% for low to moderate levels of fiber waviness and nearly 50 to 60% for severe fiber waviness.

In comparison, Adams and Hyer reported compression strength reductions from 5 to 60% for out-of-plane waviness [*19,20*]. Adams and Hyer [*20*] reported a 36% (maximum) reduction in compressive strength for their "layer wave" specimens for fiber rotations on the order of 15°. Perhaps the most obvious explanation for this disparity is the layup of the specimen used by Adams and Hyer; in their specimens, waviness persists over only 20% of the thickness, while in our investigation the waviness extends in a cooperative fashion over nearly 50% of the specimen thickness. Moreover, the effects of fiber waviness may be exacerbated

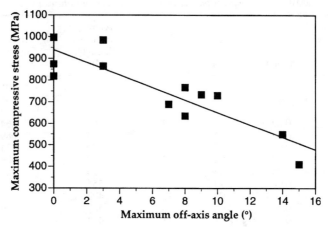

FIG. 15—*Maximum compressive stress corrected for bending.*

in the relatively high-stiffness unidirectional laminates studied in this investigation as compared to the "layer wave" crossply laminates studied by Adams and Hyer [20].

As a final check on our results, we can compare our experimental results to established theories for compressive microbuckling and kinking in fiber-reinforced composites. The earliest prediction of compressive strength, proposed by Rosen [6], neglects initial imperfections; nonetheless, his fiber-microbuckling model is still the most widely used. A quick comparison with Rosen's model will indicate the relative quality of our non-wavy, baseline specimens. Rosen's model of straight fibers embedded in an elastic matrix (continuum) employs beam theory to estimate the lowest buckling stress, which is associated with shear mode buckling

$$\sigma_c = \frac{G_m}{(1 - v_f)}$$

where G_m is the elastic shear modulus of the matrix and v_f is the fiber volume fraction of the composite. In fact, the right-hand side is just an approximation for the effective longitudinal shear modulus of the composite, G_{12}; neglecting initial imperfections, Rosen's model is interpreted as $\sigma_c/G_{12} = 1$.

However, Rosen's model has been demonstrated to grossly overpredict the compressive strength of most composites; experimental compression test data typically fall in the range $1/5 \leq \sigma_c/G_{12} \leq 1/4$. For AS4/3501-6, the shear modulus is 5.2 MPa [40], which when combined with compression test results [39] yields $0.148 \leq \sigma_c/G_{12} \leq 0.269$; a number of researchers have attributed this discrepancy between theory and experiment to the presence of fiber waviness. For T300/P1700, $G_{12} = 4.1$ GPa [20] and we measured $\overline{\sigma}_c(0) = 868$ MPa; therefore, $\sigma_c/G_{12} = 0.212$ for our non-wavy specimens, suggesting the presence of some low angle misalignment.

We can estimate the initial fiber misalignment in our non-wavy specimens using Budiansky's model for kink banding in an elastic/perfectly-plastic composite [8]. Improving on Rosen's model, he incorporated the initial misalignment angle, ϕ (as suggested by Argon [7]) as follows:

$$\sigma_c = \frac{G_{12}}{1 + \dfrac{\phi}{\gamma_y}}$$

where $\gamma_y = \tau_y/G_{12}$ represents the elastic shear strain at $\tau = \tau_y$, and τ_y is the shear yield stress of the composite. Approximating $\tau_y \approx 30$ MPa for our T300/P1700 composite [20] and substituting $\bar{\sigma}_c(0)$ to solve for the initial misalignment angle in our non-wavy specimens, we obtain $\phi = 1.5°$. This estimate corresponds to the waviness observability threshold described previously and quantifies our hypothesis that even our non-wavy specimens contain low levels of fiber misalignment.

Conclusions

Compression strength reduction due to the presence of in-plane fiber waviness is significant. These compression test results demonstrate that the strength reduction due to in-plane waviness is comparable to the stress concentration at the tab termination, even where low-angle waviness is concerned. The strength reduction becomes more pronounced with increasing levels of fiber waviness and is especially significant in unidirectional composites as compared to crossply laminates [20].

It is clear from the compression test results that the fiber waviness measurement technique [5] is sensitive to varying levels of fiber waviness and that off-axis angle can be a useful measure of fiber waviness severity. The results also demonstrate that even our nominally, non-wavy specimens most likely contain at least some low-level fiber misalignment, since the measured strengths are still only one fifth the values estimated using Rosen's classic microbuckling model [6]. Assuming an elastic/perfectly-plastic material, we have demonstrated an initial misalignment angle of 1.5°, even in our nominally straight-fiber specimens. It is noteworthy that such a small degree of initial fiber misalignment causes such a precipitous drop in the predicted compressive strength. This, coupled with the limitations of our measurement technique in this regime, is indeed a cautionary result.

This example also demonstrates that the compressive strength of T300/P1700 is significantly lower than the strength of other high-performance composite systems, e.g., AS4/3501-6. We attribute this to the relatively low shear modulus of the T300/P1700 composite, which is not widely regarded as a high-strength thermoplastic matrix material. As a followup to this investigation, we intend to explore the effect of fiber waviness in S2 glass-reinforced laminates to assess the role of fiber stiffness on the compression response, complete with a thorough shear characterization so that the experimental results can be compared with Schapery's analytical model [11].

While the WTF combined loading compression test fixture performed adequately for the wavy specimens, the non-wavy specimens exhibited an unacceptable number of tab failures (approximately 85%). However, the WTF combined loading fixture exhibited far less coupon bending and hence yielded significantly higher values of average compressive stress than reported for T300/P1700 by Naley [17] using the modified D-695 fixture. An in depth comparison of our test results with Naley's investigation using the D-695 fixture brings up another important conclusion: the bending correction described herein should be used with caution; it should be used only when the ratio R of the bending stress to average axial stress in the gage section at failure is small. The WTF combined loading test fixture clearly has potential advantages; we recommend a more thorough analysis to determine the optimal testing conditions (proportion of shear-to-end loading, as well as tab geometry and material).

Typical of kink band images found in the literature [*12*], Figs. 10, 11, and 12 show very straight fibers bounding the kink zones. However, the presence of fiber waviness produced far more fiber curvature in the kink band zone than most other researchers have observed, suggesting a more continuous yielding process in the presence of wavy fibers. This hypothesis requires further investigation of the evolution of kink banding in wavy composites. As a natural next step, we recommend the use of video interrogation to follow a band of wavy fibers through the failure sequence in real time to understand the dynamics of kink banding.

Acknowledgments

Financial support from Dr. Yapa Rajapakse of the Office of Naval Research under Grant ONR-N00014-91-J-4091 is gratefully acknowledged. Additional support is provided by Dr. Bruce Kramer of the National Science Foundation under Grant DDM-9258413. The authors would like to express their appreciation to Dr. Richard A. Schapery and Dr. Kenneth Liechti of the Aerospace Engineering and Engineering Mechanics Department, The University of Texas at Austin, for the use of their laboratory facilities and equipment. The authors would also like to acknowledge Mark Naley for his developmental work on the testing interface and software development, without which much of this testing would not have been possible; Robert Verastiqui for his help with the specimen preparation and data collection; Melanie Violette for many thoughtful discussions and especially her help in conducting the compression tests; and Danielle Kugler for her editorial assistance in preparing this manuscript.

References

[*1*] Cantwell, W. J. and Morton, J. "The Significance of Damage and Defects and Their Detection in Composite Materials: A Review," *Journal of Strain Analysis,* Vol. 27, No. 1, 1992, pp. 29–42.
[*2*] Summerscales, J., "Manufacturing Defects in Fibre-Reinforced Plastics Composites," *INSIGHT Non-Destructive Testing and Condition Monitoring,* Vol. 36, No. 12, December 1994, pp. 936–942.
[*3*] Bhalerao, M. S., "On Process-Induced Fiber Waviness in Composites: Theory and Experiments," Ph.D. dissertation, The University of Texas at Austin, Austin, TX, 1996.
[*4*] Kugler, D. and Moon, T. J., "Investigation of the Effect of Select Processing Parameters on the Development of Fiber Waviness in Thin Laminates," *Journal of Composite Materials,* in preparation.
[*5*] Joyce, P. J., Kugler, D., and Moon, T. J., "A Technique for Characterizing Process-Induced Fiber Waviness in Unidirectional Composite Laminates—Using Optical Microscopy," *Journal of Composite Materials,* pp. 1694–1727.
[*6*] Rosen, B. W., "Mechanics of Composites Strengthening," *Fiber Composite Materials,* 1964, pp. 37–75.
[*7*] Argon, A. S., "Fracture of Composites," *Treatise on Materials Science and Technology,* Vol. 1," H. Herman, Ed., Academic Press, New York, NY, 1972, pp. 79–114.
[*8*] Budiansky, B., "Micromechanics," *Computers and Structures,* Vol. 16, Nos. 1–4, 1983, pp. 3–12.
[*9*] Hahn, H. T. and Williams, J. G., "Compression Failure Mechanisms in Unidirectional Composites," *Composite Materials: Testing and Design, ASTM STP 694,* American Society for Testing and Materials, West Conshohocken, PA, 1986, pp. 118–139.
[*10*] Schapery, R. A., "Analysis of Local Buckling in Viscoelastic Composites," *IUTAM Symposium on Mechanics Concepts for Composite Material Systems,* Virginia Polytechnic Institute and State University, Blacksburg, VA, 1991.
[*11*] Schapery, R. A., "Prediction of Compressive Strength and Kink Bands in Composites Using a Work Potential," *International Journal of Solids and Structures,* Vol. 32, 1995, pp. 739–765.
[*12*] Schultheisz, C. R. and Waas, A. M., "Compressive Failure of Composites, Part I: Testing and Micromechanical Theories," *Progress in Aerospace Science,* Vol. 32, 1996, pp. 1–42.
[*13*] Waas, A. M. and Schultheisz, C. R., "Compressive Failure of Composites, Part 2: Experimental Studies," *Progress in Aerospace Studies,* Vol. 32, 1996, pp. 43–78.
[*14*] Mrse, A. and Piggott, M. R., "Relation Between Fibre Divagation and Compressive Properties of Fibre Composites," *Proceedings, 14th International SAMPE Symposium,* 1990, pp. 2236–2244.

[15] Camponeschi, E. T. Jr., "Lamina Waviness Levels in Thick Composites and Its Effect on Their Compression Strength," *Composites, Design, Manufacture, and Application*, 1991, pp. 30E1–30E13.

[16] Highsmith, A. L., Davis, J. J., and Helms, K. L. E., "The Influence of Fiber Waviness on the Compressive Behavior of Unidirectional Continuous Fiber Composites," *Composite Materials: Testing and Design, ASTM STP 1120*, American Society for Testing and Materials, West Conshohocken, PA, 1992, pp. 20–36.

[17] Naley, M., "Fabrication and Testing of Unidirectional Thermoplastic Composite Plates with Varying Degrees of Fiber Waviness," Master's thesis, The University of Texas at Austin, Austin, TX, 1994.

[18] Adams, D. O. and Hyer, M. W., "Fabrication and Compression Testing of Layer Waviness in Thermoplastic Composite Laminates," technical report, Virginia Polytechnic Institute and State University, Blacksburg, VA, 1990.

[19] Adams, D. O. and Hyer, M. W., "Characterization of the Influence of Layer Waviness on the Response of Laminates," technical report, Virginia Polytechnic Institute and State University, Blacksburg, VA, 1991.

[20] Adams, D. O. and Hyer, M. W., "Effects of Layer Waviness on Compression Loaded Thermoplastic Composite Laminates," technical report, Virginia Polytechnic Institute and State University, Blacksburg, VA, 1992.

[21] Hsiao, H. M., Daniel, I. M., and Wooh, S. C., "Effect of Fiber Waviness on the Compression Behavior of Thick Composites," *Failure Mechanics in Advanced Polymeric Composites*, ASME AMD Vol. 196, 1994, pp. 141–189.

[22] Hsiao, H. M., Wooh, S. C., and Daniel, I. M., "Fabrication Methods for Unidirectional and Cross-ply Composites with Fiber Waviness," *Journal of Advanced Materials*, Vol. 26, 1995, pp. 19–26.

[23] Slaughter, W. S. and Fleck, N. A., "Compressive Fatigue of Fibre Composites," *Journal of the Mechanics and Physics of Solids*, Vol. 41, 1993, pp. 1268–1284.

[24] Piggott, M. R. and Lam, P. W. K., "Fatigue Failure Processes in Aligned Carbon Epoxy Laminates," *Composite Materials: Fatigue and Fracture (Third Volume), ASTM STP 1110*, American Society for Testing and Materials, West Conshohocken, PA, 1991, pp. 686–698.

[25] Van Dreumel, W. H. and Kamp, J. L. M., "Non Hookean Behavior in the Fibre Direction of Carbon Fibre Composites and the Influence of Fibre Waviness on the Tensile Properties," *Journal of Composite Materials*, Vol. 13, 1977, pp. 461–469.

[26] Wang, S. J., Baptiste, D., Bompard, P., and Francois, D., "Microscopic Failure Mechanisms of an Unidirectional Glass Fiber Composite," *Fatigue and Fracture of Engineering Materials & Structures*, Vol. 14, 1991, pp. 391–403.

[27] Rai, H. G., Rogers, C. W., and Crane, D. A., "Mechanics of Curved Fiber Composites," *Proceedings, 47th Annual Forum of the American Helicopter Society*, 1991, pp. 297–304.

[28] Berg, J. S. and Adams, D. F., "An Evaluation of Composite Material Compression Test Methods," *Journal of Composites Technology and Research*, Vol. 11, No. 2, 1989, pp. 41–46.

[29] Adams, D. F. and Lewis, E. Q., "Influence of Specimen Gage Length and Loading Method on the Axial Compressive Strength of a Unidirectional Composite Material," *Experimental Mechanics*, March 1991, pp. 14–20.

[30] Adams, D. F. and Odom, E. M., "Influence of Specimen Tabs on the Compressive Strength of a Unidirectional Composite Material," *Journal of Composite Materials*, Vol. 25, 1991, pp. 774–786.

[31] Camponeschi, E. T. Jr., "Compression of Composite Materials: A Review," *Composite Materials: Fatigue and Fracture (Third Volume), ASTM STP 1110*, T. K. O'Brien, Ed., American Society for Testing and Materials, West Conshohocken, PA, 1991, pp. 550–580.

[32] Adams, D. F., personal communication, Wyoming Test Fixtures, Laramie, WY, July 1995.

[33] Haeberle, J. G. and Matthews, F. L., "The Influence of Test Method on the Compressive Strength of Several Fiber-Reinforced Plastics," *Journal of Advanced Materials (SAMPE Quarterly)*, Vol. 25, No. 1, 1993, pp. 35–45.

[34] Hsiao, H. M., Daniel, I. M., and Wooh, S. C., "A New Compression Test Method for Thick Composites," *Journal of Composite Materials*, Vol. 29, 1995, pp. 1789–1806.

[35] Joyce, P. J. and Moon, T. J., "Analysis and Testing of the Wyoming Combined Loading Compression Test Fixture with Unidirectional Composite Specimens," *Experimental Mechanics*, in preparation.

[36] Rolfes, R. L., "Test Method for Compressive Properties of Oriented Fiber Composites with the Prototype Compression Fixture," *Air Force Wright Aeronautical Laboratories Technical Memorandum, AFWAL-TM-85-222-FIBC*, August 1985.

[37] Tan, S. C., "Stress Analysis and Testing of Celanese and IITRI Compression Specimens," *Composites Science and Technology*, Vol. 44, 1992, pp. 57–70.

[*38*] Ault, M. and Waas, A., "Finite Element Analysis of a Proposed Specimen for Compression Testing of Laminated Materials," presented at the AIAA SDM Meeting, La Jolla, CA, *AIAA Paper No. 93-1518*, 1993.
[*39*] Daniels, J. A. and Sandhu, R. S., "Evaluation of Compression Specimens and Fixtures for Testing Unidirectional Composite Laminates," *Composite Materials: Testing and Design (Eleventh Volume), ASTM STP 1206*, E. T. Camponeschi, Jr., Ed., American Society for Testing and Materials, West Conshohocken, PA, 1993, pp. 103–123.
[*40*] Schapery, R. A., "Mechanical Characterization and Analysis of Inelastic Composite Laminates with Growing Damage," *Mechanics of Composite Materials and Structures*, ASME AMD-Vol. 100, pp. 1–9, 1989.

Michael K. Cvitkovich,[1] T. Kevin O'Brien,[2] and Pierre J. Minguet[3]

Fatigue Debonding Characterization in Composite Skin/Stringer Configurations

REFERENCE: Cvitkovich, M. K., O'Brien, T. K., and Minguet, P. J., **"Fatigue Debonding Characterization in Composite Skin/Stringer Configurations,"** *Composite Materials: Fatigue and Fracture, Seventh Volume, ASTM STP 1330*, R. B. Bucinell, Ed., American Society for Testing and Materials, 1998, pp. 97–121.

ABSTRACT: The objective of this work was to investigate the fatigue damage mechanisms and to identify the influence of skin stacking sequence in carbon epoxy composite bonded skin/stringer constructions. A simple four-point-bending test fixture originally designed for previously performed monotonic tests was used to evaluate the fatigue debonding mechanisms between the skin and the bonded frame when the dominant loading in the skin is flexure along the edge of the frame. The specimens consisted of a tapered flange, representing the stringer, bonded onto a skin. Based on the results of previous monotonic tests, two different skin layups in combination with one flange layup were investigated. The tests were performed at load levels corresponding to 40, 50, 60, 70, and 80% of the monotonic fracture loads. Microscopic investigations of the specimen edges were used to document the onset of matrix cracking and delamination and subsequent fatigue delamination growth. Typical damage patterns for both specimen configurations were identified. The observations showed that failure initiated near the tip of the flange in the form of matrix cracks at one of two locations, one in the skin and one in the flange. The location of the 90° flange and skin plies relative to the bondline was identified as the dominant layup feature that controlled the location and onset of matrix cracking and subsequent delamination. The fatigue delamination growth experiments yielded matrix cracking and delamination onset as a function of fatigue cycles as well as delamination length as a function of the number of cycles.

KEYWORDS: composite materials, fatigue testing, design, bond strength, skin/flange interface, secondary bonding

One of the major objectives of today's aircraft industry is the reduction of manufacturing costs without loss, or with even an increase, in quality and reliability. One of the many approaches to achieve this goal is the simplification of the production processes. Composite structures have already proven to be very useful due to their light weight. These structures are also very attractive since secondary bonding or co-curing may replace mechanical fastening methods and reduce component assembly time and cost.

As of today, little data exist in the open literature on the mechanical capabilities of bonded skin/stringer structures. Initial investigations have been performed on the effects of pressure loading on a state-of-the-art composite materials bonded fuselage panel [1]. Due to flexure

[1]National Research Council resident research associate, NASA Langley Research Center, Mail Stop 188E, Hampton, VA 23681-0001.
[2]Senior scientist, U.S. Army Research Laboratory, Vehicle Technology Center, NASA Langley Research Center, Mail Stop 188E, Hampton, VA 23681-0001.
[3]Senior tech specialist, Boeing Defense & Space Group, Helicopters Division, P.O. Box 16858, Mail Stop P38-13, Philadelphia, PA 19142-0858.

along the edge of the frame (Fig. 1), shear stresses and moments along the frame length were identified as potentially critical loading conditions very likely to result in structural failure. Experiments under monotonic loading conditions were carried out utilizing a modified frame pull-off test with specimens cut from a full-size panel. The major drawback of this test methodology is the expensive production of the pull-off specimens and the complex test setup.

Based on these results, a new test methodology for analyzing the failure mechanisms at the skin/stringer interface has recently been introduced [2,3]. Comparatively simple specimens consisting of a flange bonded onto a skin were tested in three- and four-point bending fixtures. It was shown in experiments with monotonic loading that failure initiated at the tip of the flange, identical to the failure mode in complex specimens tested in Ref 1. Depending on specimen layup, delaminations were reported to propagate at two different locations, either in the flange or in the skin [3]. A later examination of the specimens tested under monotonic loading revealed that in one specimen configuration debonding also occurred between the flange and the skin within the bondline.

To obtain a more complete understanding of skin/stringer debonding mechanisms, it is vital to perform fatigue tests along with the monotonic experiments to establish the durability of these bonded joints over the anticipated life cycle. Unfortunately, no such data are available in the literature. Therefore, the objective of this investigation was to shed more light on fatigue skin/stringer debond failure. Four-point bending fatigue tests were carried out using the same specimen configurations and fixtures used in Refs 2 and 3. Two specimen configurations with the same flange layup but with different skin layups were investigated to evaluate the influence of the ply-stacking sequence on the fracture mode near the flange tip.

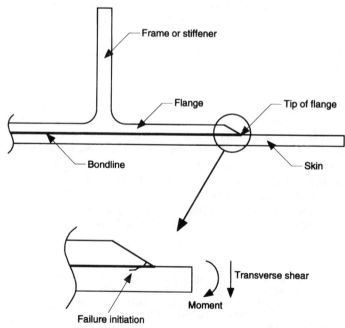

FIG. 1—*Illustration of frame/skin interface.*

Materials and Specimen Preparation

The specimens for fatigue loading were identical to the specimens used in previous monotonic tests [2,3]. They were machined from the same panels and consisted of a bonded skin and flange assembly shown in Fig. 2. To study the influence of skin layup only, two skin laminates, labeled S1 and S3, and one flange laminate, labeled F1, were combined to give two specimen configurations, A and D (Table 1). The terms S1, S3, F1, A, and D were chosen in accordance with the nomenclature used for monotonic testing [3]. Two panels of each configuration were produced. Both the skin and the flange laminates had a multidirectional layup, containing 0, 90, +45, and −45° plies. Moreover, the skin layups were chosen to given an almost identical bending stiffness, D_{11}. Laminate characteristics and 0° ply material properties are summarized in Tables 1 and 2, respectively.

Both the skin and flange were made from IM6/3501-6 graphite/epoxy prepreg tape with a nominal ply thickness of 0.188 mm. First, the flange and skin laminates were cured separately. The flange parts were then cut into 50-mm-long strips and machined with a 27° taper along the edges. Subsequently, the flange was adhesively bonded to the skin using a 177°C cure film adhesive from American Cyanamid (CYTEC 1515). A Grade 5 film was used to

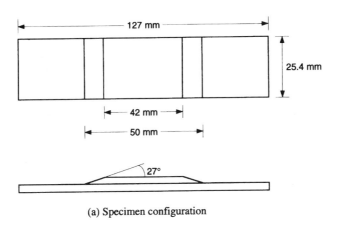

(a) Specimen configuration

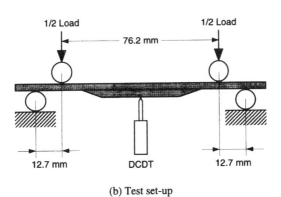

(b) Test set-up

FIG. 2—*Four-point bending specimen configuration and test setup:* (a) *specimen configuration;* (b) *test setup.*

TABLE 1—*Laminate characteristics.*

	Layup	Skin/Flange Thickness, mm	Bending Stiffness D_{11}, Nm
S1	[45/-45/0/0/45/90/-45]$_s$	2.8	112
S3	[90/45/0/0/-45/45/-45/$\overline{90}$]$_s$	3.0	117
F1	[45/90/-45/0/90]$_s$	2.0	22.5

NOTE: Configuration A = S1 + F1.
        Configuration D = S3 + F1.

yield a nominally 0.127-mm-thick bondline. However, because some of the adhesive flowed outwards during cure, the actual bondline thickness was 0.102 mm. Moreover, one panel of Configuration D showed ply waviness and ply drops on one flange side due to fabrication imperfections as shown in Fig. 3. This side represents the end of the flange panel. It was not trimmed and also not machined; therefore, the taper was mistakenly formed by terminated plies. A diamond saw was used to cut the panels into 25-mm-wide by 127-mm-long specimens. The specimen dimensions are shown in Fig. 2a.

Experimental Procedure

The same four-point bending test configuration used for monotonic testing was employed to perform the fatigue tests [2,3]. A schematic of the test fixture is shown in Fig. 2b. The bottom support had a 101.6-mm span, while the upper fixture had a 76.2-mm span. Midspan deflection was recorded using a spring-loaded direct current displacement transducer (DCDT) contacting the center of the frame flange as shown in Fig. 2b. The experiments were performed in a servohydraulic load frame in load control at a cyclic frequency of 5 Hz and an R ratio of 0.1. From the monotonic tests a load, P_{nl}, was determined at which the load versus stroke curves deviated slightly from the initial linear slope [3]. The average value of P_{nl} for the monotonic tests was found to be 1470 N for Configuration A and 1220 N for Configuration D. It was assumed that at this load level possible damage initiation may occur. When comparing different test configurations or when dealing with structural components, calculating the bending moment in the skin at the tip of the flange has been shown to be a better way of analyzing the data [1]. Since only one test configuration was used in the present investigation, the data are still displayed in terms of load. However, for reasons of comparison the bending moment at the tip of the flange may easily be calculated as $(P_{nl}/2)$ times the moment arm of 12.7 mm.

For each configuration, fatigue tests were run at load levels corresponding to 40, 50, 60, 70, and 80% of the load P_{nl}. Two tests were performed at the five levels with the exception of 40%, where only one specimen was tested. The cyclic loading was stopped at fixed intervals depending on the load level, and a photograph of the polished specimen edge was taken under a light microscope to document the occurrence and growth of matrix cracks and

TABLE 2—*IM6/3501-6 unidirectional graphite/epoxy tape
material properties.*

E_{11}, GPa	E_{22}, GPa	G_{12}, GPa	v_{12}
144.8	9.7	5.2	0.3

Ply Waviness   Flange ply drops

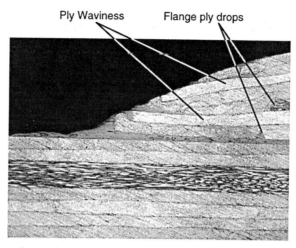

FIG. 3—*Side view of a specimen of Configuration D showing ply waviness and flange ply drops.*

delaminations. Table 3 lists the schedule used to obtain photographs for each cyclic load level. Damage was documented based on location at each of the four corners identified in Fig. 4. Also shown are cracks labeled a_1 to a_4 with respect to the four corners. The specimens were clamped into a three-point bending rig as shown in Fig. 5, and a small load was applied by hand tightening a set screw while the specimen and rig were placed under the microscope to open the matrix cracks and delaminations slightly, thereby increasing the visibility of the damage. Furthermore, at each interval the specimens were loaded monotonically to the mean load, and a plot of load versus mid-span deflection was recorded using an X-Y plotter. As damage was initiated, the specimen compliance, given by the slope of the plot, increased. Tests run at 40, 50, and 60% were terminated at 1 000 000 (one million) cycles. Unless specified otherwise in the text, tests at 70 and 80% were terminated at 100 000 cycles. To investigate the influence of the fabrication imperfections in one panel of Configuration D, four additional specimens were cut from that panel and were tested at load levels of 70 and 80% of P_{nl}.

TABLE 3—*Photographic schedule.*

Photograph taken every N cycle	Between Cycles N_1 and N_2	
	N_1	N_2
100	1	100
1 000	100	10 000
2 000	10 000	20 000
5 000	20 000	50 000
10 000	50 000	100 000
20 000	100 000	300 000
50 000	300 000	700 000
100 000	700 000	1 000 000

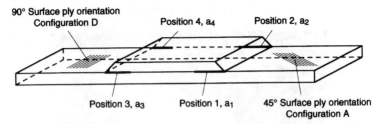

FIG. 4—*Four-point bending specimen with crack locations and surface ply orientations.*

Results and Discussion

Specimen Configuration A

In Fig. 6, results of Configuration A are summarized for the two replicate specimens at each load level as a plot of the number of cycles to the onset of matrix cracking and subsequent delamination. In some cases, a left hand arrow indicates damage initiation within the first 100 cycles. At a load level of 40%, a right hand arrow shows that no delaminations occurred within the test period of 1 000 000 cycles. The loads at onset of damage obtained from the monotonic tests are shown at the ordinate. These data points represent a load level of 100% of P_{nl}. A clear distinction between matrix cracking and delamination onset can be observed from the plot. Furthermore, a linear relationship between P_{max} and log N exists for each event, with very little scatter between the two replicates tested at each load level. The number of cycles between onset of matrix cracking and delamination covers a little over one order of magnitude for all load levels investigated with the exception of the 80% load data. A linear extrapolation of the fatigue data for matrix cracking onset and delamination onset suggests that no time delay exists between matrix cracking and delamination formation during monotonic loading. This is also consistent with the results at the highest fatigue load level (80%), where both events occurred within the same observation period.

Typical damage patterns observed in specimens of Configuration A are shown in Fig. 7. These drawings are based on the microscopic investigation performed during the tests. As

FIG. 5—*Three-point bend rig with specimen in place.*

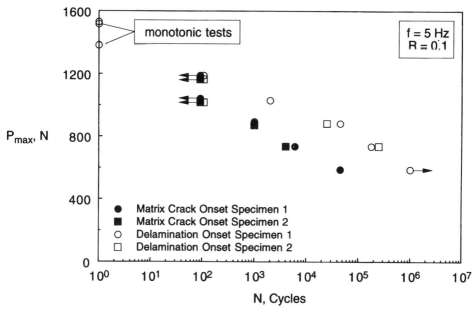

FIG. 6—*Maximum cyclic load as a function of the number of cycles to matrix cracking and subsequent delamination onset for Configuration A.*

shown in Fig. 7a, initial matrix cracks formed first at Corners 2 and 3, typically in the 90° flange ply as described in Ref 3. They initiated further matrix cracks in the lower 45° flange ply. Subsequently, delaminations (labeled "B" for bondline failures) formed from these matrix cracks. These "B" delaminations ran to one side of the interface between the bondline and the composite, usually the top skin ply interface. The first delamination always corresponded to the first matrix crack that formed. They were followed by delamination between plies (labeled "P") at Corners 1 and 4. In some specimens, the "P" delaminations initiated in the −45°/90° flange ply interface with no matrix crack, whereas in other specimens they initiated from matrix cracks in the 90° flange ply. Both scenarios are displayed in Fig. 7b. "P" delaminations always resulted in a delamination running in the 90°/45° flange ply interface. Each flange side (Side 1 = Corners 1 and 2, Side 2 = Corners 3 and 4) consisted of a "P" and a "B" delamination. At each flange side, the "P" delaminations started later than the corresponding "B" delaminations, but almost immediately equaled or exceeded them in length. As the "P" delaminations grew, they would tend to arrest and form new matrix cracks branching into the bottom 45° flange ply (see Fig. 7b). These matrix cracks always stopped at the bondline. After the matrix crack had formed, the delamination would start to grow again. Delaminations arrested only beyond the tapered region of the flange. No branching into the skin ply or into the bondline was observed. As an example, Figs. 8 and 9 show micrographs of a Configuration A specimen tested at 80% of P_{nl} exhibiting typical "B" and "P" damage patterns.

In Figs. 10 to 17, the results are shown as plots of delamination length (see Fig. 7 for definition) versus the number of cycles for each load level for the Configuration A specimens with delaminations. At a maximum load, P_{max}, of 80% of P_{nl} (see Fig. 10 and Fig. 11), matrix cracks and delaminations had formed within the first 100 cycles in both specimens. The tests were terminated at crack lengths of about 20 mm, corresponding to almost total

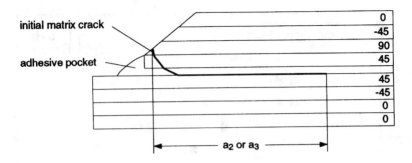

(a) "B"-delaminations: positions 2 and 3

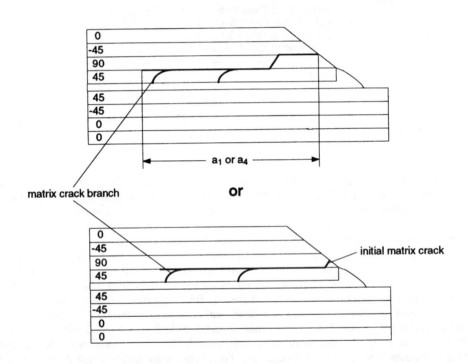

(b) "P"-delaminations: positions 1 and 4

FIG. 7—*Typical damage patterns in specimens of Configuration A: (a) "B" delaminations: Positions 2 and 3; (b) "P" delaminations: Positions 1 and 4.*

debonding of the flange from the skin. At P_{max} of 70% of P_{nl} (see Fig. 12 and Fig. 13), matrix cracks also formed within the first 100 cycles in both specimens. As delaminations grew from all four corners, delaminations on Side 1 tended to slow down, while delaminations on Side 2 continued to grow until the tests were terminated. At P_{max} of 60% of P_{nl} (see Fig. 14 and Fig. 15), matrix cracks formed in all specimens within the first 1000 cycles. Delaminations on Side 2 grew faster than delaminations on Side 1. At P_{max} of 50% of P_{nl}

Delamination   Matrix cracks

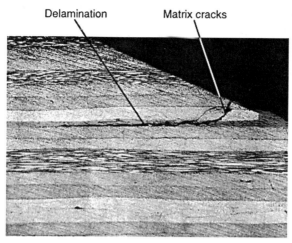

FIG. 8—*Side view of a failed specimen of Configuration A at a load level of 80% of* P_{nl} *showing a type "B" delamination between skin and flange at the bondline at Corner 2.*

(see Fig. 16 and Fig. 17), matrix cracks formed first between 4000 and 6000 cycles in the 90° flange ply closest to the skin, initiating a single delamination in the bondline of each specimen after 100 000 cycles. In one specimen, the delamination formed at Corner 3, while in the second specimen it formed at Corner 2. At P_{max} of 40% of P_{nl}, only one specimen was tested. A single transverse matrix crack formed at 45 000 cycles at Corner 3 in the 90° flange ply closest to the skin. No delaminations formed within one million cycles.

Matrix crack   Delamination   Matrix crack branch

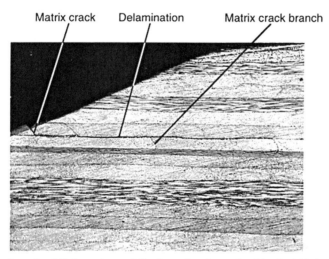

FIG. 9—*Side view of a failed specimen of Configuration A at a load level of 80% of* P_{nl} *showing a type "P" delamination in the 90°/45° flange interface at Corner 4.*

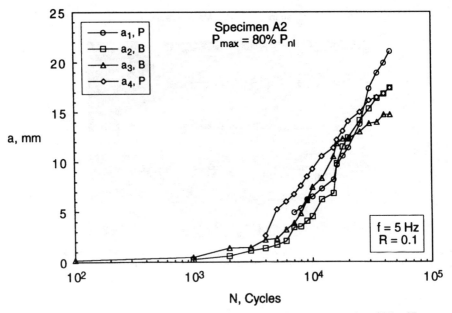

FIG. 10—*Delamination length versus number of cycles for Specimen A2 at 80% of* P_{nl}.

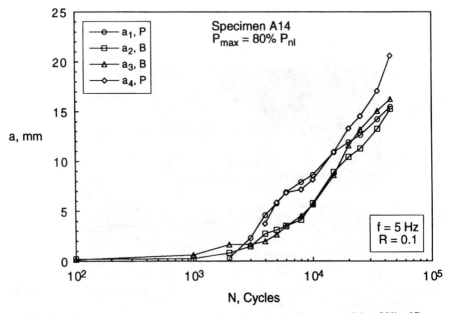

FIG. 11—*Delamination length versus number of cycles for Specimen A14 at 80% of* P_{nl}.

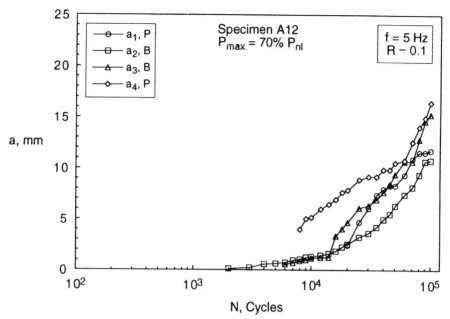

FIG. 12—*Delamination length versus number of cycles for Specimen A12 at 70% of* P_{nl}.

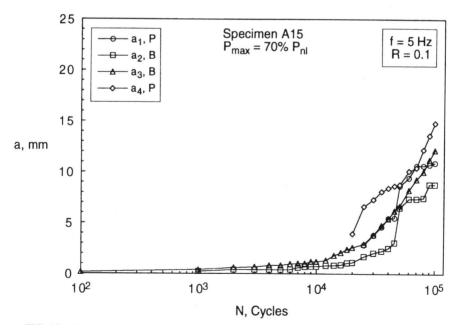

FIG. 13—*Delamination length versus number of cycles for Specimen A15 at 70% of* P_{nl}.

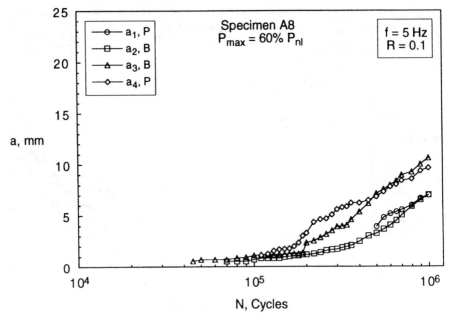

FIG. 14—*Delamination length versus number of cycles for Specimen A8 at 60% of* P_{nl}.

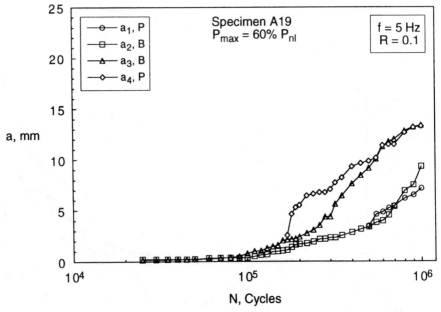

FIG. 15—*Delamination length versus number of cycles for Specimen A19 at 60% of* P_{nl}.

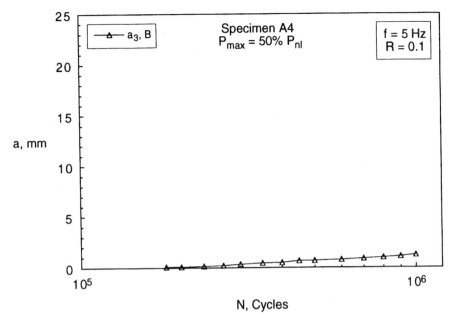

FIG. 16—*Delamination length versus number of cycles for Specimen A4 at 50% of* P_{nl}.

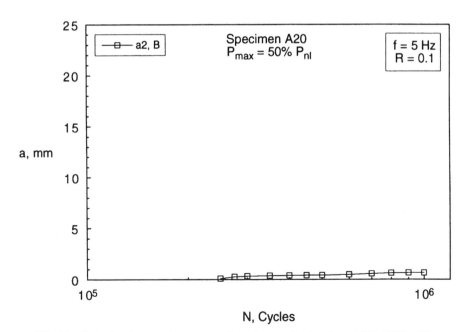

FIG. 17—*Delamination length versus number of cycles for Specimen A20 at 50% of* P_{nl}.

Specimen Configuration D

In Fig. 18, results of Configuration D are summarized as a plot of the number of cycles to the onset of matrix cracking and subsequent delamination. At a load level of 80%, a left hand arrow indicates matrix cracking within the first 100 cycles. At other load levels, a right hand arrow shows that no delaminations or matrix cracks occurred within the test period of either 100 000 or 1 000 000 cycles. The loads at onset of damage obtained from the monotonic tests are shown at the ordinate. These data points represent a load level of 100% of P_{nl}. In contrast to Configuration A, no clear differentiation between matrix cracking and delamination onset is seen, as the scatter for each event is significantly higher. Hence, the data overlap and no delay between those two events is apparent. This overlap is more pronounced for load levels of 40, 50, and 60%. The data scatter and the random manner of delamination growth also reflects the higher scatter compared to Configuration A observed in monotonic tests. At the investigated load levels, the presence of an imperfection in the form of a flange ply drop does not seem to have any influence on either event. This is shown in Fig. 19, where the data from Fig. 18 are replotted for onset to matrix cracking only to discriminate between ideal specimens and specimens with flange ply drop. In one of the two specimens tested at 50 and at 40%, both matrix cracking and delamination formation could not be observed within the test duration.

Unlike the previous specimen Configuration A that was tested, delaminations in specimens of Configuration D propagated in a more random manner. The only consistency found in the damage patterns is related to the four flange corners (see drawings in Fig. 20).

Transverse matrix cracks typically formed in the 90° skin ply closest to the flange at Corner 1 in all specimens and at Corner 4 in specimens with no flange ply drop ("ideal") (see top section of Fig. 20a). Subsequently, these 90° skin ply cracks initiated matrix cracks

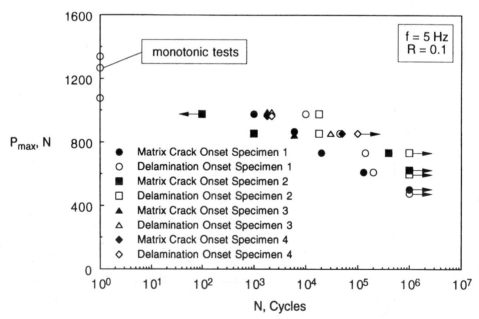

FIG. 18—*Maximum cyclic load as a function of the number of cycles to matrix cracking and subsequent delamination onset for Configuration D.*

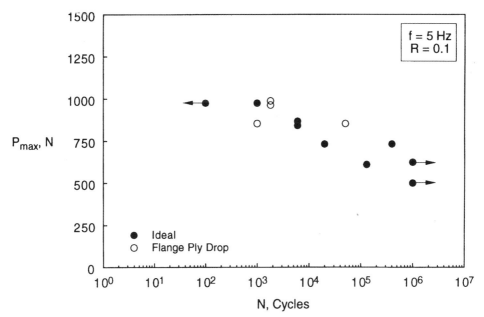

FIG. 19—*Maximum cyclic load as a function of the number of cycles to matrix cracking onset for Configuration D.*

in the 45° ply below, which in turn initiated delaminations in the 45°/0° interface. Once the delamination started to grow, the adhesive pocket also cracked. However, in the specimens with internal ply drops on one flange side (see Fig. 3 and bottom section of Fig. 20a), matrix cracks at Corner 4 formed first in the ply drop region of the bottom 45 or 90° flange ply and in the bondline and then initiated further matrix cracks and delaminations as described above. The majority of these delaminations stayed in the 45°/0° interface. In some specimens, however, those delaminations started to split into the two 0° plies at longer crack lengths. These events are labeled "split" in Fig. 20.

At Corners 2 and 3, no consistent correlation between the flange imperfections and the damage patterns was observed. Typically, some transverse matrix cracks formed in the 90° skin ply closest to the flange, initiating a delamination in the 90°/45° interface as depicted in the top section of Fig. 20b. The other matrix cracks formed either in the flange ply drop region (not shown) or in the 90° flange ply closest to the skin, initiating further matrix cracks in the bottom 45° flange ply shown in the bottom section of Fig. 20b. Subsequently, a short delamination formed in the bondline, creating another matrix crack in the 90° top skin ply. At the end of this transverse crack a delamination formed in the 90°/45° interface. At this point, all delaminations started to grow in a similar manner. As they propagated, they would tend to arrest and form new matrix cracks branching into the top 45° skin ply. These matrix cracks always stopped at the 45°/0° interface. After the matrix crack had formed, the delamination would start to grow again. However, delamination arrest was not related to any position within the laminate. As the delaminations grew further, they started to branch into the lower 45°/0° interface (labeled "branch" in Fig. 20), along with subsequent crack splitting into the two 0° plies. Branching and splitting were either detected within the same observation cycle or occurred within a few thousand cycles of each other. As an example, micrographs of a specimen tested at 80% of P_{nl} display typical damage patterns in Figs. 21

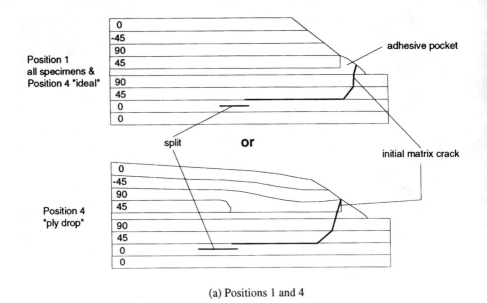

(a) Positions 1 and 4

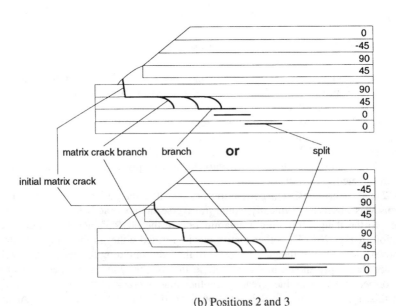

(b) Positions 2 and 3

FIG. 20—*Typical damage patterns in specimens of Configuration D: (a) Positions 1 and 4; (b) Positions 2 and 3.*

and 22. In Figs. 23 to 31, the results are shown as plots of delamination length versus the number of cycles for each load level for the specimens with delaminations.

At P_{max} of 80% of P_{nl} (see Figs. 23 to 26), matrix cracks formed first between 100 and 2000 cycles. Subsequently, delaminations formed from these matrix cracks. The first delam-

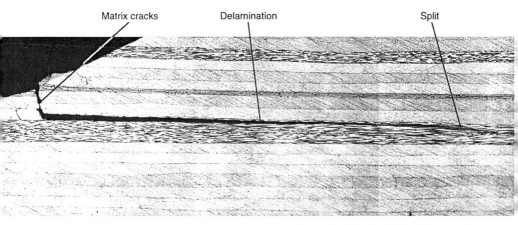

FIG. 21—*Side view of a failed specimen of Configuration D at a load level of 80% of* P_{nl} *showing a delamination in the 45°/0° skin interface at Corner 1.*

ination always corresponded to the first matrix crack that formed. Only in one specimen (D16) were no branching and splitting found, which is most likely due to the comparatively very short crack lengths at test termination.

At P_{max} of 70% of P_{nl} (see Figs. 27 to 29), matrix cracks formed first between 1000 and 50 000 cycles. In some cases, delaminations formed from these matrix cracks. For the "ideal" specimens without a flange ply drop, the first delamination always corresponded to the first matrix crack that formed. In one specimen with a flange ply drop, a matrix crack occurred at 1000 cycles but did not result in a delamination for the time span investigated. Instead, at 18 000 cycles a matrix crack and a delamination formed at a different corner. In the second specimen with a flange ply drop, a single matrix crack occurred at 50 000 cycles. No other matrix cracks and no delamination formed in this specimen before the test was terminated.

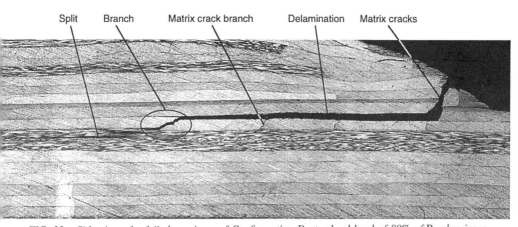

FIG. 22—*Side view of a failed specimen of Configuration D at a load level of 80% of* P_{nl} *showing a delamination in the 90°/45° skin interface branching into the 45°/0° interface at Corner 3.*

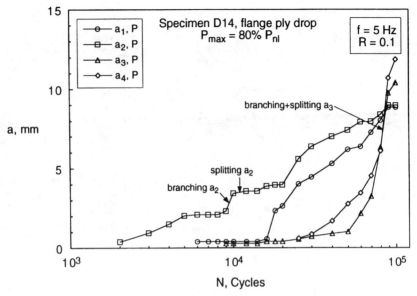

FIG. 23—*Delamination length versus number of cycles for Specimen D14 at 80% of* P_{nl}.

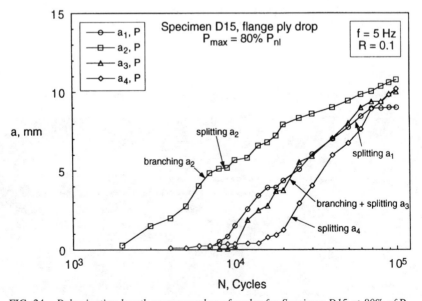

FIG. 24—*Delamination length versus number of cycles for Specimen D15 at 80% of* P_{nl}.

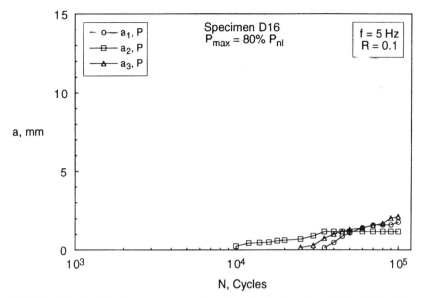

FIG. 25—*Delamination length versus number of cycles for Specimen D16 at 80% of* P_{nl}.

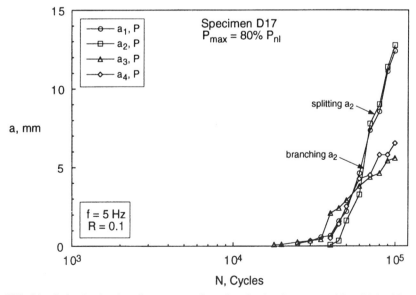

FIG. 26—*Delamination length versus number of cycles for Specimen D17 at 80% of* P_{nl}.

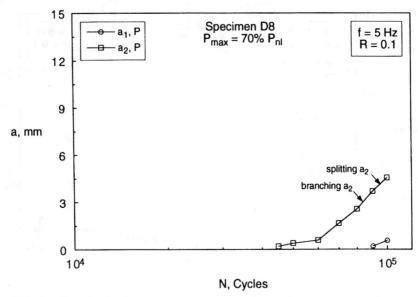

FIG. 27—*Delamination length versus number of cycles for Specimen D8 at 70% of* P_{nl}.

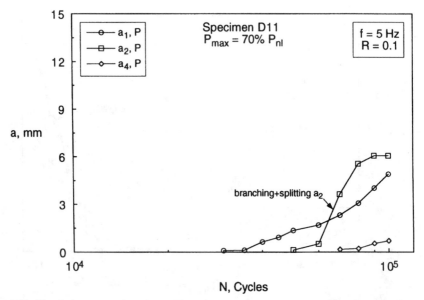

FIG. 28—*Delamination length versus number of cycles for Specimen D11 at 70% of* P_{nl}.

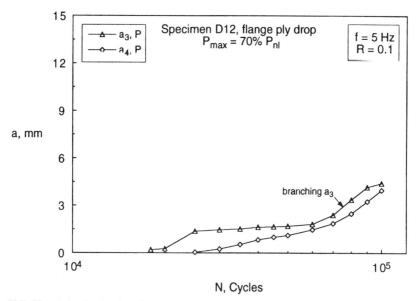

FIG. 29—*Delamination length versus number of cycles for Specimen D12 at 70% of* P_{nl}.

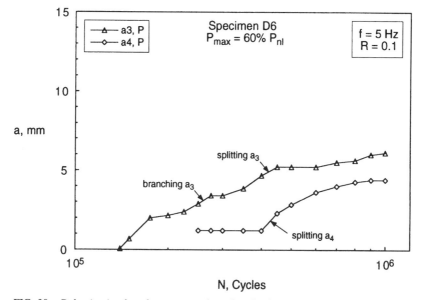

FIG. 30—*Delamination length versus number of cycles for Specimen D6 at 60% of* P_{nl}.

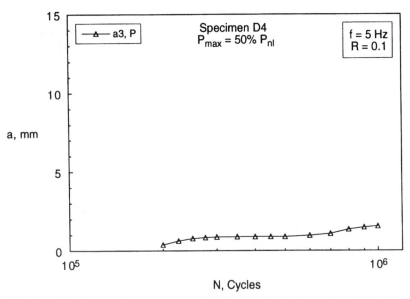

FIG. 31—*Delamination length versus number of cycles for Specimen D4 at 50% of* P_{nl}.

At P_{max} of 60% of P_{nl} (see Fig. 30), matrix cracks formed first between 20 000 and 400 000 cycles. In one of the two specimens tested at this load level, delaminations formed from these matrix cracks. The first transverse matrix crack formed at Corner 4. Subsequently, a matrix crack formed at Corner 3 and initiated a delamination in the 90°/45° interface at 140 000 cycles. At 250 000 cycles another matrix crack formed at Corner 4 in the 45° ply below the top 90° skin ply. It instantly initiated a delamination in the 45°/0° interface. In the other specimen, a matrix crack formed at 400 000 cycles but did not result in a delamination.

At P_{max} of 50% of P_{nl} (see Fig. 31), a single matrix crack and a delamination formed only in one of the two specimens within the period investigated. In this specimen, the only matrix crack occurred at 130 000 cycles at Corner 3 in the 90° skin ply closest to the flange. It resulted in a delamination in the 90°/45° interface, forming at 200 000 cycles. No matrix cracks and delaminations were detected in the second specimen. At P_{max} of 40% of P_{nl}, no matrix cracks and delaminations formed within one million cycles.

Configuration Comparison

In Fig. 32, the two specimen configurations are compared with respect to the onset of matrix cracking. Again, left hand arrows indicate damage within the first 100 cycles, while right hand arrows show that no matrix cracks were observed within the test period of one million cycles. Matrix cracking typically occurred earlier in Configuration A, where 90° plies in the skin were located further away from the bondline than in Configuration D. This indicates that the location of the 90° skin and flange plies relative to each other and to the bondline is the dominant layup feature that controls the location and onset of matrix cracking and subsequent delamination. While specimens of Configuration A show very little scatter, the data for Configuration D displays significant variability at each load level. It is believed

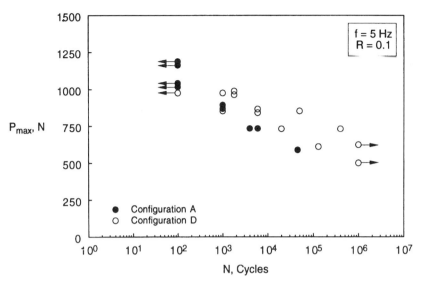

FIG. 32—*Comparison of the maximum cyclic load as a function of the number of cycles to matrix crack onset.*

that this scatter is related to the transverse tension strength of the surface 90° skin ply. This large variability has also been found in 90° flexural fatigue tests performed in Ref *4*.

Another way of comparing the results is to show the number of cycles to delamination onset once a matrix crack has already formed. This can be established by simply subtracting the number of cycles to onset of matrix cracking from the number of cycles to delamination onset. The combined data for Configurations A and D are shown in Fig. 33. As pointed out before, in some cases both events occurred within the first 100 cycles or could not be detected until the tests were terminated. Hence, as no complete information of these damage stages was obtained, these data points have been excluded from the graph. Right hand arrows indicate that matrix cracking occurred within one million cycles, whereas delaminations did not occur. Although both configurations have been tested at different absolute load levels, there is little difference in the fatigue response for delamination onset between the two configurations once matrix cracks are formed. As noted before, with decreasing load levels the number of cycles to delamination onset is shifted towards higher values. The scatter for the onset of delamination once a matrix crack has formed in specimens of Configuration D is smaller than observed for matrix cracking alone, indicating that delamination onset is not so sensitive to the skin layup.

Summary and Conclusions

This paper addresses the fatigue debonding behavior of multidirectional composite skin/stringer configurations. Two different specimen configurations, A and D, were tested at various load levels in four-point bending to investigate the influence of skin-stacking sequence. The specimen edges were examined at discrete time intervals under a light microscope to study the location of damage onset and subsequent damage progression. The location of the 90° skin and flange plies relative to each other and to the bondline was identified as the

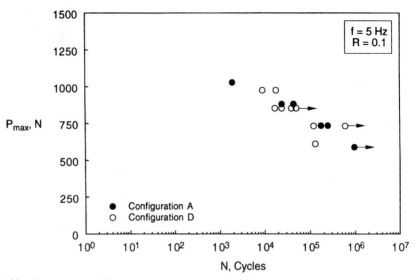

FIG. 33—*Comparison of the maximum cyclic load as a function of the number of cycles to delamination onset once a matrix crack is present.*

dominant layup feature that controls the location and onset of matrix cracking and subsequent delamination.

In specimens of Configuration A, delaminations always initiated from a matrix crack in the flange near the flange tip and then grew at slightly different rates depending on delamination location (between the flange or skin plies or at the bondline). All three failure modes have also been observed in monotonic experiments previously performed. For specimen Configuration D, matrix cracks formed either in 90° plies in the skin, as reported for monotonic tests, or in the flange. In both events, damage initiation was again limited to an area near the tip of the flange. Subsequently, delaminations always grew in the skin, eventually exhibiting branching into lower interfaces and splitting within 0° plies at longer crack lengths. Flange ply drops due to fabrication imperfections observed in some specimens resulted in only slightly different damage patterns.

As expected, damage onset and propagation was shifted towards higher lifetimes as load levels decreased. At 40% of the monotonic failure initiation load, no damage progression could be observed in either configuration within the test period of one million cycles. Again, specimens cut from laminates with flange ply drops due to manufacturing imperfections did not exhibit different fatigue crack growth behavior. Matrix cracking typically occurred earlier in Configuration A, where 90° plies in the skin where located further away from the bondline than in Configuration D. However, a comparison of both configurations shows that the number of cycles to delamination onset once matrix cracks are present does not strongly depend on skin layup. The comparatively high scatter in the fatigue data for specimen Configuration D and the random manner of delamination growth observed for this configuration also reflect the scatter observed in monotonic tests. It is believed that this scatter is related to the low transverse tension strength of the surface 90° skin ply as pointed out in Ref 3. This large variability has also been found in 90° flexural fatigue tests [4]. Finally, it can be concluded that when designing for bending stiffness, the skin and flange stacking sequences with the 90° plies located as far away from the bondline as possible should be the most durable in terms of matrix cracking and delaminations formation and growth.

Acknowledgments

This work was performed while the first author held a National Research Council Research Associateship supported by the U.S. Army Research Laboratory Vehicle Technology Center at NASA Langley Research Center.

References

[1] Minguet, P. J., Fedro, M. J., O'Brien, T. K., Martin, R. H., Ilcewicz, L. B., Awerbuch, J., and Wang, A., "Development of a Structural Test Simulating Pressure Pillowing Effects in a Bonded Skin/Stringer/Frame Configuration," *Proceedings,* Fourth NASA/DoD Advanced Composites Technology Conference, Salt Lake City, UT, June 1993.

[2] Minguet, P. J. and O'Brien, T. K., "Analysis of Test Methods for Characterizing Skin/Stringer Debonding Failures in Reinforced Composite Panels," *Composite Materials: Testing and Design, Twelfth Volume, ASTM STP 1274,* August 1996, pp. 105–124.

[3] Minguet, P. J. and O'Brien, T. K., "Analysis of Composite Skin/Stringer Failure Using a Strain Energy Release Rate Approach," *Proceedings,* Tenth International Conference on Composite Materials (ICCM X), Vancouver, British Columbia, Canada, August 1995, Vol. 1, pp. 245–252.

[4] Peck, A. W., "Transverse Tension Fatigue Characterization of Composite Materials," presented at the Seventh ASTM Symposium on *Composite Materials: Fatigue and Fracture,* (ASTM STP 1330), in press.

Reji John,[1] Jay R. Jira,[1] and James M. Larsen[1]

Effect of Stress and Geometry on Fatigue Crack Growth Perpendicular to Fibers in Ti-6Al-4V Reinforced with Unidirectional SiC Fibers

REFERENCE: John, R., Jira, J. R., and Larsen, J. M., **"Effect of Stress and Geometry on Fatigue Crack Growth Perpendicular to Fibers in Ti-6Al-4V Reinforced with Unidirectional SiC Fibers,"** *Composite Materials: Fatigue and Fracture, Seventh Volume, ASTM STP 1330,* R. B. Bucinell, Ed., American Society for Testing and Materials, 1998, pp. 122–144.

ABSTRACT: Critical turbine engine and aircraft components fabricated from continuous fiber-reinforced metal matrix composite (MMC) will experience cyclic loads during service, and many of these components typically contain crack initiators. Hence, extensive characterization of the fatigue crack growth behavior of a model MMC ($[0]_8$ SCS-6/Ti-6Al-4V) was initiated by the USAF Wright Laboratory. This paper discusses some of the results of the experimental and analytical investigation of fatigue crack propagation in $[0]_8$ SCS-6/Ti-6Al-4V. Automated fatigue crack growth tests were conducted using middle tension, M(T), specimens at 23°C with a stress ratio of 0.1. During some of the tests, the crack opening displacement profile was measured to verify the stress distributions predicted by fiber-bridging models. The results are also compared with those available for SM1240/Ti-6Al-4V under tension and bending fatigue loading. This study showed that the shear lag model assuming a constant value of τ can be used to predict bridged crack growth perpendicular to fibers in SCS-6/Ti-6Al-4V and SM1240/Ti-6Al-4V over a wide range of stress levels and under tension and bending fatigue loading conditions. The predictions of partially bridged and unbridged crack growth, crack opening displacements, and slip lengths correlated well with the data. The value of the fiber/matrix interfacial shear stress, τ, was the same for SCS-6/Ti-6Al-4V and SM1240/Ti-6Al-4V, implying that the bridging mechanism of the SM1240 fiber is identical to that of the SCS-6 fiber at room temperature. The results also indicate that the onset of fiber failure could be predicted using the bundle strength as the critical value.

KEYWORDS: bridging stress-intensity factor, center crack, crack growth rate, crack opening displacement, cyclic loading, fatigue crack growth, fiber bridging, life prediction, metal matrix composite, shear lag model, titanium matrix composite

Titanium alloy matrix composites (TMC) are targeted for use in many aerospace applications that require high specific strength and stiffness at elevated temperatures [1,2]. Critical turbine engine and aircraft components fabricated from continuous fiber-reinforced TMC will experience cyclic loads during service, and any of these components typically contain crack initiators. Hence, many investigations [3–25] have been conducted to characterize the fatigue crack growth behavior of titanium matrix composites such as SCS-6/Ti-24Al-11Nb, SCS-6/TIMETAL®21S, SM1240/Ti-6Al-4V, SM1240/TIMETAL®21S, SCS-6/Ti-15V-

[1] Materials and Manufacturing Directorate, Air Force Research Laboratory (AFRL/MLLN), Wright-Patterson Air Force Base, OH 45433-7817. The first author is also with the University of Dayton Research Institute, Dayton, OH 45469-0128.

3Cr-3Al-3Sn, and SCS-6/Ti-6Al-4V. The matrix crack growth in these composites is dominated typically by fiber bridging, resulting in decreasing crack growth rate with increasing crack length [3–22]. This behavior has been modeled generally as fully bridged crack growth using various fiber-bridging models such as shear lag models [4–7,11,13–15,18–25], fiber pressure models [7], and constant-bridging stress models [6,14,15]. Most of the available data have been restricted to one or two stress levels. Jira and Larsen [12] and Larsen et al. [13] provided extensive data under a wide range of stress levels for crack growth from holes in [0]$_8$ SCS-6/Ti-24Al-11Nb and various layups of SCS-6/TIMETAL®21S, respectively. Larsen et al. [13,26] showed that the shear lag model could be used to predict the crack growth in various layups of SCS-6/TIMETAL®21S over a wide range of stress levels, and at room and elevated temperatures. The fiber/matrix interfacial stress, τ, in the shear lag models was found to be independent of applied stress and decreased with increasing temperature [13,27]. The effect of temperature on τ is similar to the results reported by Bakuckas and Johnson [8,25] and Connell and Zok [21].

All these investigations essentially determined τ as a fitting parameter based on the measured crack length versus cycles response and the constituent properties. As shown by Bakuckas and Johnson [8] and Larsen et al. [13], the value of τ deduced from crack growth data depends on the fiber-bridging model, and the assumed relationship between the effective stress intensity factor range at the matrix crack tip in the composite and the neat (fiberless) matrix. Many attempts [4–8,13–16,18,20] have been made to relate the τ required to predict the crack growth behavior to that obtained from push-out tests. Unfortunately, the wide range of values reported for τ appears to indicate the futility of such a correlation. Cox [19], in discussing the limitations and applicability of the existing fiber-bridging models, suggested the deduction of the bridging law (and associated parameters) directly from critical experiments. This is consistent with deducing τ as a fit parameter. Thus, determination of the parameters required for developing a comprehensive life prediction model requires extensive testing and critical measurements.

An extensive characterization of the fatigue crack growth behavior of a model TMC was initiated by the USAF Wright Laboratory under the Metal Matrix Composite (MMC) Life Prediction Cooperative Program. The model TMC system chosen by the Cooperative was [0]$_8$ SCS-6/Ti-6Al-4V. The test program included a wide range of loading conditions such as low to high stress levels, stress ratios from 0.1 to 0.7, and temperatures from 23 to 316°C. This paper discusses the results from the tests on [0]$_8$ SCS-6/Ti-6Al-4V conducted under tension fatigue loading at room temperature with a stress ratio of 0.1. These results are compared with the results available in the literature for [0]$_4$ SCS-6/Ti-6Al-4V [6] and [0]$_6$ SM1240/Ti-6Al-4V [20,21] under tension loading and [0]$_6$ SM1240/Ti-6Al-4V under three-point bend loading [9]. The ability of a shear lag model to predict the crack growth in these composites under a wide range of stress levels is also discussed.

Experimental Procedure

Material

The composite consisted of the Ti-6Al-4V matrix reinforced with continuous SiC fibers (SCS-6™). The composite panels were manufactured by Textron Specialty Materials Division (SMD) by hot isostatic pressing alternating layers of woven fiber mat and matrix foil. The Textron SCS-6 fiber is produced by chemical vapor deposition on a carbon monofilament core. The fiber has a nominal diameter of 142 μm and possesses a double-pass, carbon-rich outer coating. The fiber mat was woven with a fiber spacing equivalent to 129 fibers per 25.4 mm. The thickness of the 8-ply panels ranged from 1.82 to 1.94 mm. Knowing the thickness of the composite and the fiber spacing, the volume faction of the fibers, V_f, can be calculated as

$$V_f = \frac{\pi R_f^2}{B_p s} \tag{1}$$

where

R_f = fiber radius,
B_p = average ply thickness = B/n,
B = thickness of composite,
n = number of plies, and
s = fiber spacing (= 25.4/129 = 0.197 mm).

Thus, V_f for the specimens ranged from 0.32 to 0.35. Following consolidation, the composite was aged under vacuum for 8 h at 621°C to stabilize the matrix prior to testing. The properties of the constituent materials can be found in Refs 28 and 29.

In this paper, the crack growth behavior of [0]$_8$ SCS-6/Ti-6Al-4V is compared with that obtained from [0]$_4$ SCS-6/Ti-6Al-4V [6] and [0]$_6$ SM1240/Ti-6Al-4V [9,20,21]. The loading conditions, geometric parameters, material constituents, and stress levels are listed in Table 1. All these tests were conducted at R = 0.1 in laboratory air. Note that the stress levels shown in Table 1 are the maximum far-field stresses for the center-cracked geometry and the maximum stresses in the outer layer for the bending geometry.

SCS-6 and SM1240 are essentially SiC fibers with carbon and tungsten cores, respectively. The SCS-6 and SM1240 fibers are coated with two layers of carbon-rich coating (≈ 1 μm thick/layer). SM1240 has an additional coating of TiB_2 on the outer layer. The presence of the carbon layers promotes the fiber/matrix interfacial sliding required to generate fiber bridging across the crack. After consolidation, significant residual stresses exist in SCS-6/Ti-6Al-4V [28] and SM1240/Ti-6Al-4V due to the mismatch of the coefficient of thermal expansion (CTE) between the fibers and the matrix. A finite-difference-based nonlinear code, FIDEP2 [29], was used to estimate the residual stresses in these composites. For $V_f \approx 0.32$, the radial residual stress in matrix at the fiber/matrix interface ≈ -260 and -235 MPa for SCS-6/Ti-6Al-4V and SM1240/Ti-6Al-4V, respectively. Since the fibers contain similar carbon coatings and are subjected to similar radial residual stresses, the fiber/matrix frictional shear stresses opposing the crack opening in the matrix may be expected to be similar.

Testing

A schematic of the center-cracked-tension specimen, M(T), is shown in Fig. 1. Typically, width, W, was 19.0 mm; thickness, B, was 1.82 to 1.94 mm; and initial notch to width ratio, $2a_0/W$, was 0.11. The dimensions of the specimens corresponding to the tested stress levels are shown in Table 1. The initial notches, created using electrical discharge machining (EDM), were ≈ 0.3 mm wide. The notches were oriented perpendicular to the fibers. The specimens were ≈ 152 mm long with a clear distance of ≈ 102 mm between the clamped ends. Automated fatigue crack growth tests were conducted with the stress ratio R (= minimum stress/maximum stress) of 0.1 and a loading frequency of 1 Hz. During the tests, crack extension from the machined notch was monitored using the direct current electric potential (DCEP) technique. Periodic optical measurements of the four crack tips enabled verification and subsequent correction of the DCEP data. During some of the tests, the crack

TABLE 1—*Fatigue crack growth data used for analysis.*[a]

Fiber	Matrix	Layup	V_f	Geometry	Maximum Stress, MPa	Stress Ratio	Width, W (mm)	Unbridged Notch/Width	Reference
SCS-6[c]	Ti-6Al-4V	[0]_8	0.35 0.33 0.33	Center-cracked in tension, M(T)	400 400 700	0.1 0.1 0.1	19.0 19.1 18.9	0.11[e] 0.11 0.11	This study
SCS-6[c]	Ti-6Al-4V	[0]_4	≈0.42	Center-cracked in tension, M(T)	171	0.1	24.0	0.25[e]	Davidson [5]
SM1240[d] (Sigma)	Ti-6Al-4V	[0]_6	≈0.32	Center-cracked in tension, M(T)	844 889	0.1	15.0	0.20	Zok, Connell and Du [20]
SM1240[d] (Sigma)	Ti-6Al-4V	[0]_6	≈0.32	Center-cracked in tension, M(T)	222 278 333 389 444	0.1	15.0	0.20	Connell and Zok [21]
SM1240[d] (Sigma)	Ti-6Al-4V	[0]_6	≈0.31	Single edge in bending, SE(B)	330[b] 404[b] 453[b]	0.1	4.5	0.24	Cardona, Barney and Bowen [9]

[a] All the data were obtained at room temperature (23 to 25°C) in laboratory air.
[b] Maximum stress calculated at the outermost surface for a three-point bend configuration.
[c] SCS-6 is a SiC fiber (diameter = 142 μm) with a carbon core and dual coatings of carbon on the surface. Manufactured by Textron Specialty Materials Division (SMD), Lowell, MA.
[d] SM1240 is a SiC fiber (diameter = 100 μm) with a tungsten core and dual coatings of carbon and boron-rich TiB$_2$ on the surface. Manufactured by BP Metal Composites Limited, Hampshire, U.K.
[e] Crack opening displacement profile measured during crack growth.

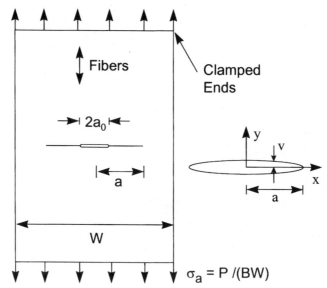

FIG. 1—*Schematic of center-cracked-tension specimen, M(T), used for testing* $[0]_8$ *SCS-6/Ti-6Al-4V.*

opening displacement (COD) profile was measured along both cracks on one side of the specimen using the laser interferometric displacement gage (IDG) system [30].

In addition to the tests on composites, fatigue crack tests were also conducted on the neat (fiberless) Ti-6Al-4V, which was consolidated from matrix alloy foils into 1.44-mm-thick panels using the composite material processing and aging conditions. The single-edge-cracked geometry with clamped ends [31,32] was used to conduct threshold-type and constant-load tests at $R = 0.1$ and 0.7. All the tests were conducted with a loading frequency of 1 Hz. During these tests, crack length and closure behavior were determined from crack mouth opening compliance measurements made using an extensometer.

Experimental Results

Crack Growth in Neat Ti-6Al-4V

The crack growth rate (da/dN) versus effective stress-intensity factor range ($\Delta K_{m,e}$) behavior of neat Ti-6Al-4V is shown in Fig. 2. The data from the tests at $R = 0.1$ exhibited the closure stress-intensity factor, K_{cl} of ≈ 0.3 to 0.5 K_{max}, where K_{max} = maximum stress intensity factor. Thus, $\Delta K_{m,e} = K_{max} - K_{cl}$. The tests at $R = 0.7$ did not exhibit closure. As shown in Fig. 2, the closure-corrected data obtained at $R = 0.1$ correlated well with the data obtained at $R = 0.7$. These data can be represented by

$$\log\left(\frac{da}{dN}\right) = C_1 \left(\sinh[C_2\{\log(\Delta K_{m,e}) + C_3\}]\right) + C_4 \tag{2}$$

where $C_1 = 1.4474$, $C_2 = 2$, $C_3 = -1.0822$ and $C_4 = -7.1261$, $\Delta K_{m,e}$ is in MPa$\sqrt{\text{m}}$, and da/dN is in m/cycle. This equation, which is valid for growth rates ranging from 5×10^{-10} to 3×10^{-6} m/cycle, is shown as a solid line in Fig. 2. Three other lines corresponding to the matrix behavior assumed by Davidson [6], Zok et al. [20], and Connell and Zok [21]

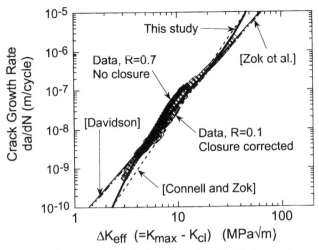

FIG. 2—*Crack growth rate behavior of neat Ti-6Al-4V at room temperature.*

are also shown in Fig. 2. These Paris-law-type equations are close to the data obtained during this study for neat Ti-6Al-4V. Davidson's [6] equation is based on test data from recrystallization annealed (RA) Ti-6Al-4V. The equation by Connell and Zok [21] was deduced by fitting the fatigue crack growth data from $[0]_6$ SM1240/Ti-6Al-4V.

Crack Growth in $[0]_8$ SCS-6/Ti-6Al-4V

The crack growth response obtained from specimens tested at 400 and 700 MPa are shown in Figs. 3 and 4, respectively. The specimens tested at 400 MPa exhibited decreasing crack

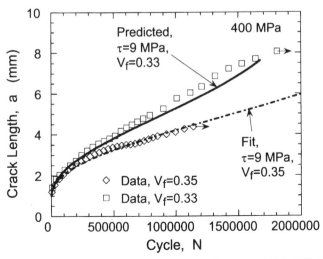

FIG. 3—*Comparison of predicted and measured crack growth response in $[0]_8$ SCS-6/Ti-6Al-4V tested under tension fatigue loading at σ_{max} = 400 MPa. Prediction assumes fully bridged crack growth.*

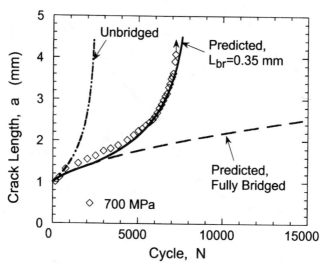

FIG. 4—*Comparison of predicted and measured crack growth response in* $[0]_8$ *SCS-6/Ti-6Al-4V tested under tension fatigue loading at* σ_{max} = 700 MPa.

growth rate (da/dN = slope of crack length versus cycles data) with increasing cracking length, consistent with the phenomenon of fully bridged crack growth [4]. One specimen was stopped after ≈1.2 × 10^6 cycles, and the other was tested until the crack tips were close to the edges of the specimen. As shown in the figure, the V_f of these two specimens are different because they were obtained from different plates. There is a minor difference in the crack growth response for the two specimens, i.e., difference of ≈1 mm after 10^6 cycles. Not surprisingly, the specimen with the lower V_f shows a faster crack growth rate. Figure 4 shows the data obtained at 700 MPa. The crack growth response is distinctly different from that shown in Fig. 3. The specimen failed at ≈7200 cycles, which is three orders of magnitude lower than the cycle count achieved at 400 MPa without failure.

Comparison of Crack Growth in $[0]_8$ *SCS-6/Ti-6Al-4V and* $[0]_6$ *SM1240/Ti-6Al-4V*

The crack growth behavior of $[0]_8$ SCS-6/Ti-6Al-4V is compared with that of $[0]_6$ SM1240/Ti-6Al-4V [9,20,21] in Figs. 5 and 6. As shown in Table 1, V_f for these specimens ranges from 0.31 to 0.35. Figure 5 compares the crack growth rate (da/dN) versus applied stress intensity factor range (ΔK_a) obtained at similar maximum stress levels of ≈400 MPa under tension fatigue loading and bending fatigue loading. Interestingly, under tension fatigue loading, the data from $[0]_8$ SCS-6/Ti-6Al-4V are close to those obtained from $[0]_6$ SM1240/Ti-6Al-4V. This similarity is also seen in the crack extension versus cycles response as shown in Fig. 6. The similarity in the crack growth behavior implies that, at room temperature, the bridging contribution of the SCS-6 fibers is similar to that of SM1240 fibers. This can possibly be attributed to the presence of similar carbon layers on the fibers and near-identical radial compressive stresses at the fiber/matrix interface (also discussed in section "Material"). The crack growth behavior of neat Ti-6Al-4V (Eq 2) is also shown in Fig. 5. The crack growth rate in the composite is ≈2 to 3 orders of magnitude slower than in neat Ti-6Al-4V, highlighting the significantly higher damage tolerance of the composite.

The data from SM1240/Ti-6Al-4V [9] subjected to bending fatigue loading are also shown in Figs. 5 and 6. During the initial stages, i.e., when $\Delta K_a \leq 40$ MPa√m, the crack growth

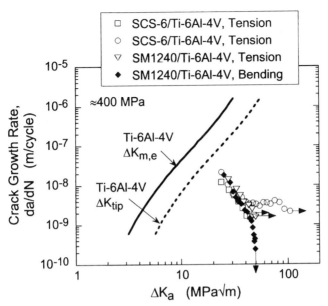

FIG. 5—*Crack growth rate versus applied stress-intensity factor range behavior of* $[0]_8$ *SCS-6/Ti-6Al-4V and* $[0]_6$ *SM1240/Ti-6Al-4V at maximum stress level of* ≈400 *MPa.*

rates are similar to those under tension fatigue loading. With an increase in crack length, the crack growth rate under bending decreases rapidly and appears to exhibit near-crack arrest conditions at $\Delta K_a \approx 50$ MPa√m. In contrast, under tension fatigue loading, the crack growth rate is ≈ an order of magnitude higher at the same ΔK_a, and for $\Delta K_a \geq 50$ MPa√m the crack growth rate is nearly constant. Similar differences in crack growth rate

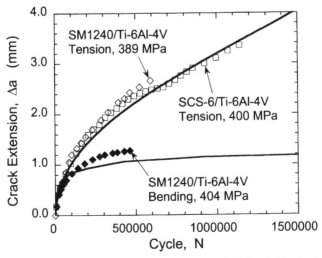

FIG. 6—*Effect of type of loading on crack extension from an initial unbridged notch in* $[0]_8$ *SCS-6/Ti-6Al-4V and* $[0]_6$ *SM1240/Ti-6Al-4V. (Data = symbols; Predictions = lines.)*

behavior can be seen in the data reported by Ghosn et al. [7] for SCS-6/Ti-15V-3Cr-3Al-3Sn. Figure 6 clearly shows the difference in crack extension from the notch under tension and bending fatigue loading. Under bending, the crack extended only ≈1.23 mm [9] prior to the near-arrest condition, while under tension the crack continued to extend at a near-constant rate. The distinct difference in the crack growth behavior of the composite under different types of loading implies that the use of a threshold ΔK, ΔK_{th}, type approach for design should account for the loading condition.

The crack growth behavior of the composites at high stress levels is compared with that of neat Ti-6Al-4V in Fig. 7. Also shown is the predicted behavior of the crack tip in the composite given by Eq 3.

$$\Delta K_{tip} = F_s \, \Delta K_{m,e}, \text{ where } F_s = \frac{E_c}{E_m} \tag{3}$$

where ΔK_{tip} = ΔK at the crack tip in the composite, E_c = elastic modulus of the composite in the direction of loading, E_m = elastic modulus of the matrix, and F_s = shielding factor [13]. Equation 3 was first proposed by Marshall, Cox, and Evans [34] to predict matrix cracking in ceramic matrix composites. Larsen et al. [13] showed that Eq 3 can be used to predict the unbridged crack growth behavior in SCS-6/TIMETAL®21S. Equation 3 implies that the damage tolerance of the composite is higher than that in the matrix even when the fibers are not bridging the crack. This can be attributed to the shielding effect of the fiber in reducing the effective stress in the matrix. Typically, $E_c/E_m \approx 2$ at room temperature. Figure 7 shows that Eq 3 predicts the crack growth rate behavior of SM1240/Ti-6Al-4V at 844 and 889 MPa satisfactorily. This correlation between Eq 3 and the composite data is consistent with crack growth without fiber bridging as reported by Zok et al. [20]. The data from SCS-6/Ti-6Al-4V at 700 MPa show an initial region of bridged crack growth followed by increasing crack growth rate. Since the da/dN of this specimen is slower than that given by Eq 3, partial fiber bridging may be active during crack growth. The analysis and prediction

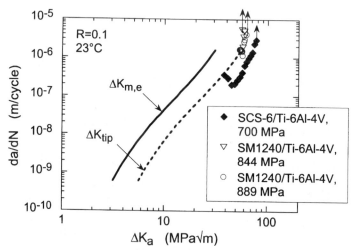

FIG. 7—*Crack growth rate versus applied stress-intensity factor range behavior of* [0]_8 *SCS-6/Ti-6Al-4V and* [0]_6 *SM1240/Ti-6Al-4V at high stresses.*

of the effect of stress and type of loading on the crack growth behavior of SCS-6/Ti-6Al-4V and SM1240/Ti-6Al-4V are discussed in the following section.

Analytical Procedure

The result of fiber bridging is the reduction of the effective ΔK at the matrix crack tip in the composite. This phenomenon is typically modeled as

$$\Delta K_{\text{tip}} = \Delta K_a - \Delta K_b \tag{4}$$

where $\Delta K_{\text{tip}} = \Delta K$ at the crack tip in the composite, ΔK_a = applied stress-intensity factor range, and ΔK_b = bridging stress-intensity factor range. ΔK_b is calculated using the weight function method given by Eq 5.

$$\Delta K_b = \int_{a_0}^{a} \Delta \sigma_b \, h(x, a) \, dx \tag{5}$$

where $\Delta \sigma_b$ = bridging stress range, $h(x, a)$ = load-independent weight function, and a_0, x, and a are defined in Fig. 1. $\Delta \sigma_b$ is generated by the fibers bridging along the crack wake. Various expressions for $\Delta \sigma_b$ are available in the literature [4–7,14–18,34,35]. The shear lag model, developed by Marshall, Cox, and Evans [34] provided an expression for σ_b as given by Eq 6.

$$\sigma_b = \beta \sqrt{v} \tag{6}$$

where

$$\beta = \left[\frac{4V_f^2 E_c E_f \tau}{(1 - V_f) E_m R_f} \right]^{1/2} \tag{7}$$

where R_f = fiber radius, E_m = elastic modulus of matrix, E_f = elastic modulus of the fiber, E_c = modulus of composite = $E_m(1 - V_f) + E_f V_f$, τ = frictional shear stress at the fiber/matrix interface, v = half-crack opening displacement as defined in Fig. 1, and other parameters were defined earlier. For cyclic loading, $\Delta \sigma_b$ is defined as [36]

$$\Delta \sigma_b = \beta \sqrt{2 \Delta v} \tag{8}$$

where Δv = half-crack opening displacement range. Using the weight function method, Δv can also be calculated as

$$\Delta v(x) = \frac{1}{E_0} \int_x^a \left[\int_0^{\bar{a}} h(\bar{x}, \bar{a}) \{\Delta \sigma_a(\bar{x}) - \Delta \sigma_b(\bar{x})\} \, d\bar{x} \right] h(x, \bar{a}) \, d\bar{a} \tag{9}$$

where $\Delta \sigma_a(x)$ = is the applied stress range, and E_0 = effective elastic modulus for a transversely isotropic material [37]. In the above calculations, $\Delta \sigma_b = 0$ for $x < a_0$. Knowing $\Delta \sigma_b$, the maximum stress in the fibers bridging the crack can be calculated as

$$\sigma_{f,\max} = \frac{\Delta\sigma_b}{V_f(1 - R)} \qquad (10)$$

Thus, for a given crack length, knowing the constituent properties, τ and ΔK_a, the quantities ΔK_b and ΔK_{tip} can be calculated using Eqs 5–9. Knowing ΔK_{tip} and using Eqs 2 and 3, the crack growth da/dN can be calculated at the current crack tip. Hence, using crack length or cycle increments, Eqs 2–9 can be used to predict the crack growth behavior of the composite. Equations 8 and 9 imply that an iterative scheme is required to predict the crack growth under cyclic loading. To overcome this problem, Cox and Lo [38] developed simplified nondimensionalized solutions for ΔK_b. ΔK_b was expressed as

$$\Delta K_b = \Delta K_a \, \xi \left(\psi, \frac{2a_0}{W}, \frac{2a}{W} \right) \qquad (11)$$

where

$$\psi = \frac{2\Delta\sigma_a \pi E_0}{4W\beta^2} \qquad (12)$$

and ξ is expressed as a canonical function [38]. Equation 11 can be used in place of Eqs 5–9, thus increasing the computation speed significantly. Cox and Lo [38] provided expressions for ξ for the center-cracked, single-edge-cracked, and double-edge-cracked geometries subjected to far-field tension loading. During this study, Eqs 2–4, 11, and 12 were used to predict the crack growth behavior of the M(T) geometry. The rigorous method using Eqs 2–9 was used to predict the crack-opening displacement profiles for specific crack lengths for the M(T) geometry and to predict the da/dN versus ΔK_a behavior of the SE(B) geometry.

The above equations are applicable for analyzing fully bridged crack growth behavior with all the fibers intact during the entire crack growth, i.e., a_0 does not change during the crack growth. The bridged length, L_{br}, for a crack extending from the initial unbridged notch is defined as

$$L_{br} = a - a_0 \qquad (13)$$

During fully bridged crack growth, L_{br} increases linearly with an increase in crack extension, $\Delta a = a - a_0$. For unbridged crack growth, $L_{br} = 0$. When $L_{br} = $ constant, $a_0 = a - L_{br}$, i.e., the unbridged notch size is increased during the analysis, thus simulating a constant bridged length behind the crack tip. This assumption implies that as the crack advances, the fibers are continuously failing at a distance equal to L_{br} behind the crack tip. As discussed later, a constant value of L_{br} was used to predict the crack growth at high stress levels in SCS-6/Ti-6Al-4V. The available solutions [39] for the M(T) and SE(B) geometries were used to calculate ΔK_a. The method proposed by Wu and Carlsson [40] were used to derive the weight functions for the M(T) and SE(B) geometries. The weight function used for the M(T) geometry is also reported in Ref 17. The constituent properties used in the analysis is reported in Table 2. The predictions of crack growth in SCS-6/Ti-6Al-4V and SM1240/Ti-6Al-4V are discussed next.

TABLE 2—*Constituent properties.*

Property	Ti-6Al-4V	SCS-6™	SM1240™
Modulus, E (GPa)	117	393	400
Poissons's Ratio, ν	0.31	0.17	0.17

Predictions of Crack Growth in SCS-6/Ti-6Al-4V and SM1240/Ti-6Al-4V

Bridged Crack Growth Under Tension Fatigue Loading

Using the simplified analytical procedure described in the previous section, the crack growth response of $[0]_8$ SCS-6/Ti-6Al-4V was predicted for the specimen tested at 400 MPa with $V_f = 0.35$ for various values of τ. As shown in Fig. 3, the prediction using $\tau = 9$ MPa correlated well with the data. Using this value of τ, the crack growth response was predicted for the specimen with $V_f = 0.33$. The prediction agrees well with the data up to $\approx 1.8 \times 10^6$ cycles. The analysis was stopped when the ratio $2a/W$ exceeded 0.8. Hence, the shear lag model is able to predict the minor differences in crack growth behavior due to small changes in V_f. Using the same value of τ, the crack growth behavior of $[0]_4$ SCS-6/Ti-6Al-4V tested at 171 MPa by Davidson [6] was predicted as shown in Fig. 8. The prediction agrees well with the limited data, confirming the applicability of the proposed analytical procedure for very low stress levels. Note that the crack extension was only ≈ 0.4 mm from the notch. Figure 9 shows our predictions of crack extension in $[0]_6$ SM1240/Ti-6Al-4V reported by Connell and Zok [21] for stress levels ranging from 222 to 444 MPa. These predictions were performed using the same value of $\tau = 9$ MPa. The good agreement between the predicted and measured crack growth response for SM1240/Ti-6Al-4V confirms the earlier observation (Figs. 5 and 6) that at room temperature SM1240/Ti-6Al-4V and SCS-

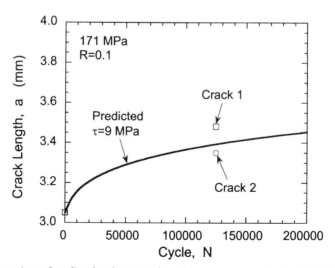

FIG. 8—*Comparison of predicted and measured crack growth response in $[0]_4$ SCS-6/Ti-6Al-4V tested under tension fatigue loading at $\sigma_{max} = 171$ MPa. Prediction assumes fully bridged crack growth. Data from Davidson [6].*

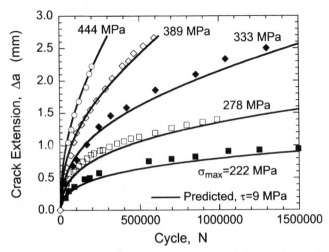

FIG. 9—*Comparison of predicted (lines) and measured (symbols) crack growth response in* $[0]_6$ *SM1240/Ti-6Al-4V tested under tension fatigue loading at maximum stress,* $\sigma_{max} = 222$ *to 444 MPa. Prediction assumes fully bridged crack growth and* $\tau = 9$ *MPa. Data from Connell and Zok [21].*

6/Ti-6Al-4V show similar crack growth behavior. Note that τ obtained from a single specimen was used to predict the trends in Figs. 3, 8, and 9. The value of τ used in this study is close to the value of 10 MPa reported by Connell and Zok [21].

Bridged Crack Growth Under Bending Fatigue Loading

Using the weight function for the SE(B) geometry [40], the crack growth behavior of $[0]_6$ SM1240/Ti-6Al-4V under bending fatigue loading was predicted with $\tau = 9$ MPa. The predictions are compared with the data from Cardona et al. [9] in Figs. 6 and 10. The stress levels shown in Figs. 6 and 10 for bending fatigue loading correspond to the maximum stress at the bottom edge of the "uncracked" composite. The predictions were performed assuming that the cracks were fully bridged during crack growth. The predictions agree well with the data obtained at maximum stresses of 330 and 404 MPa during the initial growth region. In Fig. 10, the predictions underestimate the crack growth rate at growth rates $< 10^{-9}$ m/cycle, i.e., under near-arrest conditions. As shown in Fig. 2, the matrix crack growth behavior was not adequately characterized for growth rates $< 10^{-9}$ m/cycle. Hence, crack growth predictions under near-arrest (near-threshold) conditions may not be accurate. As discussed earlier, the crack is arrested soon after initiation under bending fatigue loading, while the crack grows at a constant rate under tension fatigue loading. This trend is predicted accurately by the proposed analytical procedure as shown in Fig. 6.

At the highest stress level (453 MPa), the prediction assuming fully bridged crack growth agrees with data up to $\Delta K_a \approx 40$ MPa$\sqrt{m}$. The composite exhibits near-constant growth rate when $\Delta K_a \geq 40$ MPa$\sqrt{m}$. Based on observations under tension fatigue loading [12–15], we can conclude that the change in crack growth behavior could be attributed to the onset of noncatastrophic fiber failure. This behavior is discussed later in terms of predicted fiber stresses.

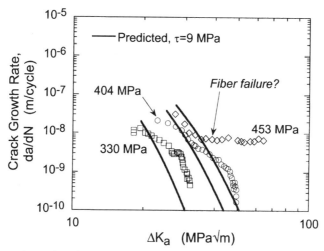

FIG. 10—*Comparison of predicted and measured crack growth response in* [0]₆ *SM1240/Ti-6Al-4V tested under bending fatigue loading at* σ_{max} = 330 to 453 MPa. Prediction assumes fully bridged crack growth. Data obtained from Cardona, Barney, and Bowen [9].*

Partially Bridged and Unbridged Crack Growth Under Tension Fatigue Loading

The crack growth behavior of SCS-6/Ti-6Al-4V at 700 MPa was predicted using three assumptions: (1) unbridged crack growth with τ = 0, (2) partially bridged crack growth with τ = 9 MPa and L_{br} = 0.35 mm, and (3) fully bridged crack growth with τ = 9 MPa. These three predictions are compared with the data in Fig. 4. The prediction assuming partially bridged crack growth agrees well with the data. The specimen failed after ≈7200 cycles, which is approximately three times the life predicted (≈2500) for unbridged crack growth. Note that the bridged length, L_{br} = 0.35 mm, is ≈ twice the fiber spacing. Hence, even a few fibers bridging behind the crack tip increases the crack growth life significantly.

Figure 11 compares the predictions with the data for unbridged crack growth in SM1240/Ti-6Al-4V at stress levels of 844 and 889 MPa. The predictions, using Eq 3 and assuming τ = 0, agree well with the data. Note that Side 1 and Side 2 refer to the crack length measurements on either side of the machined notch. The scatter in the data can be attributed to the very small crack extensions that can be expected under such high stress levels. The predictions ignoring the shielding effect of the fibers, i.e., assuming ΔK_{tip} = $\Delta K_{m,e}$, i.e., F_s = 1, are also shown in Fig. 11. As shown in Fig. 11, if unreinforced matrix properties were used, the predicted life would be ≈25 cycles and 10 cycles for the tests conducted at 844 and 889 MPa, respectively. But, as predicted using Eq 3 and confirmed by the data, the actual lives were about 5 to 10 times higher. Evidently, the assumption of F_s = 1 grossly underestimates the damage tolerance of the unbridged composite. Hence, Eq 3 should be used to predict accurately the unbridged crack growth in the composite [13,41].

Crack Opening Displacements (CODs) and Slip Lengths Along the Crack

The above predictions for SCS-6/Ti-6Al-4V and SM1240/Ti-6Al-4V are based on τ = 9 MPa. This value of τ is generally lower than the values obtained from push-out tests [5–8,13–15,21,27]. As discussed earlier, Connell and Zok [21] reported a value of τ = 10 MPa

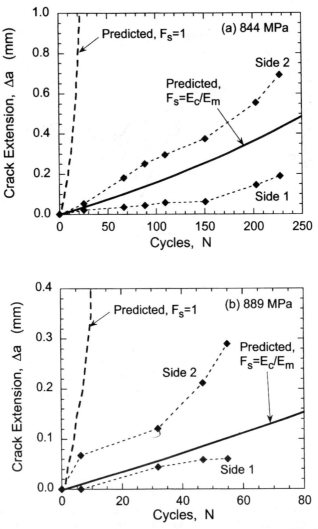

FIG. 11—*Comparison of predicted and measured crack growth response in* $[0]_6$ *SM1240/Ti-6Al-4V tested under tension fatigue loading at* (a) σ_{max} = 844 *and* (b) σ_{max} = 889 *MPa. Prediction assumes unbridged crack growth. Data obtained from Zok, Connell, and Du* [20].

for $[0]_6$ SM1240/Ti-6Al-4V based on correlation with the crack growth response and cyclic load-COD data from through-cracked (edge-to-edge) samples cut out from the interrupted fatigue tests. The hysteresis loop method proposed by Connell and Zok [21] used extensometers with gage lengths ≈10 mm to measure the crack opening displacement. This implies that the measured data could be dominated by the elastic displacement of the intact material within the gage length.

Alternate methods of more accurate measurements include *in situ* testing using special stages in a scanning electron microscope (SEM) [6,7] and direct measurements using the laser interferometric displacement gage (IDG) system [30]. We used the laser IDG system to measure the COD during some of the tests on SCS-6/Ti-6Al-4V. The laser IDG method has been successfully used to measure COD during crack growth in SCS/Ti-24Al-11Nb [14], SCS-6/TIMETAL®21S [13,15], and SM1240/TIMETAL®21S [18]. This method uses indents that are placed within about 50 μm from the crack surface. Hence the measured values are dominated by the COD at the location. As the crack grows, additional indents can be added and COD profiles obtained as functions of crack lengths and cycles.

The half-COD range (Δv) profile obtained from $[0]_8$ SCS-6/Ti-6Al-4V tested at 400 MPa is shown in Fig. 12. The different symbols correspond to data from each side of the machined notch. The displacements range from ≈ 0.5 μm near the crack-tip to ≈ 3.2 μm near the machined notch-tip. Such small displacements highlight the need to measure COD accurately and as close to the crack surface as possible. These values are similar to those reported for SCS-6/Ti-6Al-4V [6], SCS-6/Ti-24Al-11Nb [14], SCS-6/Ti-15V-3Cr-3Al-3Sn [7], and SM1240/TIMETAL®21S [18]. As discussed earlier, the rigorous method of analysis was used to predict the half-COD range using $\tau = 9$ MPa. The predictions agree well with the measured profile as shown in Fig. 12. The bridging model was also used to predict the half-COD profile for $[0]_4$ SCS-6/Ti-6Al-4V [6] corresponding to the data shown in Fig. 8. The COD data reported in Ref 6 correspond to the maximum stress. The residual COD (under zero load) was also reported in Ref 6 as a function of cycles, maximum stresses, and crack length. Using the maximum COD, COD_{max}, and residual COD, COD_r, the half-COD range, Δv, for the test conducted at $R = 0.1$ was obtained as

$$\Delta v = \frac{COD_{max} - COD_r}{2(1 - R)} \tag{14}$$

The data deduced using Eq 14 agree well with the prediction as shown in Fig. 13. The good correlation between the measured and predicted COD profiles as shown in Figs. 12 and 13 validates $\Delta \sigma_b$ and ΔK_b predicted by the shear lag model.

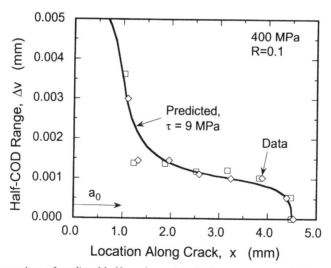

FIG. 12—*Comparison of predicted half-crack opening displacement range with that measured during crack growth in $[0]_8$ SCS-6/Ti-6Al-4V tested at $\sigma_{max} = 400$ MPa. The different symbols correspond to data from each side of the machined notch.*

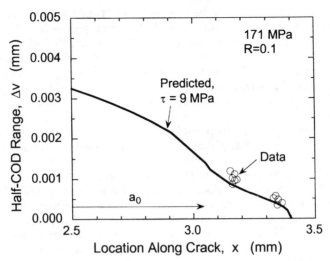

FIG. 13—*Comparison of predicted half-crack opening displacement range with that measured during crack growth in* [0]₄ *SCS-6/Ti-6Al-4V tested at σ*ₘₐₓ *= 171 MPa. Data from Davidson* [6].

The shear lag model can also be used to predict the debond or slip length along the fiber, l_s, the region along which τ is active. Marshall et al. [34] derived an expression for l_s as

$$l_s = \sqrt{\frac{R_f E_f E_m (1 - V_f)}{\tau E_c}} \sqrt{\Delta v} \tag{15}$$

The slip length predicted by Eq 15 is shown in Fig. 14 along with the data from Davidson

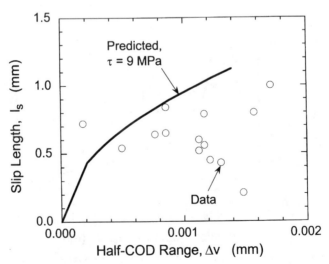

FIG. 14—*Comparison of predicted slip length with that measured during crack growth in* [0]₄ *SCS-6/Ti-6Al-4V. Data from Davidson* [6].

[6]. The data [6] shown in Fig. 14 were based on the damaged regions on the fiber coating visible after the matrix was removed. Reliable measurements of slip length are extremely difficult, but the damaged regions on the coating indicate the minimum lengths along which the fiber/matrix slip was active. All the data available in the literature (for example, Ref 13) indicate that such damaged zones are visible only behind the crack tip, consistent with the assumption that the fiber/matrix slip is active only along the crack. Hence, the data [6] shown in Fig. 14 are expected to indicate only the range of values to be expected. As shown in Fig. 14, the predicted $l_s \approx 1$ mm for $\Delta v \approx 1$ μm is consistent with the data.

Maximum Fiber Stresses at the Machined Notch Tip

The failure of the composite is governed by the strength of the fiber. Hence, during bridged crack growth analysis, the stresses in the fibers bridging the crack should be monitored. The maximum fiber stress, $\sigma_{f,max}$ occurs at the machined notch tip, i.e., at $x = a_0$ [34,35]. The predicted $\sigma_{f,max}$ for SCS-6/Ti-6Al-4V and SM1240/Ti-6Al-4V subjected to tension fatigue loading is shown in Fig. 15 for tests conducted at $\sigma_{a,max}$ = 400, 700, and 844 MPa. As expected, $\sigma_{f,max}$ increases with increase in crack extension, Δa, and the rate of increase of $\sigma_{f,max}$ depends on the applied stress and crack extension. For the specimen tested at $\sigma_{a,max}$ = 400 MPa, $\sigma_{f,max}$ never exceeded ≈2300 MPa, consistent with the bridged crack growth behavior shown in Fig. 3. When $\sigma_{a,max}$ = 700 MPa, $\sigma_{f,max}$ increases rapidly with increasing crack extension. As shown in Figs. 4 and 7, this specimen showed a sudden change in the crack growth rate at $\Delta a \approx 0.6$ mm. This corresponds to $\sigma_{f,max} \approx 2750$ MPa, which is lower than the reported mean fiber strength of 3700 MPa for SCS-6 fibers [42]. Using a Weibull distribution to represent the strength variation, Ashbaugh [43] reported a bundle strength, $\sigma_{f,uB} \approx 2600$ to 3200 MPa for the SCS-6 fibers extracted from SCS-6/Ti-6Al-4V composites. Interestingly, the critical $\sigma_{f,max}$ for the test at 700 MPa is within this range.

The predicted $\sigma_{f,max}$ for SM1240/Ti-6Al-4V tested at 844 MPa is also shown in Fig. 15. Soon after crack initiation, $\sigma_{f,max}$ increases rapidly to stress levels >3000 MPa, which exceeds the bundle strength (≈2800 MPa) [21] of the SM1240 fibers. In fact, $\sigma_{f,max}$ is already

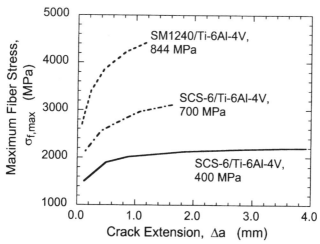

FIG. 15—*Predicted variation of maximum fiber stress during crack growth in* $[0]_8$ *SCS-6/Ti-6Al-4V and* $[0]_6$ *SM1240/Ti-6Al-4V tested under tension fatigue loading.*

close to the reported mean tensile strength ($\approx$3300 MPa) [42] of the SM1240 fibers. Hence, the shear lag model correctly predicts the unbridged crack growth behavior of SM1240/Ti-6Al-4V for stress levels $\geq$844 MPa as shown in Fig. 11.

Figure 16 shows $\sigma_{f,max}$ predicted for SM1240/Ti-6Al-4V under bending fatigue loading for different maximum stresses. For the tests conducted at 330 and 404 MPa, $\sigma_{f,max}$ was less than the bundle strength of 2800 MPa, consistent with the fully bridged crack growth behavior shown in Figs. 6 and 10. When the stress level was increased to 453 MPa, a change in the crack growth trend was observed at $\Delta K_a \approx$ 40 MPa$\sqrt{m}$ as indicated by the bold arrow in Fig. 10. This value of ΔK_a corresponds to a crack extension of $\approx$0.7 mm. Interestingly, as shown in Fig. 16, $\Delta a \approx$ 0.7 mm corresponds to predicted $\sigma_{f,max} \approx$ 2800 MPa, which is equal to the bundle strength [21] of the SM1240 fibers. Hence, the shear lag model correctly predicts the onset of fiber failure and increase in crack growth rate in SM1240/Ti-6Al-4V under bending fatigue loading. Comparing Figs. 15 and 16, we can conclude that, for similar stress levels and crack lengths, the bridging fiber stresses under bending fatigue are $\approx$30 to 35% higher than that under tension fatigue loading.

Discussion

Surprisingly, the data at room temperature indicate that the crack growth behavior of SCS-6/Ti-6Al-4V is similar to that of SM1240/Ti-6Al-4V. As discussed earlier, this similarity may be attributed to the nearly identical stiffnesses of the fibers, dual carbon coatings on the fibers, and radial compressive stresses at the fiber/matrix interface. Since the processing conditions can be expected to be different, the crack growth behavior of the matrix in these composites could be different. Connell and Zok [21] represented the matrix crack growth behavior as

$$\frac{da}{dN} = \eta(\Delta K_{m,e})^{3.8} \qquad (16)$$

Using τ and η as fitting parameters, Connell and Zok [21] determined various combinations

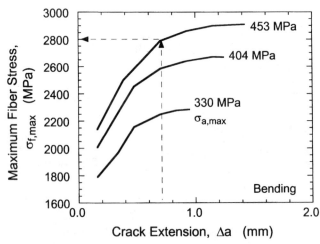

FIG. 16—*Predicted variation of maximum fiber stress during crack growth in* [0]$_6$ *SM1240/Ti-6Al-4V tested under bending fatigue loading.*

of τ and η that successfully predicted the crack behavior of SM1240/Ti-6Al-4V, i.e., the data shown in Fig. 9. As discussed earlier, cyclic load tests were conducted on through-cracked specimens and τ deduced from the hysteresis loops. Thus, the corresponding value of η was also determined. Using this value of η, the deduced matrix behavior was plotted in Fig. 2. As seen in the figure, the deduced crack growth behavior of the matrix in SM1240/Ti-6Al-4V is close to the neat Ti-6Al-4V obtained during this study. Thus, the similarity between the fibers and the matrix enabled us to successfully predict the crack growth behavior of SCS-6/Ti-6Al-4V and SM1240/Ti-6Al-4V using the same value of τ.

The value of τ used in this study compares well with that reported by Connell and Zok [21], but it is lower than the results generally reported from push-out tests. The effect of fatigue loading has been suggested as a possible reason for the low value of τ [44]. An important point to be noted is that all models and experiments indicate that the slip length ≈ 1 to 2 mm at room temperature. This implies that the correct value of τ can be deduced only if a push-out or pull-out test is conducted on specimens in which the fatigued slip zone is interrogated. In general, such specimens (from within 1 to 2 mm from the crack surface) are nearly impossible to obtain. Most of the reported data are based on specimen sections that were located at distances >2 to 5 mm from the crack surfaces. Hence, it is not surprising that the push-out results tend to yield higher values of τ. Alternate methods include cyclic-load tests with COD measurements, from which τ can be deduced. As shown during this study (Fig. 12) and by Larsen et al. [13], the COD measurements near the crack surface yield values of τ that correlate well with the crack growth behavior.

During this study and in Ref 13, we used Eq 3 to relate ΔK_{tip} at the crack tip in the composite to the neat matrix behavior. This study showed that the use of Eq 3 along with the shear lag model of Marshall et al. [34] is able to predict successfully the crack growth behavior *and* related parameters such as crack opening displacements, slip length, and fiber stresses. Larsen et al. [13] and John et al. [14,15] showed that the crack growth behavior can also be predicted assuming $F_s = 1$ with higher values of τ. But the resulting predictions of the COD profile are lower than those measured during the crack growth tests.

Cox [19] suggested that the COD data from small crack extensions may not be sensitive enough to deduce the applicability of the shear lag models. Tests on very large (and expensive) specimens are required to verify the bridging relationship given by Eq 6. As shown during this study, and in Refs 13 and 41, Eq 3 can also be used to predict the unbridged crack growth in the composites.

The results from this study indicate that the use of the bundle strength of fibers extracted from the composite can be used to predict the onset of fiber failure. We should note that the use of $F_s = 1$ will necessitate the use of higher-strength values. The analyses of Bakuckas and Johnson [8,25], Larsen et al. [13], John et al. [14,15], and Connell and Zok [21] support the use of a fiber strength that is lower than the mean strength based on tests on fibers prior to consolidation. In typical composites, depending on the layup, ≈ 8 to 30 fibers can be expected to exist adjacent to the crack tip within a fiber spacing. This number, coupled with the expected degradation of the fiber/matrix interface during fatigue, could justify the use of the bundle strength to predict fiber failure.

Summary

This study showed that the shear lag model assuming a constant value of τ can be used to predict bridged crack growth perpendicular to fibers in SCS-6/Ti-6Al-4V and SM1240/Ti-6Al-4V over a wide range of stress levels and under tension and bending fatigue loading conditions. The model predicted partially bridged and unbridged crack growth, as well as crack opening displacement and slip length measurements. Analysis of the bending fatigue loading data confirmed that under bending fatigue loading the crack growth arrested soon after initiation, in contrast to the continued growth under tension fatigue loading. The bridging fiber stresses under bending fatigue are ≈30 to 35% higher than those under tension fatigue loading. The value of τ was the same for SCS-6/Ti-6Al-4V and SM1240/Ti-6Al-4V, implying that the bridging mechanism of the SM1240 fiber is identical to that of the SCS-6 fiber at room temperature. The results also indicate that the onset of fiber failure could be predicted using the bundle strength as the critical value.

Acknowledgments

This research was conducted at the Materials and Manufacturing Directorate, Air Force Research Laboratory (AFRL/MLLN), Wright-Patterson Air Force Base, OH 45433-7817. R. John was supported under on-site contract number F33615-94-C-5200. The authors gratefully acknowledge the assistance of R. Kleismit and A. F. Lackey in conducting the experiments.

References

[1] Larsen, J. M., Williams, K. A., Balsone, S. J., and Stucke, M. A., "Titanium Aluminides for Aerospace Applications," *High Temperature Aluminides and Intermetallics*, S. H. Whang, C. T. Liu, D. P. Pope, J. O. Stiegler, Eds., The Minerals, Metals & Materials Society, Warrendale, PA, 1990, pp. 521–556.

[2] Larsen, J. M., Russ, S. M., and Jones, J. W., "An Evaluation of Fiber-Reinforced Titanium Matrix Composites for Advanced High-Temperature Aerospace Applications," *Metallurgical and Materials Transactions A*, Vol. 26A, 1995, pp. 3211–3223.

[3] Bain, K. R. and Gambone, M. L., "Fatigue Crack Growth of SCS-6/Ti-64 Metal Matrix Composite," *Fundamental Relationships Between Microstructure and Mechanical Properties of Metal Matrix Composites*, P. K. Liaw and M. N. Gungor, Eds., The Minerals, Metals, and Materials Society, Warrandale, PA, 1990, pp. 459–469.

[4] McMeeking, R. M. and Evans, A. G., "Matrix Fatigue Cracking in Fiber Composites," *Mechanics of Materials*, Vol. 9, No. 3, 1990, pp. 217–227.

[5] Sensmeier, M. D. and Wright, P. K., "The Effect of Fiber Bridging on Fatigue Crack Growth in Titanium Matrix Composites," *Fundamental Relationships Between Microstructure and Mechanical Properties of Metal Matrix Composites*, The Metallurgical Society, Inc., Warrendale, PA, 1990, pp. 441–450.

[6] Davidson, D. L., "The Micromechanics of Fatigue Crack Growth at 25°C in Ti-6Al-4V Reinforced with SCS-6 Fibers," *Metallurgical and Materials Transactions A*, Vol. 23A, 1992, pp. 865–879.

[7] Ghosn, L. J., Telesman, J., and Kantzos, P., "Specimen Geometry Effects on Fiber Bridging in Composites," *FATIGUE 93, Vol. 2*, J.-P. Bailon and I. J. Dickson, Eds., Engineering Materials Advisory Services, Ltd., U.K., 1993, pp. 1231–1238.

[8] Bakuckas, J. G., Jr. and Johnson, W. S., "Application of Fiber Bridging Models to Fatigue Crack Growth in Unidirectional Titanium Matrix Composites," NASA Technical Memorandum 107588, NASA Langley, Hampton, VA, July 1992.

[9] Cardona, D. C., Barney, C., and Bowen, P., "Modeling and Prediction of Crack Arrest in Fiber Reinforced Composites," *Life Prediction Methodology for Titanium Matrix Composites, ASTM STP 1253*, W. S. Johnson, J. M. Larsen, and B. N. Cox, Eds., American Society for Testing and Materials, West Conshohocken, PA, 1996, pp. 164–181.

[10] Bowen, P., "Characterization of Crack Growth Resistance Under Cyclic Loading in the Presence of an Unbridged Defect in Fiber-Reinforced Titanium Metal Matrix Composite," *Life Prediction Methodology for Titanium Matrix Composites, ASTM STP 1253*, W. S. Johnson, J. M. Larsen, and B. N. Cox, Eds., American Society for Testing and Materials, West Conshohocken, PA, 1996, pp. 461–479.

[11] Bao, G. and McMeeking, R. M., "Fatigue Crack Growth in Fiber-Reinforced Metal-Matrix Composites," *Acta Metallurgica et Materialia*, Vol. 42, No. 7, 1994, pp. 2415–2425.

[12] Jira, J. R. and Larsen, J. M., "Fatigue of Unidirectional SCS-6/Ti-24Al-11Nb Composite Containing a Circular Hole," *Metallurgical and Materials Transactions A*, Vol. 25A, 1994, pp. 1413–1424.

[13] Larsen, J. M., Jira, J. R., John, R., and Ashbaugh, N. E., "Crack-Bridging Effects in Notch Fatigue of SCS-6/TIMETAL21S Composite Laminates," *Life Prediction Methodology for Titanium Matrix*

Composites, ASTM STP 1253, W. S. Johnson, J. M. Larsen, and B. N. Cox, Eds., American Society for Testing and Materials, West Conshohocken, PA, 1996, pp. 114–136.

[14] John, R., Jira, J. R., Larsen, J. M., and Ashbaugh, N. E., "Analysis of Bridged Fatigue Cracks in Unidirectional SCS-6/Ti-24Al-11Nb Composite," *FATIGUE 93, Vol. 2,* J.-P. Bailon and I. J. Dickson, Eds., Engineering Materials Advisory Services, Ltd., U.K., 1993, pp. 1091–1096.

[15] John, R., Kaldon, S. G., and Ashbaugh, N. E., "Applicability of Fiber Bridging Models to Describe Crack Growth in Unidirectional Titanium Matrix Composites," *Titanium Metal Matrix Composites II, WL-TR-93-4105,* P. R. Smith and W. C. Revelos, Eds., Wright-Patterson Air Force Base, OH, 1993, pp. 233–250.

[16] John, R., Stibich, P. R., Johnson, D. A., and Ashbaugh, N. E., "Bridging Fiber Stress Distribution During Fatigue Crack Growth in [0]₄ SCS-6/TIMETAL®21S," *Scripta Metallurgica et Materialia,* Vol. 33, No. 1, 1995, pp. 75–80.

[17] Buchanan, D. J., John, R., and Johnson, D. A., "Determination of Crack Bridging Stresses from Crack Opening Displacement Profiles," *International Journal of Fracture,* Vol. 87, No. 2, 1997, pp. 101–117.

[18] Zheng, D. and Ghonem, H., "High Temperature/High Frequency Fatigue Crack Growth in Titanium Metal Matrix Composites," *Life Prediction Methodology for Titanium Matrix Composites, ASTM STP 1253,* W. S. Johnson, J. M. Larsen, and B. N. Cox, Eds., American Society for Testing and Materials, West Conshohocken, PA, 1996, pp. 137–163.

[19] Cox, B. N., "Life Prediction for Bridged Fatigue Cracks," *Life Prediction Methodology for Titanium Matrix Composites, ASTM STP 1253,* W. S. Johnson, J. M. Larsen, and B. N. Cox, Eds., American Society for Testing and Materials, West Conshohocken, PA, 1996, pp. 552–572.

[20] Zok, F. W., Connell, S. J., and Du, Z.-Z., "Fatigue Maps for Titanium Matrix Composites," *Life Prediction Methodology for Titanium Matrix Composites, ASTM STP 1253,* W. S. Johnson, J. M. Larsen, and B. N. Cox, Eds., American Society for Testing and Materials, West Conshohocken, PA, 1996, pp. 432–460.

[21] Connell, S. J. and Zok, F. W., "Measurement of the Cyclic Bridging Law in a Titanium Matrix Composite and It's Application to Simulating Crack Growth," *Acta Materialia,* Vol. 45, No. 12, 1997, pp. 5203–5211.

[22] Walls, D. P., Bao, G., and Zok, F. W., "Mode I Fatigue Cracking in a Fiber Reinforced Metal Matrix Composite," *Acta Metallurgica et Materialia,* Vol. 41, No. 7, 1993, pp. 2061–2071.

[23] Nguyen, T.-H.B and Yang, J.-M., "Elastic Bridging for Modeling Fatigue Crack Propagation in a Fiber-Reinforced Titanium Matrix Composite," *Fatigue and Fracture of Engineering Materials and Structures,* Vol. 17, No. 2, 1994, pp. 119–131.

[24] Herrmann, D. J. and Hillberry, B. M., "A New Approach to the Analysis of Unidirectional Titanium Matrix Composites with Bridged and Unbridged Cracks," *Engineering Fracture Mechanics,* Vol. 56, No. 5, 1997, pp. 711–726.

[25] Bakuckas, J. G., Jr. and Johnson, W. S., "A Methodology to Predict Damage Initiation, Damage Growth, and Residual Strength in Titanium Matrix Composites," *Life Prediction Methodology for Titanium Matrix Composites, ASTM STP 1253,* W. S. Johnson, J. M. Larsen, and B. N. Cox, Eds., American Society for Testing and Materials, West Conshohocken, PA, 1996, pp. 497–519.

[26] Larsen, J. M., Jira, J. R., John, R., and Blatt, D., "Temperature Dependent Crack Bridging Effects in SCS-6/TIMETAL®21S Composite," *Materials Science and Engineering,* accepted for publication, 1997.

[27] Hutson, A., John, R., and Jira, J. R., "The Effect of Temperature on Fiber/Matrix Interface Sliding in SCS-6/TIMETAL®21S," *Scripta Materialia,* to be submitted for publication, May 1998.

[28] Nicholas, T. and Kroupa, J. L., "Micromechanics Analysis and Life Prediction of Titanium Matrix Composites," *Journal of Composites and Technology and Research,* Vol. 20, No. 2, April 1998, pp. 101–105.

[29] Coker, D., Boller, F., Kroupa, J. L., and Ashbaugh, N. E., *FIDEP2: User Manual for Micromechanical Models for Thermoviscoplastic Behavior of Metal Matrix Composites,* University of Dayton Research Institute, Dayton, OH, to be published, 1998.

[30] Sharpe, W. N., Jr., Jira, J. R., and Larsen, J. M., "Real-Time Measurement of Small-Crack Opening Behavior Using an Interferometric Strain/Displacement Gage," *Small-Crack Test Methods, ASTM STP 1149,* American Society for Testing and Materials, West Conshohocken, PA, 1992, pp. 92–115.

[31] John, R., Kaldon, S. G., Johnson, D. A., and Coker, D., "Weight Function for a Single Edge Cracked Geometry with Clamped Ends," *International Journal of Fracture,* Vol. 72, No. 2, 1995, pp. 145–158.

[32] John, R. and Rigling, B., "Effect of Height to Width Ratio on K and CMOD Solutions for Single Edge Cracked Geometry with Clamped Ends," *Engineering Fracture Mechanics,* Vol. 60, No. 2, 1998, pp. 147–156.

[33] Miller, M. S. and Gallagher, J. P., "An Analysis of Several Fatigue Crack Growth Rate (FCGR) Descriptions," *Fatigue Crack Growth Measurement and Data Analysis, ASTM STP 738,* American Society for Testing and Materials, West Conshohocken, PA, 1981, pp. 205–251.

[34] Marshall, D. B., Cox, B. N., and Evans, A. G., "The Mechanics of Matrix Cracking in Brittle-Matrix Fiber Composites," *Acta Metallurgica,* Vol. 33, No. 11, 1985, pp. 2013–2021.

[35] McCartney, L. N., "Mechanics of Matrix Cracking in Brittle-Matrix Fibre-Reinforced Composites," *Proceedings of Royal Society of London A,* Vol. A409, 1987, pp. 329–350.

[36] Cox, B. N. and Marshall, D. B., "Concepts for Bridged Cracks in Fracture and Fatigue," *Acta Metallurgica et Materialia,* Vol. 42, 1994, pp. 341–363.

[37] Cox, B. N. and Marshall, D. B., "Crack Bridging in the Fatigue of Fibrous Composites," *Fatigue and Fracture of Engineering Materials and Structures,* Vol. 14, No. 8, 1991, pp. 847–861.

[38] Cox, B. N. and Lo, C. S., "Simple Approximations for Bridged Cracks in Fibrous Composites," *Acta Metallurgica et Materialia,* Vol. 40, No. 7, 1992, pp. 1487–1496.

[39] Tada, H., Paris, P. C., and Irwin, G. R., *The Stress Analysis of Cracks Handbook,* Del Research Corporation, St. Louis, MO, 1985.

[40] Wu, X.-R. and Carlsson, A. J., *Weight Functions and Stress Intensity Factor Solutions,* Pergamon Press, Elmsford, NY, 1991.

[41] John, R., Larsen, J. M., and Jira, J. R., "Effective Stress Intensity Factor at the Matrix Crack Tip During Fatigue Crack Growth in a Metal Matrix Composite," *Scripta Metallurgica et Materialia,* to be submitted for publication, 1998.

[42] Petitcorps, Y. L., Lahaye, M., Pailler, R., and Naslain, R., "Modern Boron and SiC CVD Filaments: A Comparative Study," *Composites Science and Technology,* Vol. 32, 1988, pp. 31–55.

[43] Ashbaugh, N. E., Metzcar, J., and Rosenberger, A. H., "Deformation and Rupture Model of [0] Metal Matrix Composite Under Sustained Load," to be submitted for publication, 1998.

[44] Kantzos, P., Eldridge, J., Koss, D. A., and Ghosn, L. J., "The Effect of Fatigue Loading on the Interfacial Shear Properties of SCS-6/Ti-Based MMCs," *Intermetallic Composites II,* D. B. Miracle, D. L. Anton, and J. A. Graves, Eds., MRS Proceedings, Pittsburgh, PA, Vol. 273, 1992, pp. 135–142.

Ann W. Peck[1]

An Experimental Investigation of Transverse Tension Fatigue Characterization of IM6/ 3501-6 Composite Materials Using a Three-Point Bend Test

REFERENCE: Peck, A. W., "An Experimental Investigation of Transverse Tension Fatigue Characterization of IM6/3501-6 Composite Materials Using a Three-Point Bend Test," *Composing Materials: Fatigue and Fracture, Seventh Volume, ASTM STP 1330*, R. B. Bucinell, Ed., American Society for Testing and Materials, 1998, pp. 145–161.

ABSTRACT: As composites are introduced into more complex structures with out-of-plane loadings, a better understanding is needed of the out-of-plane, matrix-dominated failure mechanisms. This work investigates the transverse tension fatigue characteristics of IM6/3501 composite materials. To test the 90° laminae, a three-point bend test was chosen, potentially minimizing handling and gripping issues associated with tension tests. Static testing of 50 specimens of nine different-sized configurations produced a mean transverse tensile strength of 61.3 MPa (8.0 ksi). The smallest configuration (10.2 mm wide, span-to-thickness ratio of 3) consistently exhibited transverse tensile failures. A volume scale effect was difficult to discern due to the large scatter in the data. Static testing of 10 different specimens taken from a second panel produced a mean transverse tensile strength of 82.7 MPa (12.0 ksi). Weibull parameterization of the data was possible, but due to variability in raw material and/or manufacturing, more replicates are needed for greater confidence. Three-point flex fatigue testing of the smallest configuration was performed on 59 specimens at various levels of the mean static transverse tensile strength using predominantly an *R* ratio of 0.1 and a frequency of 20 Hz. A great deal of scatter was seen in the data. The majority of specimens failed near the center loading roller. To determine whether the scatter in the fatigue data is due to variability in raw material and/or the manufacturing process, additional testing should be performed on panels manufactured from different sources.

KEYWORDS: composite materials, graphite/epoxy, transverse tensile strength, flexural fatigue, Weibull statistics

A great deal of research has been performed characterizing the in-plane fiber-dominated properties, under both static and fatigue loading, of advanced composites materials. This understanding is imperative wherever composite materials are used in order to gain a better understanding of the reliability of structures when designing for maximum efficiency in weight, volume, and payload. However, as composites are introduced into more complex loading states, there is a need to better understand some of the out-of-plane and matrix-dominated phenomena that have not received as much attention. For instance, consider the design of bonded composite airframe structure where repeated, cyclic out-of-plane bending may occur. Two such scenarios where this loading may take place are in a compressively

[1]Assistant professor, Department of Mechanical Engineering, P.O. Box 3295, University of Wyoming, Laramie, WY, 82071-3295.

loaded post-buckled panel or a full-scale pressurized fuselage such as the one investigated by the NASA Advanced Composite Technology (ACT) program. In the latter case, as a result of the internal pressurization within each panel bay, the skin will bulge or "pillow" as shown in Fig. 1. These out-of-plane deformations create local bending moments along the skin-stiffener and skin-frame interfaces, which in turn create shear and peel stresses along the various bondlines [1].

Recent tests characterizing skin/stringer debond failures in reinforced composite panels where the dominant loading in the skin is flexure along the edge of the frame indicate that failure initiates as transverse matrix cracks either in the skin or the flange near the flange tip [2]. When failure initiated in the skin, transverse matrix cracks formed in the surface ply closest to the flange and either initiated delaminations or created matrix cracks in the next lower ply, which in turn initiated delaminations. When failure initiated in the flanges, transverse cracks formed in the flange angle ply closest to the skin and initiated delaminations. In no configuration did failure propagate through the adhesive bond layer. This is a significant finding in that the limiting component in this particular bonded structure under internal pressurization appears not to be the adhesive, but rather the particular skin and flange laminates, and in particular the choice and location of the angle plies. The failure initiation site corresponded well with the site of maximum transverse tension stress, not the site of the maximum interlaminar tension site. For the examined skin/flange configuration, the maximum transverse tension stress at failure correlated well with the transverse tension strength of the composite [2]. Therefore, it is important to understand how the transverse strength of the composite degrades under repeated cyclic loading.

An extensive literature search on the topic of transverse tension fatigue revealed little in the published literature on this particular topic. Perhaps this is a result of the focus, to-date on in-plane loaded structures, such as composite wing skins, where matrix cracking in fatigue may be a rare occurrence at the relatively low operating strains dictated by low-velocity impact concerns, and where matrix cracks are fairly benign if they do occur. However, for the out-of-plane loads experienced in a composite fuselage, the onset of matrix cracking under repeated pressurization may trigger a catastrophic failure. Hence, the need for a transverse tension fatigue characterization is now apparent. It should be noted that there has been extensive work, both theoretical and experimental, focusing on transverse crack formation in laminates under fatigue loading [3]. The laminates studied have generally consisted of 90° plies embedded within a general composite laminate that usually also included 0° and ±45° plies. The crack formation and corresponding failure stress in the 90° plies of the laminates

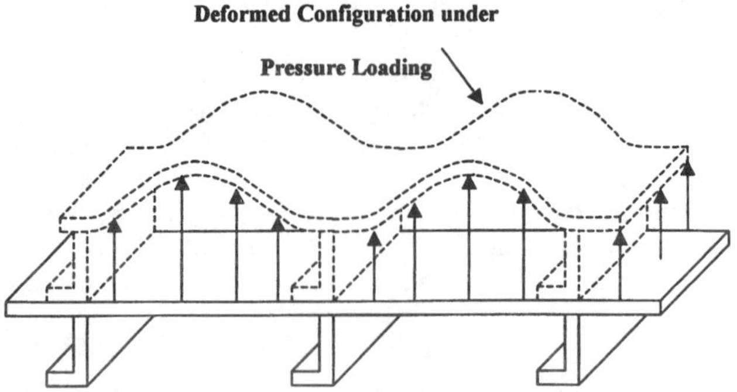

Deformed Configuration under

Pressure Loading

FIG. 1—*Illustration of pressure pillowing in a pressurized fuselage.*

differ substantially due to the constraining effect of the non-90° plies. This study will focus on the failure behavior of unconstrained 90° plies.

When testing 90° laminae, several issues must be addressed. The first is the concern over handling and gripping sensitivities. The handling and gripping issues may be minimized by testing in three-point bending as opposed to testing in tension. This has an additional benefit in that relatively small specimen sizes may be used, thereby reducing the amount of material needed for a single replicate. A small specimen size has the added benefit of minimizing internal heating due to the cyclic loading and, hence, allowing higher frequencies with shorter testing times to be performed. This can become significant for a test that potentially may become a standard for fatigue characterization. Adams, King, and Blackketter [4] evaluated the transverse flexure test using unsized AU4 and AS4 fibers, sized AS4 fibers, and EPON 828 resin. Span-to-thickness ratios of the specimens ranged form 4.0 to 16.7. All the specimens produced tensile failures and yielded similar strengths. The transverse flexure tests produced higher values of the transverse tensile strength than standard transverse tensile tests by as great as a factor of 2.5, as demonstrated by the AS4/EPON 828 combination.

The transverse tension strength of graphite epoxy composites has been shown to exhibit a volume dependence due to the inherent flaws in the microstructure. O'Brien and Salkepar investigated the volume scale effect using AS4/3501-6 90° tension specimens [5]. They found that the transverse tensile strength of a composite laminate depends on the volume of material stressed. The dependence reflects the presence of inherent flaws in the lamina's microstructure. As the volume of material under tension increases, transverse tensile strength decreases. The probability distribution function of the strength as defined by Weibull [6] is

$$P(\sigma) = 1 - \exp(\sigma/\sigma_c)^m \qquad (1)$$

where m is the shape parameter, indicating the measure of scatter in the data, and σ_c is the location parameter, similar to the mean of a normal distribution. Specimens of different volumes can then be compared via a Weibull scaling law for static loading that states

$$\frac{(\sigma_{ult})_1}{(\sigma_{ult})_2} = - \left(\frac{V_2}{V_1}\right)^{1/m} \qquad (2)$$

where $(\sigma_{ult})_1$ and $(\sigma_{ult})_2$ are the different strengths associated with the different specimen volumes V_1 and V_2, and m is a material constant found experimentally.

The primary objective of this study was to experimentally investigate the transverse strength of a composite under repeated cyclic loading. In this study, multiple 90° laminates of different specimen sizes were tested in flexure under static loading to first determine an optimum specimen configuration with *optimum* being defined as the smallest-sized configuration producing transverse tensile failures consistently. Using the optimum-sized configuration, multiple 90° laminates were tested in flexure under cyclic loading to first investigate the behavior of transverse tensile strength under fatigue loading. The data were then used to evaluate the validity of the Weibull scale law for fatigue loading. Additional tests were conducted at different frequencies to determine if the transverse tension fatigue strength exhibits a frequency dependency.

Materials and Specimen Preparation

Two 30.48-cm by 45.72-cm (12-in. by 18-in.) panels of unidirectional prepreg IM6/3501-6 graphite/epoxy material were layed up and cured in an autoclave according to the man-

TABLE 1—*Static test specimen configurations and number of test samples*

Nominal Width, w (mm)[a]	Span-to-thickness Ratio, $(S/t)^{b,c}$			
	3	4	6	8
10.2 (0.40)	A (6)	D (10)	G (2)	K (3)
12.7 (0.50)	B (10)	E (9)	...	...
15.2 (0.60)	C (6)	F (10)	...	M (3)

[a] Number in parentheses is value in inches where 1 in. = 25.4 mm.
[b] Letter indicates specimen configuration.
[c] Number in parentheses is number of samples tested.

ufacturer's specifications. The nominal cured ply thickness is 0.188 mm (0.0074 in.). Each panel was constructed of 36 plies.

The two panels were ultrasonically C-scanned after manufacturing to assess the integrity of the manufacturing process. Each panel exhibited patches of possible voids, located near the borders of the panels. The first panel exhibited a greater amount of non-uniformity than the second panel. Using the C-scans as a guide, specimens were cut from the regions of the panels exhibiting the most uniform C-scan patterns.

Nine different specimen configurations were used in the static study. Table 1 indicates the different widths (w) and span-to-thickness (S/t) ratios examined. Different-sized configurations were tested, as an initial objective of this study was to determine the minimum-sized specimen configuration that would consistently generate a transverse tensile failure under three-point bending.

The first set of static specimens was cut from Panel 1 using a diamond-wheel saw blade. Edges were not polished before testing so as not to bias results by reducing edge flaws or possibly internal flaws. Static and fatigue specimens from Panel 2 were cut using a water-cooled aluminum oxide abrasive wheel (180 grit), then edge ground to their final dimensions. Specimen sizes were determined per a pre-established cutting plan, which purposely distributed the different configuration sizes throughout the panel so as not to concentrate a single configuration to one area of the panel. The specimen cross-sectional dimensions were measured using ball-point calipers at the center locations along the specimen length. A single length measurement was taken along the centerline of the specimen.

Table 2 shows the average nominal ply thicknesses of the static and fatigue specimens. The average laminate thickness for the specimens tested from a particular panel and under a particular loading (static versus fatigue) was divided by the number of plies to obtain a

TABLE 2—*Nominal ply thickness and fiber volume fractions, IM6/3501-6 graphite-epoxy.*

	Panel 1		Panel 2			
	Static[a,b]		Static[a,b]		Fatigue[a,b]	
Nominal Ply Thickness, t (mm)	0.188 (0.0074)	[0.99]	0.191 (0.0075)	[1.0]	0.191 (0.0075)	[1.02]
Estimated V_f, %	58.0	...	57.9	...	57.7	...
Measured V_f, %	59.4	[0.84]	59.8	[1.31]	...	...

[a] Number in parentheses is value in inches where 1 in. = 25.4 mm.
[b] Number in brackets is coefficient of variation.

nominal ply thickness (t). The product of the nominal ply thickness and the manufacturer's supplied fiber density (1.7325 g/cm^3) was divided into the assumed fiber aerial weight (190 g/m^2) for IM6/3501-6 to obtain estimated fiber volume fractions. These estimated fiber volume fractions are compared with experimentally measured fiber volume fractions in Table 2.

Visual examination of the specimens revealed pitting in the majority of the fatigue specimens. In most instances, the pits were approximately pinpoint in size and shallow in depth. In a few instances, the pits were larger in size. The pits were located within approximately the upper 25% of the laminate thickness as measured from the upper surface of the specimen, which corresponds to the bottom surface of the manufactured panel. The number of pinpoint-sized pits in a particular specimen was of the order of 20 to 30. The pits were distributed along the length of the specimens. Because the pits were located on the compressive side of specimens when loaded, they were not believed to have a significant effect on the results.

Experimental Procedure

Fiber volume measurements were performed per the following ASTM standards: *D 3171: Fiber Volume by Acid Digestion, D 792: Density by Water Displacement,* and *D 2734: Void Content.*

Static transverse tension tests were performed using the three-point flex procedure as specified in *ASTM D 790-92, Standard Test Method for Flexural Properties of Unreinforced and Reinforced Plastics and Electrical Insulating Materials.* The fixture was equipped with 6.35-mm (0.25-in.)-diameter loading roller and 3.175-mm (0.125-in.)-diameter support rollers. Span-to-thickness (S/t) ratios of 3, 4, 6, and 8 were used as required for the various static tests. The support rollers were located using the center loading roller as a datum. The specimens were loaded into the fixture with the midpoint of the specimen aligned under the loading roller. Adjustments to the centering of the specimen were made by eye. As the specimens were fairly short in length, there was little extension of an individual specimen beyond the support rollers. All specimens were tested with the compression face of the specimen corresponding to the bottom of the panel. The fixture was mounted on a 22.24-kN (5000-lb) MTS brand servo-hydraulic universal test frame with an MTS TestStar digital controller under ambient laboratory conditions, as shown in Fig. 2. The test frame has a current certification of calibration providing traceability to NIST standards. Static tests were performed at a constant ram speed of 1.27 mm/min (0.05 in./min) with a stroke range set at 12.7 mm (0.5 in.). After an individual specimen failure, the specimen was removed and the moment arm distance from the centerline of the edge support roller to the location of tensile failure, defined as l, measured. The bending moment was calculated as the product of half the failure load P and the moment arm (l). The nominal transverse tensile stress was calculated as

$$\sigma = \frac{(\frac{1}{2}P)l}{(\frac{1}{12})wt^3}$$ (3)

using the nominal cross-sectional dimensions w and t. Local cross-sectional dimensions were taken close to the failure location. Figure 3 provides a schematic of the test setup with relevant dimensions. The distance S represents the distance from the centerline of the center loading roller to the location of tensile failure as measured along the bottom surface of the specimen.

FIG. 2—*Three-point bend test setup.*

Fatigue tests were performed at a variety of frequencies and amplitudes under load control. The specimens were mounted in the same machine under the same procedure as described above for the static specimens. Fatigue tests were performed at a constant ram speed of 1.27 mm/min (0.05 in./min) with a stroke range set at 12.7 mm (0.5 in.). The average ultimate transverse tensile strength (UTS) obtained from the Panel 2 "A" configuration static specimens (w = 10.2 mm and S/t = 3) was used to calculate the load levels for fatigue. The majority of fatigue tests were run under the conditions R = 0.1, where R equals the ratio of the minimum applied load to the maximum applied load. Tension-tension fatigue tests were performed at 75, 80, 85, and 95% UTS levels with a frequency of ω equal to 20 Hz. Additional tests were performed at ω equal to 5 Hz to examine the effect of frequency. A few tests were run at R = 0.5 to examine the effect of load ratio. Tests were stopped at 10^5 cycles if no failure occurred. Several tests were run beyond that limit and stopped at 10^6 cycles if no failure occurred. The same operator conducted all the fatigue tests.

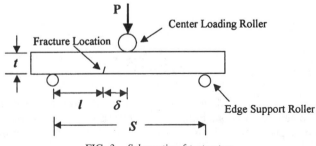

FIG. 3—*Schematic of test setup.*

Experimental Results

Material Properties of Fiber Volume

The fiber, resin, and void volume fractions were determined according to ASTM specifications assuming a resin density of 1.262 g/cm³ and fiber density of 1.7325 g/cm³. Tests were performed on three specimens from each of the two panels. The measured fiber volume values are given in Table 2 with the coefficient of variation values shown in brackets. The estimated fiber volume fractions are slightly lower than the measured values. The high variability in laminate thickness contributes to the deviation in the estimated and measured fiber volume fraction values.

Transverse Tensile Static Strength Measurements

A summary of the results of the static tests is seen in Table 3. Individual specimen data (static loading) are found in Table 4. The nominal strengths are listed for the different configurations using the nominal cross-sectional dimensions of the individual specimens. Strengths are also listed for the different configurations using the local cross-sectional dimensions measured as described in the Experimental Procedure section. The coefficients of variation of the strength for the different configurations are also given. The number of specimens tested per configuration is given in Table 1.

All the specimens exhibited a transverse tensile failure regardless of the width or span-to-thickness ratio. For the specimens of S/t equal to 3, the nominal transverse tensile strength increased as the width increased. For the specimens of nominal width 10.2 mm (0.4 in.), there was little difference in strength except when S/t equaled 6. Only two specimens of that particular configuration were tested. The coefficient of variations (CV) for the different configurations ranged from 7.8%, corresponding to Configuration G ($S/t = 6$, $W = 10.2$

TABLE 3—*Summary of nominal and local transverse tensile strengths in MPa for 90° bend tests, IM6 /3501-6 graphite-epoxy.*

	NOMINAL TRANSVERSE TENSILE STRENGTH Panel 1				Panel 2
	S/t				S/t
Width, w (mm)	3	4	6	8	3
10.2	56.3 [15.9]	59.6 [13.7]	68.9 [15.7]	59.5 [9.98]	91.7 [8.9]
12.7	58.5 [15.6]	63.1 [18.8]	...	...	82.7 [7.1]
15.2	67.5 [23.6]	67.5 [14.3]	...	52.4 [6.58]	...
	LOCAL TRANSVERSE TENSILE STRENGTH Panel 1				Panel 2
	S/t				S/t
Width, w (mm)	3	4	6	8	3
10.2	57.7 [13.3]	60.3 [31.5]	69.6 [8.35]	60.2 [9.95]	91.0 [8.8]
12.7	59.1 [15.9]	63.7 [28.6]	...	...	82.0 [6.8]
15.2	67.9 [33.1]	68.2 [13.6]	...	52.9 [12.6]	...

[a] Number in brackets is coefficient of variation.

TABLE 4—*Individual specimen data for transverse tensile strength, 90° bend tests.*

PANEL 1 RESULTS

Specimen ID No.	(S/t)	Nominal Width, w (mm)	Thickness, t (mm)	Local Width, w (mm)	Thickness, t (mm)	Failure Load, P, (N)	Moment Arm, l (mm)	Nominal Stress, σ (MPa)	CV, %	Local Stress, σ (MPa)	CV, %
SA4	3	10.26	6.83	10.24	6.82	1108	9.19	63.8		64.2	
SA5	3	10.26	6.76	10.27	6.75	894	11.73	67.2		67.2	
SA7	3	10.26	6.93	10.26	6.47	925	7.33	41.2		47.4	
SA8	3	10.29	6.91	10.26	6.89	970	9.73	57.6		58.1	
SA9	3	10.29	6.78	10.26	6.77	943	8.36	50.0		50.2	
SA10	3	10.31	6.73	10.26	6.71	974	9.31	58.2		58.9	
						969		56.3	15.9	57.7	13.4
SB1	3	12.47	6.83	...	...	1259	10.54	68.4		...	
SB2	3	12.47	6.78	...	...	1232	6.99	45.0		...	
SB3	3	12.80	6.88	...	...	1303	9.31	60.0		...	
SB4	3	12.78	6.83	12.78	6.82	1210	8.07	49.1		49.3	
SB5	3	12.80	6.76	12.78	6.73	1139	9.65	56.4		57.0	
SB6	3	12.80	6.76	12.83	6.76	1343	9.22	63.6		63.5	
SB7	3	12.80	6.93	12.80	6.91	1183	10.98	63.3		63.8	
SB8	3	12.80	6.88	12.80	6.88	1214	11.96	71.8		71.9	
SB9	3	12.80	6.78	12.80	6.76	1299	9.56	63.3		63.7	
SB10	3	12.80	6.91	12.78	6.89	1112	8.14	44.4		44.8	
						1229		58.5	15.6	59.1	15.9
SC1	3	15.19	6.81	15.19	6.80	1730	12.90	95.1		95.2	
SC2	3	15.21	6.78	15.18	6.79	1512	6.03	39.1		39.1	
SC3	3	15.34	6.88	15.34	6.87	1481	10.02	61.3		61.4	
SC4	3	15.16	6.86	15.14	6.82	1290	10.41	56.5		57.1	
SC5	3	15.14	6.73	15.16	6.72	1174	11.61	59.6		59.7	
SC6	3	15.34	6.86	15.29	6.84	1801	12.53	93.9		94.8	
						1498		67.6	23.6	67.9	33.1
SD1	4	10.13	6.78	...	...	752	12.24	59.2		...	
SD2	4	10.13	6.81	...	...	841	12.07	64.8		...	
SD3	4	10.13	6.78	...	...	863	9.44	52.4		...	
SD4	4	10.29	6.83	10.21	6.82	743	14.73	68.4		69.0	
SD5	4	10.21	6.93	10.19	6.92	974	13.69	81.5		82.1	
SD6	4	10.21	6.83	10.20	6.85	721	12.44	56.4		56.1	
SD7	4	10.21	6.73	10.20	6.72	707	9.94	45.6		45.7	
SD8	4	10.21	6.73	10.21	6.71	934	14.14	85.7		86.2	
SD9	4	10.26	6.76	10.20	6.76	543	12.19	42.4		42.6	
SD10	4	10.24	6.81	10.21	6.79	538	11.73	39.9		40.3	
						761		59.6	13.7	60.3	31.5
SE1	4	12.93	6.78	12.88	6.75	596	11.58	34.8		35.3	
SE3	4	12.93	6.81	12.90	6.78	1094	13.92	76.3		77.0	
SE4	4	12.80	6.86	12.80	6.83	1063	17.41	92.2		93.0	
SE5	4	12.75	6.93	12.72	6.91	850	14.38	59.8		60.4	
SE6	4	12.73	6.86	12.73	6.84	987	10.11	50.0		50.3	
SE7	4	12.75	6.73	12.76	6.69	752	11.28	44.0		44.5	
SE8	4	12.78	6.73	12.73	6.70	898	13.39	62.3		63.1	
SE9	4	12.78	6.76	12.75	6.75	1014	13.97	72.9		73.1	
SE10	4	12.75	6.81	12.76	6.80	850	17.68	76.3		76.4	
						900		63.2	18.8	63.7	28.6

TABLE 4—*Continued*

		Nominal		Local				Nominal		Local	
Specimen ID No.	(S/t)	Width, w (mm)	Thickness, t (mm)	Width, w (mm)	Thickness, t (mm)	Failure Load, P, (N)	Moment Arm, l (mm)	Stress, σ (MPa)	CV, %	Stress, σ (MPa)	CV, %
SF1	4	15.29	6.81	15.29	6.77	1050	13.09	58.2		58.8	
SF2	4	15.32	6.81	15.32	6.79	1317	14.96	83.3		83.7	
SF3	4	15.32	6.78	15.34	6.78	1339	10.29	58.7		58.6	
SF4	4	15.34	6.73	15.32	6.72	1054	14.58	66.4		66.7	
SF5	4	15.19	6.96	15.18	6.91	1156	15.01	70.8		71.9	
SF6	4	15.16	6.83	15.16	6.81	885	14.58	54.7		55.0	
SF7	4	15.19	6.73	15.14	6.71	1245	12.01	65.2		65.9	
SF8	4	15.16	6.73	15.13	6.67	1183	15.30	79.0		80.6	
SF9	4	15.16	6.78	15.14	6.75	947	17.20	70.1		70.9	
SF10	4	15.16	6.81	15.13	6.79	987	16.51	69.6		70.0	
						1116		67.6	14.3	68.2	13.6
SG1	6	10.11	6.93	10.08	6.90	480	24.56	72.8		73.7	
SG2	6	10.11	6.73	10.08	6.73	507	19.65	65.3		65.5	
						494		69.0	15.7	69.6	8.35
SK1	8	10.29	6.88	10.29	6.86	356	30.05	65.8		66.3	
SK2	8	10.29	6.71	10.25	6.67	329	25.07	53.5		54.3	
SK3	8	10.29	6.76	10.25	6.74	311	29.88	59.4		60.0	
						332		59.6	10.0	60.2	9.95
SM1	8	15.19	6.91	15.19	6.87	587	22.92	55.7		56.4	
SM2	8	15.24	6.71	15.19	6.68	498	25.98	56.7		57.3	
SM3	8	15.21	6.76	15.19	6.74	409	25.40	44.9		45.3	
						498		52.4	6.58	53.0	12.6
PANEL 2 RESULTS											
PSA11	3	10.21	6.71	10.21	6.76	1379	9.98	89.9		88.6	
PSA12	3	10.19	6.83	10.21	6.83	1339	9.83	83.0		82.8	
PSA13	3	10.26	6.86	10.21	6.91	1659	9.93	102.4		101.4	
PSA14	3	10.24	6.86	10.21	6.88	1592	9.91	98.3		97.8	
PSA15	3	10.19	6.83	10.24	6.85	1579	8.64	86.0		85.3	
						1510		91.9	8.91	91.2	8.84
PSB11	3	12.67	6.65	12.71	6.67	1744	9.73	90.7		90.0	
PSB12	3	12.70	6.78	12.71	6.82	1993	7.90	80.8		79.9	
PSB13	3	12.70	6.81	12.71	6.83	1966	8.51	85.3		84.6	
PSB14	3	12.70	6.81	12.71	6.83	1837	8.61	80.6		80.0	
PSB15	3	12.70	6.81	12.73	6.78	1552	9.47	75.0		75.4	
						1818		82.5	7.10	82.0	6.78

mm) to 33% for Configuration C ($S/t = 3$, $W = 15.2$ mm). The average CV for the nominal transverse tensile strength for Panel 1 was 22.1%. Because of the large scatter in strength of the different-sized specimens from Panel 1, specimens for the smaller configurations (A and B) were tested from the second panel. The overall nominal transverse tensile strength was considerably higher (87.5 MPa) in Panel 2 compared to Panel 1 (61.3 MPa). Five specimens of Configuration A and B were each tested. The CVs were considerably smaller, with an overall average CV of 8.0% for Panel 2 static transverse tensile strength.

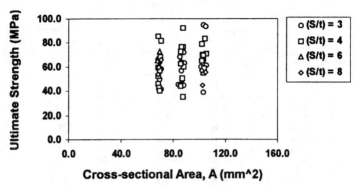

FIG. 4—*Effect of nominal cross-sectional area on transverse tensile strength, Panel 1.*

Figures 4 and 5 show the static strengths for the various specimens from Panel 1 as a function of the nominal cross-sectional area and the ratio of S/t. Each plot shows a great deal of scatter for a given width or S/t ratio. Assuming a volume dependency on the strength, the strength should decrease with increasing cross-sectional area for a given width. This trend is not seen in Fig. 5. This does not imply there is no volume dependency. Rather, because of the unusual amount of scatter, a volume dependency may be obscured.

Figure 6 shows the probability distribution of the static transverse tensile strength based on the data from Panel 1. When generating the probability distribution of the strength using Weibull scaling laws, it is best to compare only specimens of similar configurations. However, from a statistical point of view, it is also desirable to have a large number of specimens to evaluate. As each specimen failed in the same mode—transverse tensile failure, it is reasonable, albeit not optimal, to evaluate all of the specimens ($n = 59$) from Panel 1 as one data set. Using a linear regression technique, the Weibull parameters were found to be $m = 5.32$ and $\sigma_c = 66.9$ MPa (9.71 ksi). When considering the specimens with a S/t ratio of 3 ($n = 22$) and 4 ($n = 29$) individually, the Weibull parameters were found to be $m = 5.00$, $\sigma_c = 65.7$ MPa (9.53 ksi) and $m = 4.90$, $\sigma_c = 69.1$ MPa (9.90 ksi), respectively. The results of the static transverse tension strengths are comparable with those found by other

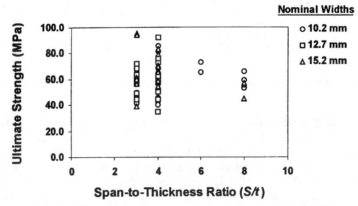

FIG. 5—*Effect of S/t ratio on transverse tensile strength, Panel 1.*

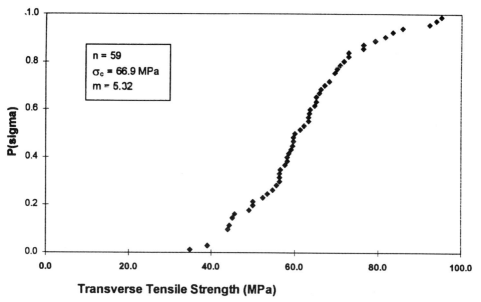

FIG. 6—*Probability distribution of the static transverse tensile strength, Panel 1.*

investigators for a similar material system tested under three-point bending but with thinner laminates and larger S/t ratios; O'Brien and Salkepar, evaluating 33 AS4/3501-6 specimens, found m and σ_c to be 7.63 and 61.1 MPa (8.87 ksi), respectively [5]. The larger m value associated with our data is consistent with the greater scatter of the data. A plot of the probability distribution for the ten specimens from Panel 2 is shown in Fig. 7 along with the Weibull parameters. The values of m and σ_c were found to be 13.0 and 92.2 MPa (13.38 ksi), respectively.

Transverse Tensile Fatigue Strength Measurements

The results of the fatigue testing can be seen in Fig. 8. Individual specimen data (fatigue loading) are found in Table 5. Forty-eight different specimens were tested under the conditions of $w = 20$ Hz, $R = 0.1$. At a frequency of $w = 5$ Hz, six and five specimens were tested at $R = 0.1$ and $R = 0.5$ values, respectively. Many tests were stopped at 10^5 and 10^6 cycles. The number in parentheses represents the number of specimens tested at $\omega = 20$ Hz, $R = 0.1$ that were stopped at 10^5 and 10^6 cycles. The number of tests stopped at 10^5 or 10^6 cycles as well as a fuller description of the testing parameters are given in a box for those specimens that were tested at conditions other than $\omega = 20$ Hz, $R = 0.1$.

Figure 9 shows four typical specimen failures along with the number of cycles-to-failure for the particular specimens. The specimens all exhibited transverse tensile failures on the lower surface. The final fracture surface exhibited a variety of patterns with the crack "kicking" left or right from the initial failure site. Regardless of whether the final fracture pattern through the thickness of the specimen was nearly vertical or angled, the moment arm l was measured as the distance from the centerline of the edge support roller to the location of tensile failure along the bottom surface of the specimen. The distance δ represents the distance from the centerline of the center loading roller to the location of tensile failure as measured along the bottom surface of the specimen.

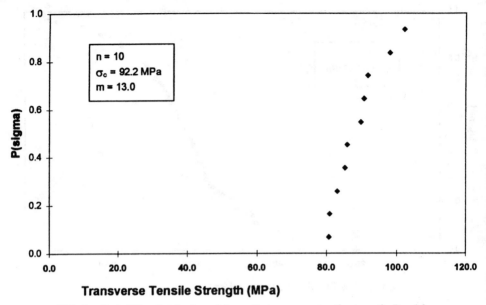

FIG. 7—*Probability distribution of the static transverse tensile strength, Panel 2.*

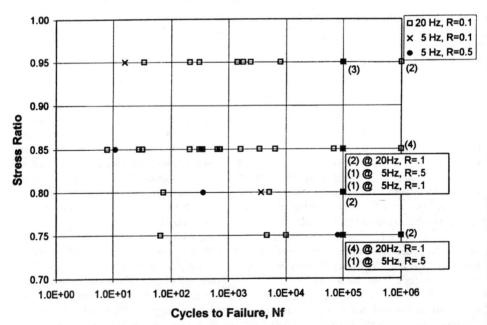

FIG. 8—*Transverse tension S-N curve under three-point bending (number of tests stopped at 10^5 or 10^6 cycles for w = 20 Hz, R = 0.1 given in parentheses; number of tests stopped at 10^5 or 10^6 cycles for conditions other than w = 20 Hz, R = 0.1 given in box).*

TABLE 5—*Individual specimen data for transverse tension fatigue, 90° bend tests.*

Specimen ID No.	Frequency, ω, (Hz)	R	Load Ratio	(S/t)	Nominal Width, w (mm)	Local Width, w (mm)	Local Thickness, t (mm)	Span, S (mm)	Failure Location, δ (mm)	Cycles-to-Failure, Nf
FA12	20	0.1	0.95	3	10.16	10.21	6.81	20.42	1.49	1 420
FA16	20	0.1	0.95	3	10.16	10.21	6.76	20.27	0.88	210*
FA20	20	0.1	0.95	3	10.16	10.21	6.88	20.65	...	1 000 000*
FA23	20	0.1	0.95	3	10.16	10.19	6.78	20.35	...	1 000 000*
FA27	20	0.1	0.95	3	10.16	10.21	6.93	20.80	...	100 000*
FA28	20	0.1	0.95	3	10.16	10.19	6.91	20.73	0.60	7 920
FA31	20	0.1	0.95	3	10.16	10.17	6.81	20.42	...	100 000*
FA38	20	0.1	0.95	3	10.16	10.19	6.93	20.80	0.94	1 797
FA47	20	0.1	0.95	3	10.16	10.20	6.90	20.70	12.74	100 000*
FA52	20	0.1	0.95	3	10.16	10.19	6.78	20.35	1.28	34
FA62	20	0.1	0.95	3	10.16	10.26	6.83	20.50	0.54	2 417
FA63	20	0.1	0.95	3	10.16	10.24	6.81	20.42	0.67	306
FA11	20	0.1	0.85	3	10.16	10.24	6.83	20.50	0.71	1 630
FA13	20	0.1	0.85	3	10.16	10.29	6.83	20.50	...	1 000 000*
FA17	20	0.1	0.85	3	10.16	10.21	6.90	20.69	0.71	720
FA24	20	0.1	0.85	3	10.16	10.17	6.55	19.66	...	1 000 000*
FA26	20	0.1	0.85	3	10.16	10.21	6.78	20.35	...	1 000 000*
FA29	20	0.1	0.85	3	10.16	10.19	6.90	20.69	5.58	28
FA30	20	0.1	0.85	3	10.16	10.19	6.86	20.57	...	1 000 000*
FA32	20	0.1	0.85	3	10.16	10.21	6.85	20.54	1.55	352
FA33	20	0.1	0.85	3	10.16	10.19	6.81	20.42	0.77	310*
FA66	20	0.1	0.85	3	12.70	12.75	6.76	20.27	1.26	3 484
FA68	20	0.1	0.85	3	12.70	12.75	6.92	20.76	0.14	696
FA70	20	0.1	0.85	3	12.70	12.78	6.86	20.57	...	100 000*
FA72	20	0.1	0.85	3	12.70	12.76	6.78	20.35	0.63	643
FA74	20	0.1	0.85	3	12.70	12.75	6.91	20.73	0.00	100 000*
FA4	20	0.1	0.80	3	10.16	10.24	6.86	20.57	2.58	130 000
FA6	20	0.1	0.80	3	10.16	10.24	6.74	20.23	0.11	5 150
FA9	20	0.1	0.80	3	10.16	10.20	6.88	20.65	4.74	202 100
FA19	20	0.1	0.80	3	10.16	10.19	6.91	20.73	1.82	73
FA22	20	0.1	0.80	3	10.16	10.21	6.83	20.50	0.65	100 000*
FA36	20	0.1	0.80	3	10.16	10.21	6.79	20.38	...	100 000*
FA43	20	0.1	0.80	3	10.16	10.19	6.85	20.54	0.86	3 675
FA50	20	0.1	0.80	3	10.16	10.21	6.82	20.46	2.36	361
FA3	20	0.1	0.75	3	10.16	10.21	6.73	20.19	...	1 000 000*
FA8	20	0.1	0.75	3	10.16	10.17	6.90	20.69	...	1 000 000*
FA18	20	0.1	0.75	3	10.16	10.17	6.91	20.73	...	100 000*
FA21	20	0.1	0.75	3	10.16	10.19	6.83	20.50	...	100 000*
FA37	20	0.1	0.75	3	10.16	10.20	6.93	20.80	0.33	81 089
FA42	20	0.1	0.75	3	10.16	10.17	6.82	20.46	0.74	65
FA48	20	0.1	0.75	3	10.16	10.19	6.91	20.73	...	100 000*
FA51	20	0.1	0.75	3	10.16	10.20	6.78	20.35	...	100 000*
FA1	5	0.1	0.85	3	10.16	10.21	6.83	20.50	0.04	6 470
FA2	5	0.1	0.85	3	10.16	10.22	6.81	20.42	1.63	210
FA39	5	0.1	0.85	3	10.16	10.21	6.93	20.80	1.05	332
FA41	5	0.1	0.85	3	10.16	10.21	6.85	20.54	3.35	32
FA46	5	0.1	0.85	3	10.16	10.22	6.78	20.35	...	100 000*
FA53	5	0.1	0.85	3	10.16	10.22	6.86	20.57	0.43	11
FA40	5	0.5	0.75	3	10.16	10.17	6.83	20.50	...	100 000*
FA49	5	0.5	0.75	3	10.16	10.24	6.91	20.73	0.48	4 682
FA56	5	0.5	0.85	3	10.16	10.24	6.81	20.42	1.49	68 877
FA60	5	0.5	0.85	3	10.16	10.25	6.90	20.69	0.24	8
FA64	5	0.5	0.85	3	10.16	10.24	6.91	20.73	...	100 000*

NOTE: * indicates test was stopped at 10^5 or 10^6 cycles.

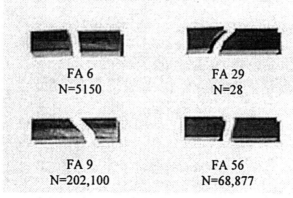

FIG. 9—*Representative transverse tension fatigue failures.*

Failure Location Influence on Strength

The transverse tensile failure location of over 80% of the specimens was within 20% of the distance from the centerline of the center loading roller to the centerline of the support rollers. Figure 10 shows the distribution of the normalized failure locations where the normalized failure location is defined as the distance from the center loading roller to the tensile failure location (δ) divided by half the span length ($0.5S$). Two specimens, including FA29 ($N = 28$ cycles), failed under the support roller. In Fig. 11, the cycles-to-failure are shown for the various strength ratios as a function of the normalized failure location.

Discussion

Testing of the optimum-sized configuration consistently yielded transverse tensile failures, yet still exhibited a great deal of scatter. It was anticipated that the smallest configuration

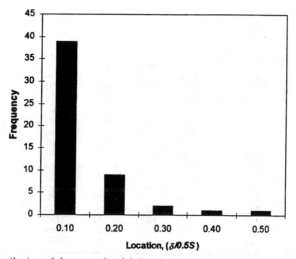

FIG. 10—*Distribution of the normalized failure location of transverse tension fatigue data.*

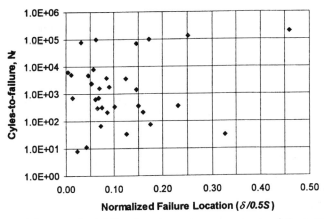

FIG. 11—*Effect of the normalized failure location on transverse tension fatigue cycles-to-failure.*

would yield transverse tensile failures as from a strength of materials approach, the ratio of transverse tensile stress to shear stress for a rectangular cross-section subjected to three-point bending is twice the span-to-thickness ratio [4]. Considering the transverse tensile strength of 90° graphite/epoxy laminates is of similar magnitude as the shear strength ($\sigma_{TT} \sim 51.7$ MPa, $\tau_{12} \sim 93$ MPa) and considerably smaller than the transverse compressive strength ($\sigma_{TC} \sim 206$ MPa), it would be reasonable to expect failure to initiate at the tensile surface [7]. It is unclear whether the large scatter in the static strengths is due to material variability, i.e., inherent presence of flaws or voids, or manufacturing variability, i.e., improper or poor processing of the panels themselves. Considering that other researchers have found considerably less scatter when testing 90° laminates of comparable material systems in three-point bending [4,5], it is probable the scatter is due to the manufacturing variability of the particular panels from which the specimens were taken. There did not appear to be a correlation between the failure loads and the number of pits present in a particular specimen. Considering the degree of scatter of transverse tensile strengths in the specimens from Panel 1, it was decided that fatigue specimens would be taken only from Panel 2, whose specimens exhibited considerably less scatter than those of Panel 1 under static loading.

Before starting the fatigue testing, certain trends were considered likely to occur. First, a steep S-N curve was considered possible. Although a high moduli graphite/epoxy material system was being tested, the fibers were oriented in the 90° orientation, rather than a 0° orientation. Therefore, the matrix would likely see more strain. As the tested laminates are composed solely of 90° plies without the benefit of any "constraining" plies to constrain matrix cracking, once the crack growth reached a characteristic damage stage, defined by stable crack growth, damage would progress quickly to failure [8]. Recognizing that the fatigue strain limit decreases with an increase of off-axis angled plies, the opening displacement mode would probably be more important than a sliding displacement occurring parallel to the fibers. Another expectation of the testing was that some specimens would fail by transverse shear failures due to the small S/t ratio. Shorter lives were also anticipated due to the testing occurring in load control; displacement and energy control loading have been shown to tend to produce longer lives [3].

The original fatigue testing was to be performed at $w = 5$ and $w = 0.1$ Hz and tests stopped if specimens did not fail after 10^6 cycles. However, initial testing at 5 Hz produced fatigue lives of 10^6 cycles, even at high load ratios. The test program was then modified so tests were performed primarily at 5 Hz and stopped at 10^5 cycles. Examining the S-N curve in Fig. 8, it can be seen that the testing produced a great deal of scatter. The number of tests stopped at 10^5 and 10^6 cycles is given in parentheses. Looking only at the highest load ratio, the fatigue lives ranged from an order of 10^1 to 10^6 cycles. A similar trend can be seen at the other load levels. The range of scatter seems to decrease slightly when looking at the lower load levels. Comparing the data at P/P_{ult} equals 0.95 and 0.75, one can see that the fatigue lives at the lower load level are generally longer than at the higher load level. Comparing the data for $R = 0.1$ at 20 and 5 Hz, there seems to be little frequency dependency. However, any dependency may be obscured by the large amount of overall scatter of the data. Comparing the fatigue data associated with a frequency of 5 Hz and $R = 0.1$ and $R = 0.5$ values, again there seems to be little distinction between the two loadings. However, it would be unwise to draw any strong conclusions from these data, as once again there is too much scatter evident.

By taking specimens solely from Panel 2, it was hoped that scatter due to manufacturing, as seen in Panel 1's static strength data, would be minimized. However, if the static results of the specimens from Panel 2 were skewed due to that particular portion of the panel being of very good quality compared to the rest of Panel 2, then the fatigue lives could exhibit substantial scatter. However, this does not seem very likely as the specimens were cut from the portion of the panel with the most uniform C-scan patterns. Other possible sources of scatter could be the lack of edge polishing of the specimens. However, this was purposely not performed so as to mimic "real-world" structures, which would likely not have their edges polished. Overall, this preliminary study shows that further tests should be performed in order to determine whether the variability of the data was due primarily to panel and/or specimen preparation or due to material variability, inherent in the 90° bend specimen.

Conclusions

Researchers have found in bonded skin/stiffener/frame composite reinforced panels that failure initiated in the adhesive pocket at the interface of the skin/stiffener and propagated as a transverse crack in skin laminate or flange laminate near the flange tip. An experimental investigation of the transverse tensile strength of IM6/3501 composite materials was performed on specimens subjected to three-point bending under static and fatigue loading. In this study, a parametric study of S/t effects was performed experimentally to determine an "optimum" configuration, where *optimum* was defined as the smallest configuration that consistently failed in transverse tension when subjected to three-point bending under static loading. The optimum configuration was found to be specimens with nominal dimensions of S/t equal to 3 and width equal to 10.2 mm (0.4 in.). Fatigue testing under three-point bending was performed on specimens of the optimum configuration. The S-N data showed a great deal of scatter. At this time, it is unclear whether the scatter in the data is due primarily to material variability (i.e., the presence of voids or flaws) or manufacturing variability (i.e., improper preparation of panels and/or specimens).

Future work should begin with tension fatigue testing under similar test conditions of 90° laminates taken from Panel 2. This would provide insight into the integrity of the three-point bend test procedure. Additional fatigue testing should also include testing under three-point bending of similar-sized specimens as the transverse bending specimens used in this study taken from panels manufactured by different sources. These tests would be performed at similar conditions as the current study. Testing of thinner laminates performed under similar conditions should also be undertaken to monitor and to assess sizing effects due to thickness of the laminate and clumping of same-angle plies. Periodic monitoring of the edges and crack growth would be beneficial in the further understanding of the fatigue growth. A comparison of these results may provide understanding into whether the scatter observed in this study is due to material variability or manufacturing variability. If the data produced less scatter, fatigue growth laws could possibly be determined.

Acknowledgments

This work was initiated as a result of the NASA-ASEE Summer Faculty Fellowship program. This work was funded in part by NASA Langley Research Center through Grant NAG-1-1773. Special thanks go to Drs. T. K. O'Brien of the Mechanics of Materials Branch, NASA LaRC, and P. J. Minguet of the Boeing Defense & Space Group, Helicopters Division, Philadelphia, PA, as well as S. and R. Coguill of the Composites Materials Research Group (CMRG) of the University of Wyoming, Laramie, WY.

References

[1] Minguet, P. J., Fedro, M., O'Brien, T. K., Martin, R., Ilcewicz, L., Awerbuch, J., and Wang, A., "Development of a Structural Test Simulating Pressure Pillowing Effects in a Bonded Skin/Stringer/Frame Configuration," *Proceedings,* Fourth NASA/DoD Advanced Composite Technology (ACT) Conference, NASA-CP 3229, Salt Lake City, UT, 1993, pp. 863–880.
[2] Minguet, P. J. and O'Brien, T. K., "Analysis of Composite Skin/Stringer Bond Failure Using a Strain Energy Release Rate Approach," *Proceedings,* 1995 ICCM Conference, Vancouver, August 1995.
[3] Stinchcomb, W. W. and Bakis, C. E., "Fatigue Behavior of Composite Laminates," *Fatigue of Composite Materials,* K. L. Reifsnider, Ed., Elsevier Science Publishers B. V., Amsterdam, Netherlands, 1991, pp. 105–158.
[4] Adams, D. F., King, T. R., and Blackketter, D. M., "Evaluation of the Transverse Flexure Test Method for Composite Materials," *Composite Science and Technology,* Elsevier Science Publishers, Ltd., England, 1990, pp. 341–353.
[5] O'Brien, T. K. and Salpekar, S. A., "Scale Effects on the Transverse Tensile Strength of Graphite/Epoxy Composites," *Composite Materials: Testing and Design (Eleventh Volume), ASTM STP 1206,* E. T. Camponeschi, Jr., Ed., American Society for Testing and Materials, West Conshohocken, PA, 1993, pp. 23–42.
[6] Weibull, W., "A Statistical Theory of the Strength of Materials," Ing. Vetenskaps Akad. Handl. (*Royal Swedish Institute Engineering Research Proceedings*), NR151, 1939.
[7] Tsai, S. W. and Hahn, H. T., *Introduction to Composite Materials,* Technomic Publishing Company, Lancaster, PA, 1980, pp. 293–294.
[8] Reifsnider, K. L., Henneke, E. G., Stinchcomb, W. W., and Duke, J. C. in *Mechanics of Composite Materials, Recent Advances,* Z. Hashin and C. T. Herakovich, Eds., Pergamon Press, New York, 1983, pp. 399–420.

Environmental Considerations

Scott Case,[1] *Nirmal Iyengar,*[1] *and Kenneth Reifsnider*[1]

Life Prediction Tool for Ceramic Matrix Composites at Elevated Temperatures

REFERENCE: Case, S., Iyengar, N., and Reifsnider, K., **"Life Prediction Tool for Ceramic Matrix Composites at Elevated Temperatures,"** *Composite Materials: Fatigue and Fracture, Seventh Volume, ASTM STP 1330,* R. B. Bucinell, Ed., American Society for Testing and Materials, 1998, pp. 165–178.

ABSTRACT: A life prediction method for ceramic matrix composites is developed. This model is based upon damage mechanics concepts included in the framework of the critical element model. One unique feature of the model is its ability to include general variations of temperature and applied loads as functions of time. A detailed description of application of the model to elevated temperature fatigue is given. In addition, a validation example is presented that includes the combined effects of rupture and fatigue. The comparison with the experimental data is shown to be good, although the result does appear to be dominated by the rupture effect.

KEYWORDS: ceramic matrix composites, life prediction, remaining strength, damage tolerance

The durability of composite materials, in particular the area of analytical modeling of damage and the resulting degradation in material properties, has received much attention in recent years [1]. Nevertheless, the prediction of the remaining strength and life in these inhomogeneous, anisotropic materials is inherently complicated—there are many different damage modes (such as delamination or matrix cracking) that may develop and interact before a final failure occurs. These damage modes that occur before final failure make it possible to design components from composite materials that are extremely damage tolerant. Due to this complex interaction of damage modes, methods of analysis typically used for engineering metals that involve the propagation of a self-similar single crack cannot be used. However, to use the damage tolerance that composite materials possess to its fullest, we must have some method to predict the stiffness, strength, and life of these materials in differing environmental conditions. In this paper, we consider the implementation of a kinetics-based approach to such a method using damage accumulation concepts. Such an approach is implemented using the critical element model proposed by Reifsnider and Stinchcomb [2].

Concepts of the Critical Element Model

The "critical element" concept [2] is based upon the assumption that the damage associated with property degradation is distributed widely within the composite laminate. In addition, it is assumed that a representative volume can be chosen such that the state of

---

[1] Materials Response Group, Engineering Science and Mechanics Department, Virginia Tech, Blacksburg, VA 24061.

stress in that volume is typical of all other volumes in the laminate, and that the details of stress distribution and damage accumulation in that volume are sufficient to describe the final failure resulting from a specific failure mode. Thus, it may be necessary to select different representative volume elements for different failure modes. We proceed by further dividing the representative volume into "critical" and "sub-critical" elements. The critical elements are selected in such a manner that their failure controls the failure of the representative volume and therefore (by definition of the representative volume) of the laminated component. The remainder of the elements in the representative volume are regarded as subcritical because their failure does not cause failure of the representative volume and, therefore, of the component. Their failure (due to such events as cracking or delamination) does, however, lead to greater stresses in the critical element that contribute to the eventual failure of the component. As an example of such a failure process, we may consider the case of tensile fatigue failure of a cross-ply polymeric composite laminate. During the fatigue process, matrix cracks develop in the 90° plies. However, these cracks do not cause failure of the laminate. They do increase the stress level in the 0° plies. But it is only when the 0° plies fail that the laminate fails. Thus, in this simple example, the 0° plies would correspond to the critical element and the 90° plies would correspond to the subcritical element. By coupling the critical element concept with damage accumulation concepts we may develop a powerful tool for the analysis of ceramic matrix composite materials under a variety of loading conditions.

Damage Accumulation Concepts

In the present analysis we will follow arguments presented by Reifsnider et al. [3] for damage accumulation in composites. We begin our analysis by postulating that remaining strength may be used as a damage metric. We next assume that the remaining strength may be determined (or predicted) as a function of load level and some form of generalized time. For a given load level, a particular fraction of life corresponds to a certain reduction in remaining strength. We claim that a particular fraction of life at a second load level is equivalent to the first if and only if it gives the same reduction in remaining strength, as illustrated in Fig. 1. In the case of Fig. 1, time t_1 at an applied stress level S_a^1 is equivalent to time t_2^0 at stress level S_a^2 because it gives the same remaining strength. In addition, the remaining life at the second load level is given by the amount of generalized time required to reduce the remaining strength to the applied load level. In this way, the effect of several increments of loading may be accounted for by adding their respective reductions in remaining strength. For the general case, the strength reduction curves may be nonlinear, so the remaining strength and life calculations are path dependent.

Our next step in the analysis is to postulate that normalized remaining strength (our damage metric) is an internal state variable for a damaged material system. This normalized remaining strength is based upon the selection of an appropriate failure criterion (such as maximum stress or Tsai-Hill), which is a scalar combination of the principal material strengths and applied stresses in the critical element. In this way we are able to consider a single quantity rather than the individual components of the strength tensor. We denote this failure function by Fa. We next construct a second state variable, the continuity [4], which we shall define to be $(1 - Fa)$ and denote by ψ. We shall attempt to define our remaining strength and life in terms of ψ. To do so, we assume that the kinetics are defined by a specific damage accumulation process for a particular failure mode and assign different rate equations to each of the processes that may be present.

As an example, let us consider a common kinetic equation (a power law) such that

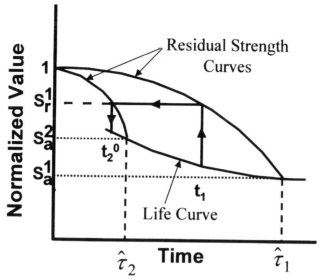

FIG. 1—*Use of remaining strength as a damage metric. Time* t_2^0 *at constant applied stress level* S_a^2 *is equivalent to time* t_1 *at constant applied stress level* S_a^1 *because it results in the same remaining strength* S_r^1.

$$\frac{d\psi}{d\tau} = A\psi^j \tag{1}$$

where j is a material constant, and τ is a generalized time variable defined by

$$\tau = \frac{t}{\hat{\tau}} \tag{2}$$

and $\hat{\tau}$ is the characteristic time for the process at hand. This characteristic time could be a creep rupture life, a creep time constant, or even a fatigue life, in which case

$$\tau = \frac{n}{N} \tag{3}$$

where n is the number of fatigue cycles, and N is the number of cycles to failure at the current applied loading conditions. Next we rearrange Eq 1 and integrate so that we have

$$\int_{\psi_0}^{\psi_i} d\psi = A \int_0^{\tau_i} (\psi(\tau))^j \, d\tau \tag{4}$$

If, for example, we set $A = -1$ and $j = 1$, we arrive at

$$\psi_i - \psi_0 = 1 - Fa_i - (1 - Fa_0) = -\Delta Fa = -\int_0^{\tau_i}(1 - Fa(\tau))d\tau \tag{5}$$

Then we define our normalized remaining strength, Fr, so that

$$Fr = 1 - \Delta Fa = 1 - \int_0^{\tau}(1 - Fa(\tau))d\tau \tag{6}$$

In such a form, the failure condition is the point at which the remaining strength equals the applied load ($Fr = Fa$).

A special case of Eq 6 is that of fatigue loading at constant amplitude in which

$$\tau = \frac{n}{N}$$

$$Fa = \frac{S_a}{S_{ult}} \tag{7}$$

$$Fr = \frac{S_r}{S_{ult}}$$

where n is the number of fatigue cycles, S_a is the constant applied stress, S_{ult} is the ultimate strength, S_r is the remaining strength after n cycles, and N is the number of cycles to failure at S_a.

Substituting into Eq 6, and making use of the fact that $Fa(\tau) = S_a/S_{ult}$ is constant, we arrive at

$$\frac{S_r}{S_{ult}} = 1 - \left(1 - \frac{S_a}{S_{ult}}\right)\left(\frac{n}{N}\right) \tag{8}$$

This form is identical to that proposed by Broutman and Sahu [5] to explain the fatigue behavior of fiberglass.

Suppose that instead of the kinetic equation given in Eq 1, we use a different form (one that is explicit in time) given by

$$\frac{d\psi}{d\tau} = \psi j \tau^{j-1} \tag{9}$$

If we integrate Eq 9 from τ_1 to τ_2, we obtain

$$\psi_2 - \psi_1 = Fr_2 - Fr_1 = \Delta Fr = -\int_{\tau_1}^{\tau_2}(1 - Fa)j\tau^{j-1}d\tau \tag{10}$$

If we set τ_1 equal to zero and Fr_1 equal to unity, we arrive at the remaining strength as a function of generalized time

$$Fr = 1 - \int_0^\tau (1 - Fa(\tau))j\tau^{j-1}d\tau \tag{11}$$

For the case in which Fa is constant, Eq 11 may be integrated analytically to yield

$$Fr = 1 - (1 - Fa)\tau^j \tag{12}$$

Equation 11 has been used with a great deal of success [6–9] for cases in which $Fa(\tau)$ is continuous. However, for cases in which $Fa(\tau)$ is not continuous (such as in the case of block loading), we use a slightly modified approach based on the idea of remaining strength as a damage metric. To explain our implementation of these ideas, we will consider the case of block loading at a level Fa_1 for a generalized time τ_1 resulting in a remaining strength Fr. This loading is followed by loading at a level Fa_2. The question then is how to determine the equivalent amount of generalized time, τ_2^0, that would have been necessary to cause an equivalent amount of damage (reduction in residual strength) at the level Fa_2. We will call a time "pseudo-time" because it does not refer to actual time. Substituting into Eq 12 and using the idea of equivalent remaining strength, we obtain

$$Fr = 1 - (1 - Fa_1)\tau_1^j = 1 - (1 - Fa_2)(\tau_2^0)^j \tag{13}$$

Equation 13 may then be solved for the pseudo-time to yield

$$\tau_2^0 = \left(\frac{1 - Fr}{1 - Fa_2}\right)^{1/j} \tag{14}$$

In order to understand such a procedure, it may be helpful to consider a specific example such as that shown in Fig. 2. This example represents creep loading at an applied stress level that corresponds to Fa_1 (that results in a characteristic time to failure, $\hat{\tau}_1$) for time t_1 followed by loading that corresponds to Fa_2 (that results in a characteristic time to failure, $\hat{\tau}_2$) for time Δt. The equivalent pseudo-time, t_2^0 (unnormalized), is given by

$$t_2^0 = \hat{\tau}_2\tau_2^0 = \left(\frac{1 - Fr}{1 - Fa_1}\right)^{1/j/\hat{\tau}_2} \tag{15}$$

Making use of Eq 10, the change in remaining strength, ΔFr, over the interval $(t_1, t_1 + \Delta t)$, which corresponds to an interval $(t_2^0, t_2^0 + \Delta t)$ in pseudo-time, is then given by

$$\Delta Fr = -(1 - Fa_2)\left\{\left[\frac{t_2^0 + \Delta t}{\hat{\tau}_2}\right]^j - \left(\frac{t_2^0}{\hat{\tau}_2}\right)^j\right\} \tag{16}$$

In general, such an approach may be carried out for any number of blocks of loading having an applied stress level Fa_i so that the remaining strength is given by

$$Fr = 1 - \sum_{i=1}^{i=N_{blocks}} \Delta Fr_i \tag{17}$$

where N_{blocks} is the number of blocks of loading.

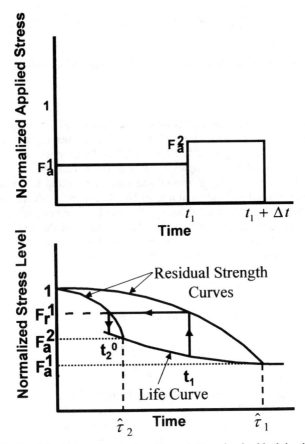

FIG. 2—*Calculation of equivalent ("pseudo") cycles for block loading.*

As an example of such an approach, we will consider a triangular waveform, as illustrated in Fig. 3. We will assume that, because of the temperatures involved, the behavior of the material is controlled by a combination of the rupture and fatigue behaviors. We begin our analysis by calculating an average value of the temperature, $\overline{T}$, and the failure function, $\overline{Fa}$ (based upon the average stresses, $\overline{\sigma_{ij}}$). A corresponding time to rupture, $\hat{\tau}$, is also determined. Using Eq 16, we calculate a remaining strength at time t_1. Next we divide the interval (t_0, t_1) into two subintervals, $(t_0, t_0 + t_1/2)$ and $(t_0 + t_1/2, t_1)$, and calculate a remaining strength at time t_1 by applying Eq 16 twice and summing the results for those respective cases. This process of subdivision and summation is repeated until the calculated value of the remaining strength at t_1 converges to some specified error tolerance.

In addition to the rupture process, we recognize that fatigue may also be contributing to the reduction in remaining strength and eventual failure of the component. To account for the fatigue process, we introduce a new generalized time, $\hat{N}$, that corresponds to the number of fatigue cycles to failure at a given failure function value. At each load reversal during the loading profile (such as occurs at t_1), we increment the number of cycles by ½ and calculate a change in remaining strength in an analogous manner to that used for the rupture process. This approach, although simplistic, does give the correct answer in the limiting cases (i.e.,

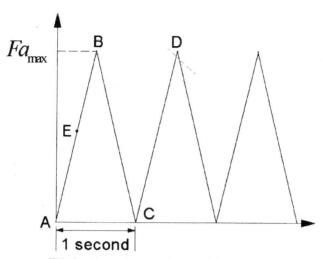

FIG. 3—*Triangular waveform used for example.*

for a constant rupture load with no failure effect, the predicted life is equal to the rupture life, and for fatigue at a constant amplitude with no rupture effect, the predicted life is equal to the fatigue life). In addition, it also allows us to preserve the sequence of the loading profile.

To further illustrate this example with specific numbers, we will estimate the fatigue life at room temperature in the form of

$$N = 10^{(Fa_{\max}-A_n/B_n)^{P_n}}$$

$$A_n = 0.95$$

$$B_n = -0.08 \tag{18}$$

$$P_n = 1.00$$

where $Fa_{\max}$ is the maximum value of the failure function experienced during a fatigue cycle, and A_n, B_n, and P_n are parameters obtained by fitting room temperature fatigue data. In addition, we will estimate the rupture life at a given temperature (in seconds) in the form of

$$\hat{\tau} = 10^{(Fa-A_t/B_t)^{P_t}}$$

$$A_t = 1.00$$

$$B_t = -0.15 \tag{19}$$

$$P_t = 1.00$$

where Fa is the instantaneous value of the failure function, and A_t, B_t, and P_t are curve-fitting parameters. Further, for the purposes of example, we will consider the value of the failure function corresponding to the applied load to vary in a triangular waveform fashion (at a constant temperature) from a value of 0 to Fa_{max} with a period of 1 s, as is illustrated in Fig. 3. To estimate the remaining strength at Point B for the case in which Fa_{max} is equal to 0.8 using the procedure described above, we follow the flowchart shown in Fig. 4. First, we calculate an average value of Fa over the interval $0 \leq t \leq 0.5$ s. This value is given by

$$\overline{Fa} = 0.4 \tag{20}$$

Substituting this value into Eq 19, we determine a rupture time that is given by

$$\hat{\tau} = 10^{\overline{Fa} - 1.00/-0.15}$$

$$= 10^{0.4 - 1.00/-0.15} \tag{21}$$

$$= 10\ 000\ s$$

Using Eq 16, the change in remaining strength over the interval is given by

$$\Delta Fr = -(1 - \overline{Fa}) \left[\frac{\Delta t}{\hat{\tau}} \right]^{j_{rupture}}$$

$$= -(1 - 0.4) \left[\frac{0.5}{10\ 000} \right]^{0.8} \tag{22}$$

$$= -2.174 \times 10^{-4}$$

so that the first estimate of the remaining strength at Point B is given by

$$Fr_B = 1 + \Delta Fr \tag{23}$$

$$= 0.999783$$

Next we divide the interval from Point A to Point B into two equally sized subintervals. The first subinterval, which ends at Point E, spans over the range $0 \leq t \leq \frac{1}{4}$. The average value of the failure function over this interval is

$$\overline{Fa_1} = 0.2 \tag{24}$$

so that the estimate of the remaining strength at Point E is given by

$$Fr_E = 1 + \Delta Fr_1$$

$$= 1 - (1 - \overline{Fa_1}) \left[\frac{\Delta t}{\hat{\tau}_1} \right]^{j_{rupture}} \tag{25}$$

$$= 1 - (1 - 0.2) \left[\frac{0.25}{2.15 \times 10^5} \right]^{0.8}$$

$$= 0.999986$$

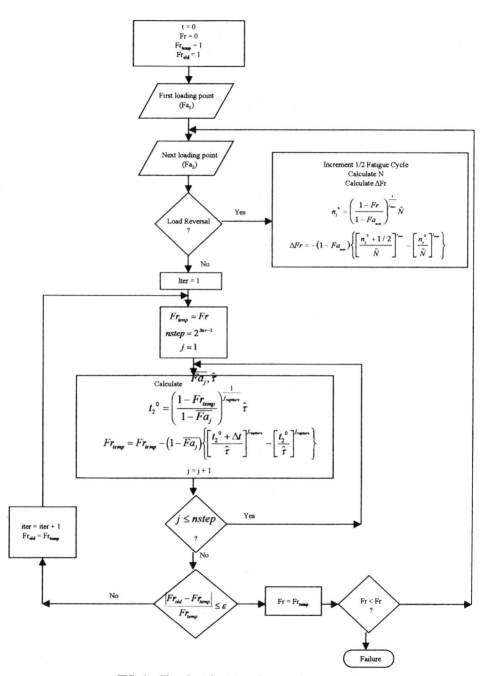

FIG. 4—*Flowchart for triangular waveform example.*

For the second subinterval (from Point E to Point B), the average value of the failure function is given by

$$\overline{FA_2} = 0.6 \tag{26}$$

The "pseudo-time" corresponding to the remaining strength at Point E and $\overline{Fa_2}$ is found by using Eq 15, so that

$$
t_2^0 = \left(\frac{1 - Fr_E}{1 - \overline{Fa_2}}\right)^{1/j_{rupture}} \hat{\tau}
$$

$$
= \left(\frac{1 - 0.999986}{1 - 0.6}\right)^{1/0.8} (464.15) \tag{27}
$$

$$
= 1.28 \times 10^{-3}
$$

The second estimate of the remaining strength at Point B is then given by

$$Fr_B = Fr_E + \Delta Fr_2$$

$$
= 0.999986 - (1 - \overline{Fa_2}) \left\{ \left[\frac{t_2^0 + \Delta t}{\hat{\tau}}\right]^{j_{rupture}} - \left(\frac{t_2^0}{\hat{\tau}_2}\right)^{j_{rupture}} \right\}
$$

$$
= 0.999986 - (1 - 0.6) \left\{ \left[\frac{1.28 \times 10^{-3} + 0.25}{464.15}\right]^{0.8} - \left[\frac{1.28 \times 10^{-3}}{464.15}\right]^{0.8} \right\} \tag{28}
$$

$$
= 0.999025
$$

The process may be repeated until the estimated remaining strength at Point B converges to within a specified error tolerance. The results for such a case are summarized in Table 1.

At this point, we may now consider the fatigue effect. Using Eq 18, the estimated fatigue life for this case is given by

TABLE 1—*Convergence of remaining strength at Point B for loading profile shown.*

Iteration	Number of Subintervals	Fr_B
1	1	0.999783
2	2	0.999025
3	4	0.998465
4	8	0.998290
5	16	0.998246
6	32	0.998235
7	64	0.998232
8	128	0.998232

$$\hat{N} = 10^{0.8 - 0.95 / -0.08}$$

$$= 75 \text{ cycles} \tag{29}$$

The number of "pseudo-cycles" that correspond to the value of Fr_B is given by

$$n_2^0 = \left(\frac{1 - Fr_B}{1 - Fa_{\max}} \right)^{1 / j_{\text{fatigue}}} \hat{N}$$

$$= \left(\frac{1 - 0.998232}{1 - 0.8} \right)^{1 / 1.2} (75) \tag{30}$$

$$= 1.458$$

where the value of the parameter j for fatigue, j_{fatigue}, has been taken to be 1.2. As a result, the remaining strength at Point B *after taking into account the fatigue effect* is given by

$$Fr_B = Fr_B^{\text{rupture}} + \Delta Fr_{\text{fatigue}}$$

$$= 0.998232 - (1 - Fa_{\max}) \left\{ \left[\frac{n_2^0 + \frac{1}{2}}{\hat{N}} \right]^{j_{\text{fatigue}}} - \left(\frac{n_2^0}{\hat{N}} \right)^{j_{\text{fatigue}}} \right\} \tag{31}$$

$$= 0.998232 - (1 - 0.8) \left\{ \left[\frac{1.458 + \frac{1}{2}}{75} \right]^{1.2} - \left(\frac{1.458}{75} \right)^{1.2} \right\}$$

$$= 0.997481$$

This process may then be repeated to calculate the remaining strength of the material at all points in its life. Failure of the material occurs when the value of the normalized remaining strength is less than the failure function. For this example, failure occurs after approximately 51 cycles. By using different values of $Fa_{\max}$, it is possible to generate a failure function (normalized applied stress) versus time-to-failure type curve for the material, as is illustrated in Fig. 5. The fatigue life curve, Eq 18 (expressed in terms of time to failure for a 1-Hz waveform) and the time to rupture curve, Eq 19, are plotted for reference.

Application to a Ceramic Matrix Composite

To illustrate the utility of the method, we consider a $[0/90]_{2s}$ 2-D woven Nicalon/enhanced silicon carbide (SiC) ceramic composite subjected to a loading profile shown schematically in Fig. 6. Due to the proprietary nature of the data, exact details pertaining to the load levels cannot be presented. Rather, all load levels will be presented in terms of the ratio S/S_{ult}, where S is the maximum applied stress and S_{ult} is the ultimate strength of the material. The values for t and σ_{ij} as shown in Fig. 6 are given in Table 2.

A number of inputs are necessary for our modeling efforts in this case. We begin by representing the characteristic rupture time, $\hat{\tau}$, at the temperature of interest in the form of

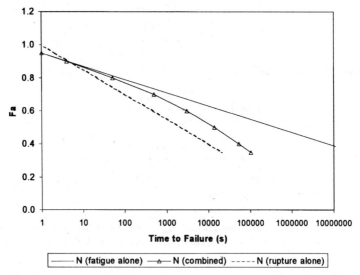

FIG. 5—*Failure function (normalized applied stress) versus time-to-failure curve for triangular wave-form example.*

$$\ln \hat{\tau} = (17.7504 - 18.8897 \cdot Fa) \qquad (32)$$

The characteristic time for the rupture process, $\hat{N}$, is represented in the form of

$$\hat{N} = 10^{(Fa-1.00/-0.025)} \qquad (33)$$

The final parameters to be specified are the values for the parameters $j_{rupture}$ and $j_{fatigue}$. For this material system, these values were determined to be as follows

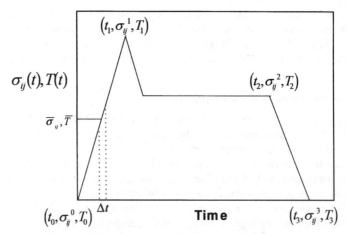

FIG. 6—*Schematic illustration of a mission-loading profile.*

TABLE 2—*Time, temperature, and stress information for mission-loading profile shown in Fig. 6. Values of σ_{ij} other than σ_{xx} are equal to zero.*

Load Step	t, s	σ_{xx}
0	0	0
1	0.50	S
2	0.67	$2S/3$
3	5400	$2S/3$
4	5401	0

$$j_{\text{fatigue}} = 1.20$$

$$j_{\text{rupture}} = 0.78$$

Using these representations, as well as representations of matrix stiffness changes (due to cracking) as a function of time and stress level, it is possible to estimate the number of repetitions of the mission-loading profile that lead to failure of the component. These predictions, as well as the corresponding experimental observations, are shown in Fig. 7 for various values of S/S_{ult}. In this case the comparison appears to be quite good, although it should be pointed out that the material behavior (at least in the analysis) is controlled largely by the rupture effect.

Conclusions

In this paper, an analysis has been developed for the prediction of the remaining strength and life of ceramic matrix composites under combined conditions such as fatigue and rupture.

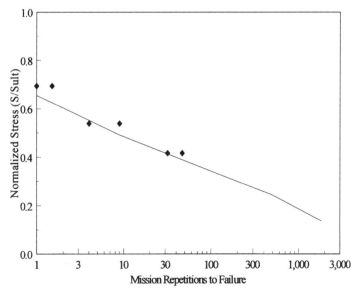

FIG. 7—*Comparison of predicted and observed life for the mission-loading profile illustrated in Fig. 6.*

This model is based upon damage mechanics concepts included in the framework of the critical element model. One important feature of the approach is that it allows the analysis of general load and temperature histories. In order to validate the analysis, a mission-loading profile at elevated temperature is considered. This profile results in a combination of fatigue and rupture effects. For the case considered, the agreement between the measured and predicted lives is quite good, although the overall material behavior in this case is dominated by the rupture effect.

References

[1] Vellios, L., Kostopoulos, V., and Paipetis, S. A., "Fatigue Effect on the Dynamic Properties of CFRP Composites," *Advanced Composites Letters,* Vol. 3, 1994, pp. 145–150.

[2] Reifsnider, K. L. and Stinchcomb, W. W., "A Critical Element Model of the Residual Strength and Life of Fatigue-Loaded Composite Coupons," *Composite Materials: Fatigue and Fracture, ASTM STP 907,* H. T. Hahn, Ed., American Society for Testing and Materials, West Conshohocken, PA, 1986, pp. 298–303.

[3] Reifsnider, K. L., Iyengar, N., Case, S. W., and Xu, Y. L., "Damage Tolerance and Durability of Fibrous Material Systems: A Micro-Kinetic Approach," *Durability Analysis of Structural Composite Systems,* A. H. Cardon, Ed., A. A. Balkema, Rotterdam, 1996, pp. 123–144.

[4] Kachanov, L. M., *Introduction to Continuum Damage Mechanics,* Martinus Nijhoff Publishers, Boston, 1986.

[5] Broutman, L. J. and Sahu, S., "A New Damage Theory to Predict Cumulative Fatigue Damage in Fiberglass Reinforced Plastics," *Composite Materials: Testing and Design (Second Conference), ASTM STP 497,* American Society for Testing and Materials, West Conshohocken, PA, 1972, pp. 170–188.

[6] Reifsnider, K. Case, S., and Iyengar, N., "Recent Advances in Composite Damage Mechanics," *Proceedings of Conference on Spacecraft Structures, Materials & Mechanical Testing,* (ESA SP-386), Noordwijk, The Netherlands, June 1996, pp. 483–490.

[7] Reifsnider, K. L., Case, S. W., and Xu, Y. L., "A Micro-Kinetic Approach to Durability Analysis: The Critical Element Method," *Progress in Durability Analysis of Composite Systems,* A. H. Cardon, H. Fukuda, and K. Reifsnider, Eds., A. A. Balkema, Rotterdam, 1996, pp. 3–12.

[8] Reifsnider, K., Iyengar, N., Case, S., and Xu, Y., "Kinetic Methods for Durability and Damage Tolerance Design of Composite Components," Keynote Address, Conference on Composite Materials, Japan Society for Mechanical Engineers, Tokyo, 26 June 1995.

[9] Reifsnider, K., Lesko, J., and Case, S., "Kinetic Methods for Prediction of Damage Tolerance of High Temperature Polymer Composites," *Composites '95: Recent Advances in Japan and the United States,* I. Kimpara, H. Miyairi, and N. Takeda, Eds., *Proceedings* Japan-U.S. CCM-VII, Kyoto, 1995, pp. 49–55.

William M. Johnston[1] and Thomas S. Gates[2]

The Effects of Stress and Temperature on the Open-Hole Tension Fatigue Behavior of a Graphite/Bismaleimide Composite

REFERENCE: Johnston, W. M. and Gates, T. S., **"The Effects of Stress and Temperature on the Open-Hole Tension Fatigue Behavior of a Graphite/Bismaleimide Composite,"** *Composite Materials: Fatigue and Fracture, Seventh Volume, ASTM STP 1330,* R. B. Bucinell, Ed., American Society for Testing and Materials, 1998, pp. 179–198.

ABSTRACT: An experimental investigation was performed to determine the behavior of an open-hole tension (OHT) graphite/bismaleimide composite specimen loaded in tension-tension fatigue under isothermal, fixed-frequency conditions. A range of stress levels and temperature levels was chosen to assess performance. These loads and elevated temperatures ranged from relatively benign conditions (low stress, room temperature) up through aggressive conditions (high stress, high temperature). Measurements were made of stiffness, edge crack density, residual strength, weight loss, and changes in glass transition temperature (T_g).

Results from this work will help explain the roles of aging and fatigue damage in the performance of graphite/bismaleimide OHT specimens as well as providing insights into the individual and synergistic contributions of each of these processes. Contributions of aging and microcrack formation during elevated temperature fatigue of OHT specimens may result in a net stiffness increase in the gage section and a decrease in residual strength. The magnitude of changes in stiffness or strength is determined by the time at temperature and the chosen stress level.

KEYWORDS: composite, open-hole tension, fatigue, damage, chemical aging, BMI, elevated temperature

For advanced polymer matrix composites (PMCs) the issue of long-term durability at elevated temperature is crucial for material selection and development of accelerated test methods and associated life prediction models. Experimental and analytical techniques must be established that provide the means for predicting long-term behavior from short-term test data. Before these predictive models and accelerated test methods can be fully developed, the individual and combined effects of elevated temperature and varied load histories must be studied.

The objective of this research was to experimentally determine the individual and synergistic responses due to mechanical fatigue loading and elevated temperature exposure. The effects of the temperature and fatigue loading on the PMC were quantified by measurements of stiffness, glass transition temperature, weight loss, edge crack density, and residual strength.

Three potential processes that contribute to degradation may occur during the fatigue/aging time: chemical aging, physical aging, and damage (microcrack) development [1–3].

[1] Old Dominion University, Norfolk, VA.
[2] NASA Langley Research Center, Hampton, VA.

Chemical aging, a thermo-irreversible process, and physical aging, a thermo-reversible process, result in the polymer matrix becoming stiffer, more brittle, and with reduced creep and stress relaxation rates. Aging in general results in a stiffness increase, and the damage accumulation through the fatigue cycles results in a stiffness decrease. These processes are not independent and can be additive or subtractive depending on the test material and the resultant performance property being considered.

Experimental Program

To develop an understanding of the individual effects of elevated temperature and fatigue loading, experimentation was divided into two parts. The first part consisted of an investigation to determine the effects of both elevated temperature and fatigue loading on the response of open-hole tension (OHT) specimens. The OHT specimen investigation varied loads and temperatures from relatively benign conditions (low stress, room temperature) up through aggressive conditions (high stress, high temperature). The isothermal aging of unloaded OHT specimens allowed comparison of changes in damage accumulation and residual properties caused by the fatigue loading and provided insight into the individual effect of temperature. The OHT fatigue investigation provided information on changes in the PMC's stiffness, damage accumulation, and residual static properties. The second portion of the experimentation provided information on the effect of unloaded isothermal aging without load on the weight loss and the glass transition temperature of the PMC. As outlined in Ref 4, the measurement of weight loss and T_g can provide information on oxidative stability and chemical aging.

Material Information

Tests were performed on a G40-800/5260 composite material system. This system consisted of a high-strength, intermediate modulus carbon fiber produced by Celion in a toughened bismaleimide (BMI) matrix that was produced by BASF. Test panels were produced at NASA Langley Research Center with a $[45/0/-45/90]_{2S}$ quasi-isotropic layup and had a nominal thickness of 2.16 mm (0.085 in.) after cure.

Test Specimen

Test panels were cured using the manufacturer's recommended procedure. As shown in Fig. 1, the panels were machined into two specimen types: open-hole tension (OHT), and dynamic-mechanical analysis (DMA). The DMA specimens were used to measure T_g and weight loss. The OHT specimens were 203.2 mm (8.0 in.) long and 38.1 mm (1.5 in.) wide with a 6.35-mm (0.25-in.) center hole. These specimens were polished on one edge to enhance visual inspection. DMA specimens were 12.7 mm (0.5 in.) wide by 50.8 mm (2.0 in.) long.

After the specimens were machined (and the OHT polished) they were dried for 24 h at 110°C and kept desiccated until test time.

Strain Field Investigations

Prior to initiating the fatigue tests on the OHT specimens, investigations were made into the effect of the center hole on the strain field of the specimens. Strain field investigations

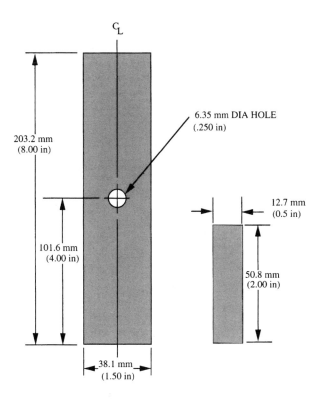

Open Hole Tension Specimen   **DMA/Weight Loss Specimen**

FIG. 1—*Specimen configurations.*

provided information on the stress concentration around the center hole and how temperature affected this strain field.

Strain Gage—Strain gages were mounted to an OHT specimen as shown in Fig. 2. In addition an extensometer, as used in the fatigue tests, was mounted on the edge to understand the relation between the strain measured at the edge and the strain across the width. The 25.4-mm-gage length extensometer was mounted centered on the 6.35-mm hole.

Static tests were performed at different test temperatures to understand the effect of temperature on the strain field. The specimen was loaded at a rate of 22.2N/s to 13 300 N (3000 lb) and back to no load. Strain measurements were taken every 20 s. Tests were performed at nine temperatures from 23 to 204°C (23, 38, 66, 93, 121, 149, 163, 177, 204°C).

Strain measured at the extensometer was found to agree to within 8% with the axial gages not mounted on the center hole (1, 4, 6) over the entire temperature range. All strain gages not near the hole (1, 4, 5, 7) were also found to be linearly proportional to strain measured by the extensometer. This demonstrated that the extensometer gave strain readings in the gage section that were representative of strains away from the center hole.

Moiré—The moiré technique was used to provide visualization of the strain field around the center hole. An array with a frequency of 2400 lines/mm was applied to the surface of the specimen. Load was applied at room temperature to the specimen using a screw-driven

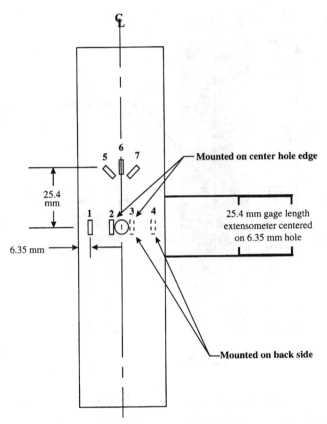

FIG. 2—*Strain gage and extensometer placement on OHT specimen used for strain field study.*

load stand. Pictures of the moiré fringe pattern were produced with a 35 mm camera that allowed calculation of the axial strain field using

$$\varepsilon = \frac{1}{f}\left(\frac{\Delta N y}{\Delta y}\right) \tag{1}$$

where

   f = frequency of array,
   $\Delta N y$ = change in fringe number (i.e., number of fringes), and
   Δy = change in axial distance (distance covered by $\Delta N y$ fringes).

   A more detailed explanation of theory and procedure can be found in references on moiré [5].
   Figure 3 shows the fringe pattern for an OHT specimen loaded to 81.1 MPa. The fringe pattern in Fig. 3 indicates that the effects of the stress concentration are minimal at the free surface and that the strain field at the edge is almost uniform. Therefore it was assumed that the extensometer, which attaches to the edge, will not be adversely affected by the stress concentration. The fringe pattern of Fig. 3 also shows that the effects of the stress concen-

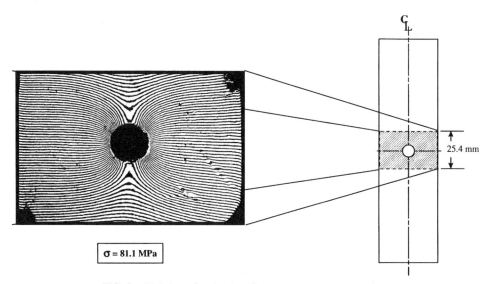

$\sigma = 81.1$ MPa

FIG. 3—*Moiré results showing the extensometer gage section.*

tration caused by the notch are all contained in the 25.4-mm (1 in.) gage section. Therefore it was further assumed that most of the damage induced by the stress concentration would occur in the gage section.

Test Procedure and Equipment

Open-Hole Tension Fatigue Testing

The tension fatigue tests were performed on a 222-kN (50-kip) capacity servo-hydraulic test stand. Load, strain, actuator displacement, and time were recorded using an automated digital data acquisition system. An environmental chamber maintained the test temperature over the entire length of the grips and specimen. A high-temperature, capacitance extensometer with a 25.4-mm (1.0-in.) gage length was centered on the 6.35-mm (0.25-in.) hole and edge mounted on the unpolished side.

All open-hole tension fatigue tests were performed under load control at a frequency of 1 Hz with a minimum/maximum stress ratio of $R = 0.1$ and a triangular waveform. A minimum of two tests were performed at each condition, and tests were concluded after 400 000 cycles. This test matrix is summarized in Table 1.

In addition to the fatigue load tests, a set of tests were performed without any fatigue loading. These tests were conducted on specimens that were aged unloaded in an oven at fatigue test temperatures for 120 h. This time was equivalent to the fatigue test time at temperature. These tests provided information on how the properties of the OHT specimens were affected by elevated temperature without load. These specimens were subjected to the same residual property tests and damage investigation as the fatigued specimens.

Fatigue loads were chosen to represent an aggressive and a moderate condition. Two maximum stresses were used, $\sigma_{max} = 119$ MPa (17.25 ksi) and $\sigma_{max} = 238$ MPa (34.5 ksi), which corresponded to 23 and 46% of ultimate strength, respectively. These stress levels also corresponded to far field strains of 0.2 and 0.4% strain, respectively.

TABLE 1—*OHT test matrix.*

	Fatigue Stress Level		
Temperature	σ_{max} = 119 MPa $R = 0.1$	σ_{max} = 238 MPa $R = 0.1$	Baseline No load Aged 120 h
23°C	2	2	2
177°C	2	2	2
191°C	2	2	2
204°C	2	2	2
218°C	2	2	2

Test temperatures were chosen to represent a range from benign to aggressively high temperatures. The higher temperatures were chosen to provide information on potential accelerated test conditions. Tests were performed at 23°C (room temperature), 177°C (350°F), 191°C (375°F), 204°C (400°F), and 218°C (425°F). The room temperature tests provided baseline information. The next highest temperature, 177°C, was chosen to correspond to the maximum recommended use temperature.

The measurement of strain and load occurred during specified cycles of the fatigue test and allowed for determination of longitudinal stiffness changes. Stiffness changes have been used as an indicator of internal damage [6,7]. For the current work the stiffness changes reflected the effect of damage and aging over the 1-in. gage section.

OHT Fatigue Stiffness—Applied load and measured strain were used to calculate an effective stiffness for the gage section. Stiffness calculations were performed by making a least-squares linear fit to the stress and strain data. Stress was calculated using the unnotched specimen area and the load from the data acquisition system. Strain was taken from the extensometer attached to the gage section. The fit was performed on the loading and unloading data between 5 and 95% of the maximum load for that fatigue cycle. To aid in comparison, fatigue stiffness data were normalized with respect to the initial value (i.e., the stiffness at cycle count, $n = 1$). The stiffness versus cycle count data were averaged for replicate tests and fit using least squares to aid in the visualization of data trends.

Plotting the normalized stiffness allowed assessment of the change in stiffness over the duration of the fatigue test as shown in Figs. 4 and 5. Both of these figures show the results for a single fatigue test conducted at a temperature of 218°C and with a maximum stress of 119 MPa. Both plots show the normalized stiffness of the gage section versus the cycle count. Figure 4 shows the data on a semi-log representation, whereas Fig. 5 shows these same data on a linear scale.

The stiffness measurements appeared to fall into three regions depending on the rate of change of stiffness. The fits to the data in the first two regions were found by performing least-squares reduction (LSR) fits to the semi-log data. The third fit was found by performing a linear LSR to the normalized stiffness versus cycle count data. The first two linear fit regions are easily visible on the semi-log plot of Fig. 4. The third linear fit region is easiest seen in Fig. 5.

OHT Damage—The end-of-test damage evaluation included two inspection procedures. The first procedure consisted of an edge inspection conducted visually with a microscope at ×400 magnification to determine the extent of matrix cracks. The number of transverse matrix cracks in each ply along the gage section was recorded. Edge damage was investigated as a possible indicator of internal damage.

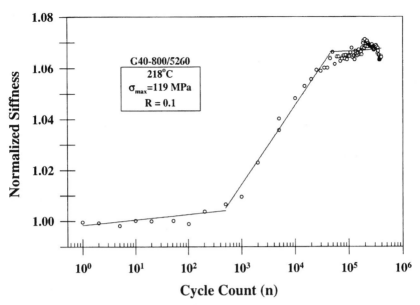

FIG. 4—*Normalized fatigue stiffness of OHT specimens plotted as a function of log of cycle count.*

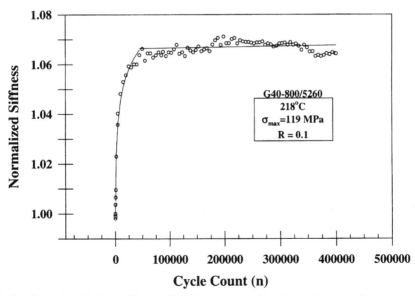

FIG. 5—*Normalized fatigue stiffness of OHT specimens plotted as a function of cycle count on a linear scale.*

In the second procedure, X-ray radiographs were made to investigate the internal damage. A zinc iodine dye penetrant was used, and the X-rays were made under a no-load condition. These radiographs were examined for indications of delamination and depth of matrix cracks.

OHT Residual Properties—After investigation for post-fatigue damage, the specimens were loaded at room temperature in static tension until failure. Measurements were made on the remaining strength and stiffness. The static tests were performed on an 88-kN (20-kip) servo-hydraulic test stand equipped with an automated data acquisition system. Tests were performed in stroke control at a ramp rate of 1 mm/min (0.039 in./min). This speed was chosen to ensure an entire test time between 1 and 10 min for all tests. Two 25.4-mm (1-in.) gage length extensometers were mounted on the edges and centered with respect to the hole to provide strain data. Load and strain were recorded at half-second intervals during the test.

Residual properties calculations consisted of ultimate strength and chord modulus. The chord modulus was calculated between the values of 1000 and 3000 $\mu\varepsilon$. Ultimate stress was calculated using the area of the unnotched cross section and the highest load carried by the laminate.

Physical Properties Testing

The effects of aging at elevated temperature on the composite material during the fatigue testing were determined through a physical property investigation. The physical properties measured consisted of weight loss and glass transition temperature.

Physical property measurements were performed after every 24 h of aging for up to five days of aging at all test temperatures. Two replicates were used for each condition. Test conditions are summarized in the test matrix shown as Table 2. DMA was used to determine T_g before and after aging. Glass transition temperature tests were run on specimens that had been dried but not exposed to any additional thermal history to provide unaged T_g measurements.

After drying and prior to testing, DMA specimens were removed from the desiccator and weighed to within ± 0.00001 g using an analytic balance. The specimens were then placed in ovens for aging. The aging ovens were opened only for the period of time required to remove the specimens and then closed immediately, causing only a minor temperature fluctuation. After completion of the aging the DMA specimens were again weighed to determine weight loss. Weight loss was calculated using

$$w_{loss} = \frac{w_{initial} - w_{final}}{w_{initial}} \times 100\% \tag{2}$$

After weight loss was measured, the DMA tests were performed to calculate T_g. DMA

TABLE 2—*Number of physical property tests at each temperature.*

Time, h	Test, DMA	Weight Loss
0	2	...
24	2	2
48	2	2
72	2	2
96	2	2
120	2	2

tests were conducted from room temperature to 400°C at a rate of 5°C/min. Glass transition temperature was taken as the peak in the tan delta curve, where tan delta is a ratio of the mechanical energy lost as heat to the mechanical energy elastically stored at a given temperature as explained by Kämpf [8].

Experimental Results

Physical Properties

Weight Loss—The weight loss results, shown in Fig. 6, indicate that weight loss occurs at all temperatures with distinct differences due to aging temperature. As temperature was increased, the weight loss increased. For the lower test temperatures (177°C and 191°C), a constant level of weight loss was obtained after the first 24 h of aging and maintained for the remaining portion of the 120 h of testing. At 204°C and 218°C, the weight loss continued to increase throughout the entire aging time.

Glass Transition Temperature—Further evidence of material changes due to aging was given by measured changes in T_g during 120 h of isothermal aging. As shown by the DMA results in Fig. 7, the T_g increased after aging at all of the fatigue test temperatures. Unfortunately, the 204 and the 177°C specimens were exposed to ambient conditions for several days prior to measuring the T_g. The moisture gain during this time may have caused the T_g of the material to decrease.

Open-Hole Tension

Damage Investigation Results—The post-test damage investigation revealed that most visible edge damage consisted of cracks in the 90° plies (Fig. 8). The extent of edge damage (crack density) in these plies is summarized in Fig. 9. This plot shows the average number

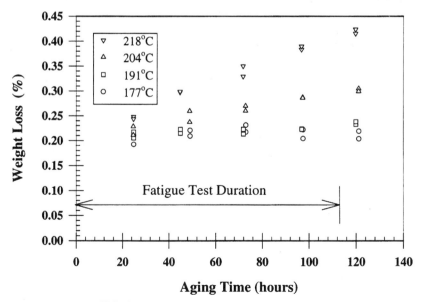

FIG. 6—*Weight loss as a function of aging time.*

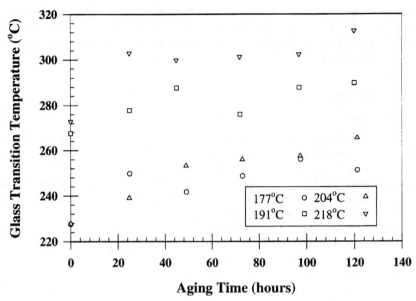

FIG. 7—T_g as measured with DMA plotted as a function of aging time.

of cracks per millimeter in 90° plies of the gage section. The data are presented for all three loading conditions (no load, 119 MPa, and 238 MPa) and all five test temperatures.

For the no-load condition, matrix cracks were not detected for aging temperatures below 191°C. Despite the absence of load, edge cracks developed at 204 and 218°C. The room-temperature specimens fatigued at the low stress level also showed little or no matrix cracking. For all three loading conditions it was observed that crack density increased with temperature. The highest crack density occurred at the highest stress level and highest temperature. At the higher temperatures (204 and 218°C), extensive matrix cracking was present for all three loading conditions.

The internal damage is illustrated in the X-ray radiographs in Figs. 10–14. Each of these figures shows the X-ray results for one test temperature. The high (a) and low (b) fatigue load tests are shown in each figure.

The specimens that experienced the higher fatigue stress showed the most extensive internal damage. Matrix cracking, axial splitting, and delamination were all visible in X-rays of the high stress case. Internal damage did not increase with the increase of test temperature as the edge damage implied. The most extensive transverse matrix cracking damage was visible in the specimen tested at room temperature. At room temperature the matrix cracks extended from the free edge towards the center hole and from the center hole towards the edge. The length of these cracks decreased as temperature increased (Figs. 10a–14a).

Radiographs of the unloaded specimens showed no internal damage. Internal damage from the low-load fatigue tests did not vary noticeably with temperature and consisted mainly of a very small area of matrix cracking and delamination around the center hole (Figs. 10b–14b). The high levels of matrix cracking visible during the optical edge investigation were not visible on the X-rays.

Fatigue Test Results—The stiffness measurements taken over the gage section during the isothermal fatigue tests provided data on the combined effects of stress level and aging at elevated temperature.

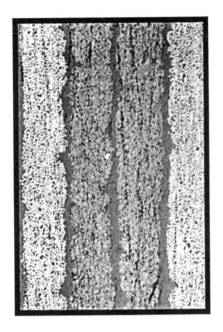 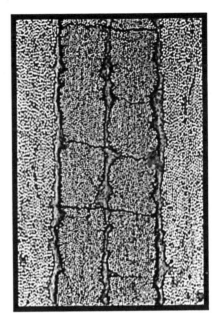

Photo Micrograph
Pretest Edge Condition

Post-Test Edge Damage
σ_{max}= 238 MPa, R =0.1
N = 400,000 cycles
T=218°C

FIG. 8—*Detail showing example of initial condition and post-fatigue damage in two central 90° plies.*

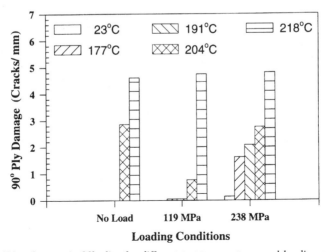

FIG. 9—*Edge damage in 90° plies for different test temperatures and loading conditions.*

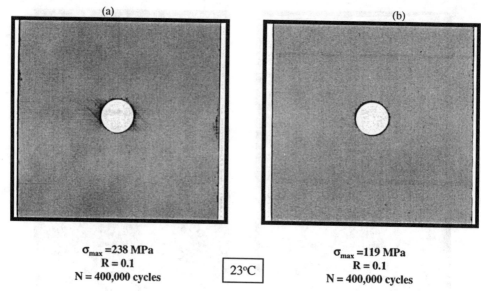

FIG. 10—*Post fatigue X-ray radiograph of 23°C fatigue test showing damage accumulation.*

The low-stress-level cases shown in Fig. 15 illustrate the dependence of stiffness on test temperature. The largest total increase in stiffness was associated with the highest test temperature. The data in Fig. 15 indicate that as the temperature increased, so did the rate of stiffness change. In a manner similar to the room temperature data of Ref 9, the lower two test temperatures showed a total net loss in stiffness over the fatigue test duration. As the

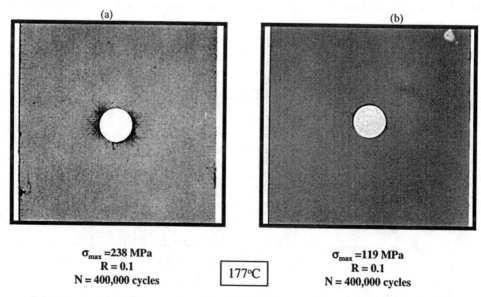

FIG. 11—*Post fatigue X-ray radiograph of 177°C fatigue test showing damage accumulation.*

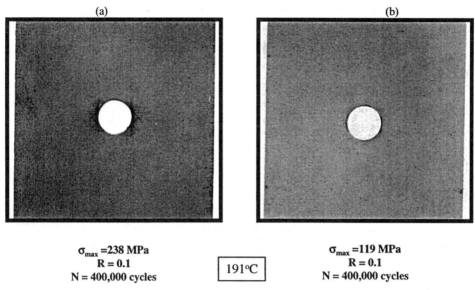

σ$_{max}$ =238 MPa
R = 0.1
N = 400,000 cycles

191°C

σ$_{max}$ =119 MPa
R = 0.1
N = 400,000 cycles

FIG. 12—*Post fatigue X-ray radiograph of 191°C fatigue test showing damage accumulation.*

temperature increased, the rate of stiffness change (after the first 1000 cycles) continued to increase. The highest rate of increase in stiffness occurred for the 218°C tests.

For the high stress case, the data given in Fig. 16 indicate that as the temperature increased so did the rate of stiffness increase. The lower two test temperatures showed a total change in stiffness similar to the low stress condition. As the temperature increased above 177°C,

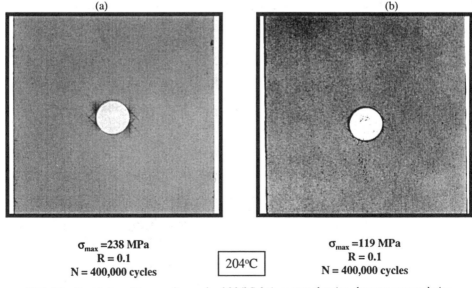

σ$_{max}$ =238 MPa
R = 0.1
N = 400,000 cycles

204°C

σ$_{max}$ =119 MPa
R = 0.1
N = 400,000 cycles

FIG. 13—*Post fatigue X-ray radiograph of 204°C fatigue test showing damage accumulation.*

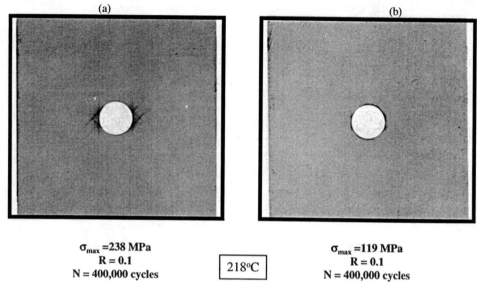

FIG. 14—*Post fatigue X-ray radiograph of 218°C fatigue test showing damage accumulation.*

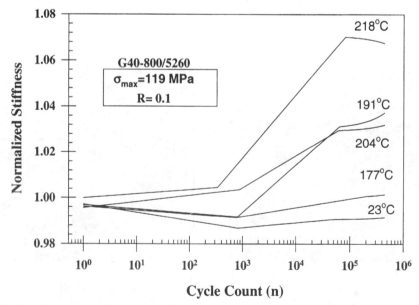

FIG. 15—*Normalized stiffness of OHT specimens during low-load fatigue tests at each test temperature.*

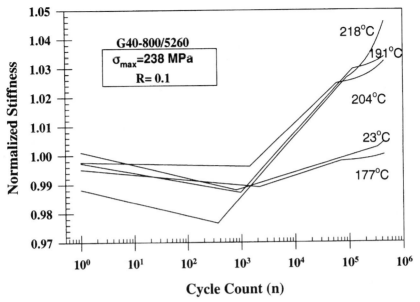

FIG. 16—*Normalized stiffness of OHT specimens during high-load fatigue tests at each test temperature.*

the rate of stiffness change continued to increase. The largest net increase (and lowest initial drop) in stiffness occurred during the 218°C test.

Figure 17 illustrates the effect of fatigue load level for the highest test temperature (218°C). Comparing the results for the two stress levels reveals that the higher stress resulted in an increased rate of stiffness loss during the initial 1000 cycles and increased rate of stiffness gain during the final portion of the fatigue test.

Residual Properties Results—Static measurements were made after the completion of the fatigue tests to determine the effects of temperature and stress level on the remaining strength and stiffness. All of these tests were conducted at room temperature and included the baseline (unfatigued) specimens for comparison.

Residual strength was taken as the maximum load-carrying capability of the laminate during the course of the static test as explained earlier in the "Test Procedure" section. Examination of the data in Fig. 18 reveals that the highest strength was associated with the room-temperature, high stress case, while the lowest strength was associated with the high-stress, next-to-highest-temperature case. Both high and low stress cases show decreasing strength with an increase in fatigue test temperature. The baseline strength values were unaffected by temperature and slightly lower than most of the strength values for the fatigue tests. These changes in strength due to stress and temperature are similar to results found by Boyd and Chang [10].

The room temperature residual stiffness in the gage section was unaffected by the fatigue stress level or fatigue temperature (Fig. 19). The results for all cases were similar to the baseline test values. The scatter in modulus from specimen to specimen masked the small changes due to aging and fatigue.

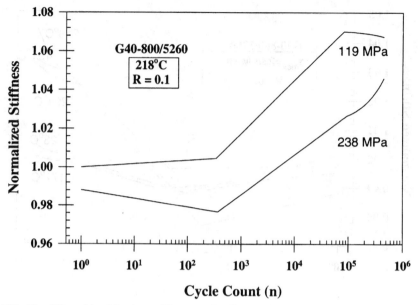

FIG. 17—*Effect of load level on stiffness results illustrated for the 218°C test temperature.*

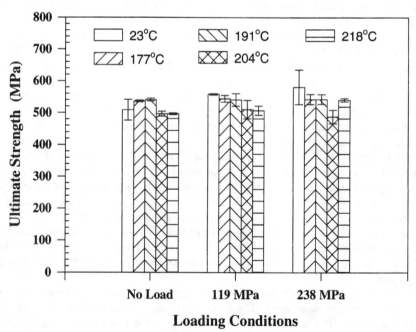

FIG. 18—*Residual tensile strength of G40-800/5260 OHT specimens after fatiguing or isothermal aging at test temperature.*

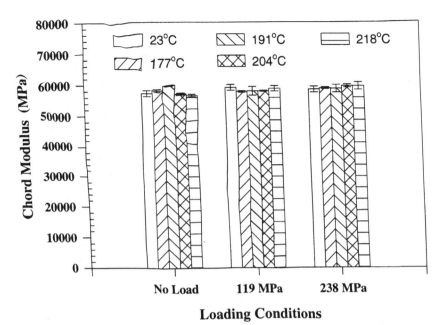

FIG. 19—*Residual chord modulus of G40-800/5260 OHT specimens after fatiguing or isothermal aging at test temperature.*

Discussion

Physical Properties

Changes in weight and T_g indicated that chemical and physical aging occurred during the elevated temperature tests. The continued weight loss and edge cracking during higher temperature exposure indicated problems with the material's thermal oxidative stability. Exposure to elevated temperature also caused time-dependent shifts in T_g, which indicated chemical aging associated with continued curing in the polymer.

Laminate Damage

Increasing the test temperature and time at temperature produced increased levels of edge damage. The edge damage measured consisted of matrix cracks in the 90° plies (visible on the edge). The level of internal damage was governed primarily by the stress level. However, increasing the test temperature decreased the length and number of internal matrix cracks in the 90° plies. The worst internal matrix cracking occurred in the high-stress, room-temperature case. These results demonstrate that for the elevated-temperature OHT test, edge damage alone cannot be used as a clear indication of the extent of internal damage.

Fatigue Stiffness

Open-hole tension fatigue tests of the composite at elevated temperatures produced stiffness changes in the gage section of the laminate that were dependent upon the relative contributions of damage development and aging. During the fatigue tests, aging of the polymer in the laminate induced a time-and-temperature-dependent increase in stiffness that competed with decreases in stiffness due to load-dependent damage.

At room temperature, aging effects were negligible and all measured stiffness losses were due to mechanical load alone. At elevated temperatures, some stiffness loss occurred during the initial 1000 cycles but was gradually overcome by a stiffness gain during aging. Raising the test temperature resulted in a higher rate of stiffness increase. Raising the load resulted in a higher rate of stiffness loss during the initial portion of the test ($n < 1000$) and also resulted in a greater rate of stiffness increase during the end of the fatigue test ($n > 100\,000$).

Up to 191°C, aging had an increased effect on stiffness as temperature was increased. This trend is easy to see on the low stress results (Fig. 15). This trend seems to change near 191 and 204°C, where the higher rate of stiffness increase occurred with the lower temperature (191°C). The 191°C data and the 204°C data for the high-load and low-load cases are plotted in Figs. 20 and 21, respectively. If the edge damage is considered, an explanation is possible. Although edge damage is not a good indicator of internal damage, it is a good indicator of thermal oxidative breakdown. At 204°C the unloaded OHT specimens displayed extensive edge cracking while the 191°C unloaded OHT specimens showed no cracks (Fig. 9). In addition, at 204°C the DMA specimens displayed continued weight loss throughout the test while the 191°C DMA specimens did not (Fig. 6). Those changes that occurred at 204°C indicated a temperature-dependent breakdown of the material that could have influenced the trend of stiffness increase.

Static Residual Properties

Fatigue test temperature and stress levels showed little effect on post-fatigue residual laminate stiffness. Changes that were observed may be due to data scatter associated with

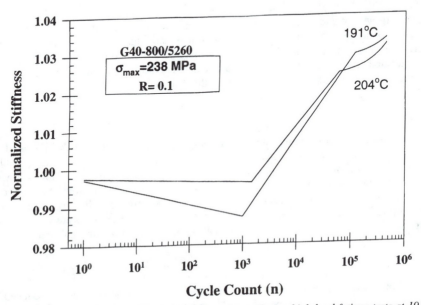

FIG. 20—*Detail of normalized stiffness of OHT specimens during high-load fatigue tests at 191 and 204°C.*

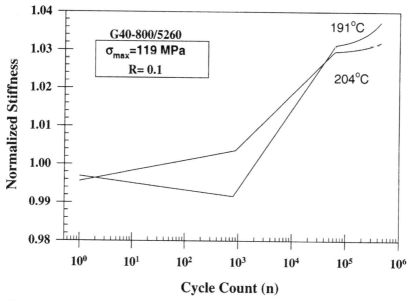

FIG. 21—*Detail of normalized stiffness of OHT specimens during the low load fatigue tests at 191 and 204°C.*

material variability. Increases in fatigue test temperature did result in measurable decreases in residual strength for the low-load case and a slight decrease for the high-load case.

Conclusions

The measured changes in weight and T_g provided evidence of chemical aging during the isothermal fatigue of G40-800/5260 composite material. The applied stress caused internal and edge damage. Increasing test temperature increased visible edge damage due to oxidative breakdown but did not increase the level of internal matrix crack damage.

Edge damage may not be an accurate measure of internal damage at elevated temperatures due to the effects of aging at the exposed surfaces. In addition, at elevated temperatures the gage section stiffness cannot be used as a strict indicator of internal damage because the stiffness gained due to aging offsets the effects of damage.

It is apparent that use of the OHT test to characterize the elevated temperature performance of G40-800/5260 requires knowledge of the relative contributions of stress and temperature to the fatigue and post-fatigue behavior. Aging during the fatigue test period causes an increase in gage section stiffness. This stiffness increase competes with decreases in stiffness due to damage. Synergistic contributions of aging and microcrack formation may result in a net stiffness increase and a decrease in residual strength, the magnitude of which is determined by the time at temperature and the chosen stress level.

References

[1] Struik, L. C. E., *Physical Aging in Amorphous Polymers and Other Materials,* Elsevier Scientific Publishing Company, New York, 1978.

[2] Wang, X. and Gillham, J. K., "Physical Aging in the Glassy State of a Thermosetting System vs. Extent of Cure," *Journal of Applied Polymer Science,* Vol. 47, 1993, pp. 447–460.
[3] Zhou, J., "A Constitutive Model of Polymer Materials Including Chemical Aging and Mechanical Damage and its Experimental Verification," *Polymer,* Vol. 34, 1993, pp. 4252–4256.
[4] Bank, L. C., Gentry, T. R., and Barkatt, A., "Accelerated Test Methods to Determine the Long-Term Behavior of FRP Composite Structures: Environmental Effects," *Journal of Reinforced Plastics and Composites,* Vol. 14, June 1995, pp. 559–587.
[5] Whitney, J. M., Daniel, I. M., and Pipes, R. B., "Experimental Mechanics of Fiber Reinforced Composite Materials," SESA Monograph No. 4, The Society for Experimental Stress Analysis Brookfield Center, CT, 1982.
[6] Talreja, R. *Fatigue of Composite Materials,* Technomic Publishing Company, Inc., Lancaster, PA, 1987.
[7] Riefsnider, K. L., Schulte, K., and Duke, J. C., "Long-Term Fatigue Behavior of Composite Materials," *Long-Term Behavior of Composites, ASTM STP 813,* T. K. O'Brien, Ed., American Society for Testing and Materials, 1983, pp. 136–159.
[8] Kämpf, G., *Characterization of Plastics by Physical Methods,* Macmillan Publishing Company, Inc., New York, 1986.
[9] Coats, T. W. and Harris, C. E., "Experimental Verification of a Progressive Damage Model for IM7/5260 Laminates Subjected to Tension-Tension Fatigue," *Journal of Composite Materials,* Vol. 29, 1995, pp. 280–305.
[10] Boyd, J. D. and Chang, G., "A Third Generation Bismaleimide Prepreg System," presented at the 35th International SAMPE Symposium, 2–5 April 1990.

Dennis J. Buchanan,[1] *Reji John,*[1] *and Kenneth E. Goecke*[1]

Influence of Temperature and Stress Ratio on the Low-Cycle Fatigue Behavior of Trimarc-1/Ti-6Al-2Sn-4Zr-2Mo

REFERENCE: Buchanan, D. J., John, R., and Goecke, K. E., **"Influence of Temperature and Stress Ratio on the Low-Cycle Fatigue Behavior of Trimarc-1/Ti-6Al-2Sn-4Zr-2Mo,"** *Composite Materials: Fatigue and Fracture, Seventh Volume, ASTM STP 1330,* R. B. Bucinell, Ed., American Society for Testing and Materials, 1998, pp. 199–216.

ABSTRACT: The results of an experimental investigation of load-controlled isothermal low-cycle fatigue behavior of a titanium matrix composite (TMC) are discussed. The TMC was composed of Ti-6Al-2Sn-4Zr-2Mo matrix (wire) reinforced with silicon-carbide (Trimarc-1)™ fibers. The composite panels were constructed by alternating layers of matrix wire and fiber using a wire-winding technique. The panels were unidirectional with a $[0]_{10}$ layup and fiber volume fraction ≈ 0.29. The longitudinal fatigue data showed good correlation with other TMC systems at both positive and negative stress ratios. The Walker equation was successful at correlating the longitudinal *S-N* data for stress ratios $R = -1.3$ and 0.1, and for predictions at $R = 0.5$ and 0.7. The maximum fiber stress versus cycles to failure for several unidirectional TMC systems at similar test conditions consolidate to a narrow band, indicating that the life is fiber dominated. The *S-N* behavior of the TMC, subjected to transverse fatigue loading, was successfully predicted using the matrix *S-N* data and a net-section model.

KEYWORDS: titanium matrix composite, isothermal, low-cycle fatigue, Walker equation, negative stress ratio, mechanical testing, Trimarc-1, Ti-6Al-2Sn-4Zr-2Mo, wire winding, longitudinal, transverse, fatigue, TMC

Engine and airframe systems of advanced aircraft require materials with high modulus, relatively low density, and elevated temperature capability. Continuous-reinforced titanium matrix composites (TMC) have been identified as a replacement material for some applications currently using conventional monolithic materials. The process of evaluating the mechanical behavior of the TMC required numerous mechanical tests to fully characterize the composite performance in tension, compression, fatigue, and creep. This paper focuses on the longitudinal and transverse fatigue results at three temperatures and two stress ratios.

Several investigators have studied fatigue damage initiation and evolution in continuous fiber-reinforced metal matrix composites (MMC) [1–11]. Many have shown that the reduction in the stiffness was a useful parameter for indication of initiation and accumulation of damage in the composite [1,2,5]. A variety of failure modes and damage accumulation theories have also been identified. A common approach that several researchers have utilized as a method to analyze the data was to calculate the stress range or maximum stress in the 0° fibers. This method has been shown to consolidate fatigue data of composites with varied

[1] Associate research engineer, research engineer, and chief mechanical technician, respectively, Materials, and Manufacturing Directorate (AFRL/MLLN), Air Force Research Laboratory, Wright-Patterson AFB, OH 45433-7817, and University of Dayton Research Institute, Dayton, OH 45469-0128.

fiber orientations and layups [1,2]. Another approach used was to compare different materials systems on a density-corrected basis [15]. This is an effective way to compare composite materials with monolithic nickel-base alloys.

Primarily, most of the fatigue research on TMC systems available in the literature has been done for positive stress ratios, R, particularly at $R = 0.1$. Several reasons exist for this, such as the difficulty of developing test fixtures for compressive loading. Another was that the low value of R provides the largest stress or strain range possible without approaching zero or compressive loads. Tensile mean stresses are representative of what a component may experience in service and tend to be the most damaging to the fatigue life and can provide a worst-case scenario. Finally, the composite panels that are available are often composed of eight or less plys, resulting in a moment of inertia that can support only compressive loads of less than half the yield strength without the benefit of antibuckling fixtures. All these reasons support the limited amount of data in the literature on negative stress ratio tests on TMC. To overcome these problems, Lerch and Halford [11] conducted tests using a 32-ply unidirectional TMC over a wide range of R (-1.0 to 0.7) for both stress and strain control at 427°C. Their results [11] showed that for a given stress or strain range the fatigue life decreases with an increase in mean stress. Lerch and Halford [11] also concluded that the Soderberg method, which normalizes the mean stress by the yield strength, was able to correlate all the data. The expense and duration of executing these sophisticated tests have limited previous investigations to focus on only one aspect of the fatigue behavior. Often the effects of stress ratio or temperature on the fatigue life are the most important. This study was initiated to investigate the combined effects of temperature and stress ratio on the fatigue life.

This paper discusses the fatigue behavior of a unidirectional reinforced TMC evaluated at two stress ratios and three temperatures. These results are compared with other TMC references in the literature.

Experimental Procedure

Material and Specimens

The TMC evaluated during this study was a Ti-6Al-2Sn-4Zr-2Mo matrix reinforced with Trimarc-1™ (SiC) fibers. The fibers were manufactured by silicon carbide deposition on a tungsten monofilament core with a subsequent deposition of approximately 3 μm of carbon in a three-layered outer coating. The composite was not constructed by the traditional foil-fiber-foil layup, but was fabricated using a wire-winding methodology. The matrix wires and the silicon carbide fibers were co-wound around a large cylinder using a small amount of binder as shown in Fig. 1. The nominal fiber and wire diameter were approximately 127 and 178 μm, respectively. The composite panels had a unidirectional fiber orientation in a 10-ply layup. The fiber spacing was approximately 5 fibers/mm, and the computed fiber volume fraction based on specimen thickness was determined to be approximately 0.29.

C-scans of the panels made prior to machining the specimens showed relatively few points of high attenuation that corresponded to exposed fibers. Hence, the specimen-machining plan was modified to avoid the regions of high attenuation and exposed fibers. Longitudinal and transverse specimens were water-jet cut from 150-mm thickness in the as-received condition. Straight-sided specimens as well as some dogbone specimens were used in this study, primarily due to the expense of manufacturing the composite panels and the large test matrix that was needed to evaluate the material behavior. Both the straight-sided and dogbone specimens had a minimum of approximately 8-mm width in the gage section, resulting in an average of 410 fibers in the gage section. The specimens were nominally 2 mm thick and

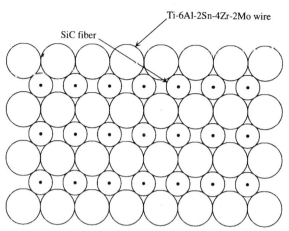

FIG. 1—*Schematic showing the co-wound SiC fibers and the titanium wires for the 10-ply layup of the unidirectional Trimarc-1/Ti-6Al-2Sn-4Zr-2Mo composite.*

119 mm long. Virgin and tested specimens revealed that fiber spacing varied from a nearly perfect rectangular array to having multiple pairs and strings of fiber touches (Fig. 2).

Mechanical Test Setup

The schematic of the mechanical test setup is shown in Fig. 3 [10]. A horizontal test frame [12,13] with commercial hydraulic wedge grips and a quartz lamp heating system [14] were used to provide uniform specimen heating for the elevated temperature tests. The quartz lamps were positioned above and below the specimen to provide a uniform heated zone. Thermal mapping demonstrated a controlled heated zone of 13 mm with less than 0.5% variation over the entire length. Three thermocouples were used inside the gage section to control the three independent temperature control zones.

The combination of straight-sided specimens and room temperature tests required silicon carbide matting and brass tabs in the grip section to reach the required test loads and to

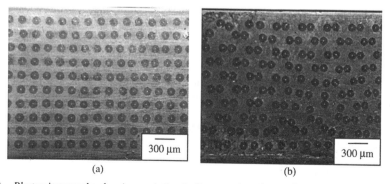

| (a) | (b) |

FIG. 2—*Photomicrographs showing variation in fiber spacing observed in Trimarc-1/Ti-6Al-2Sn-4Zr-2Mo composite specimens.*

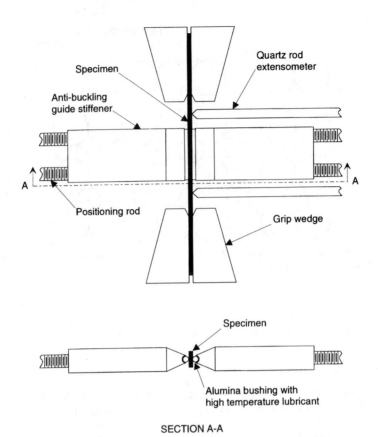

SECTION A-A

FIG. 3—*Schematic of anti-buckling guide setup for negative stress ratio tests.*

reduce the likelihood of failures in the grips. The negative stress ratio tests subjected the specimens to compressive stresses of 827 MPa, which was greater than the Euler buckling stress of 773 MPa. Therefore an anti-buckling fixture was developed to prevent buckling of the specimen during the compressive portion of the loading cycle (Fig. 3). The antibuckling guides were attached to the test frame supports with threaded rods. The threaded rods were screwed into the support to eliminate the space between the specimen and the buckling guide. A 25 by 13 by 50 mm stiffener was used to increase the moment of inertia of the buckling guides to reduce transverse deformation.

The buckling guides consisted of a cylindrical piece of alumna, 3 mm in diameter and 25 mm long, that was placed in contact with the specimen. This allowed a line contact along the length of the specimen equal to the width of the buckling guide. A point contact was not sufficient due to the large deformation that the specimen traverses from maximum to minimum load. A high-temperature lubricant was used in conjunction with the alumina rod to minimize friction between the specimen and rod. The quartz rods of the extensometer contacted the specimen on the outside of the buckling guide (Fig. 3). A commercial 12-mm gage length high-temperature extensometer was modified with bent quartz rods to achieve a 37-mm gage length between tips.

All fatigue tests followed the same procedure prior to the fatigue cycling. After the specimen was mounted in the grips, three thermocouples were attached to the specimen (Fig. 4).

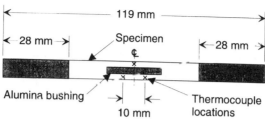

FIG. 4—*Schematic of specimen with thermocouple locations and alumina bushing for negative stress ratio tests.*

One thermocouple was attached near the top edge of the specimen at the center of the specimen's length, and two thermocouples were placed 5 mm from the center near the bottom edge. The buckling guides were than positioned in place and tightened simultaneously to minimize transverse displacement and transverse force in the system. A room temperature modulus measurement was completed at a low stress as a check on the setup and as a comparison to monotonic test results. After successful completion of the modulus check, the temperature was raised to the test temperature over a 10-min period. The specimen was held at temperature with zero applied load for another 15 min for the entire system to stabilize. Another modulus check was completed at a low stress and compared to the monotonic test results. After completion of the modulus check at test temperature, the first ten load-displacement cycles at full load were taken and recorded at 1 Hz to capture any initial damage. The load-displacement sampling interval was set to collect approximately 50 data acquisition cycles (DAC) for the entire test. The automated test control and data acquisition were achieved using personal computers.

Test Matrix

The low-cycle fatigue test matrix is shown in Table 1. The focus was to characterize the longitudinal and transverse stress-life behavior at three test temperatures: $T = 23$, 163, and 371°C. Two stress ratios, $R = -1.3$ and 0.1, were evaluated for the longitudinal specimens and one, $R = 0.1$, for the transverse specimens. At a given temperature and stress ratio, three different stress levels (σ_1, σ_2, σ_3) were chosen such that the cycles to failure (N_f) were approximately 400 000, 100 000, and 50 000 cycles, respectively. Two to six tests were conducted at each stress level.

TABLE 1—*Test matrix for load-controlled isothermal fatigue tests on Trimarc-1/Ti-6Al-2Sn-4Zr-2Mo composite.*

		Stress Levels, σ_1 (MPa)								
		23°C			163°C			371°C		
Orientation	Stress Ratio	σ_1	σ_2	σ_3	σ_1	σ_2	σ_3	σ_1	σ_2	σ_3
$[0]_{10}$	-1.3	398	530	636	398	530	636	398	530	636
$[0]_{10}$	0.1	724	827	900	636	827	900	636	827	900
$[90]_{10}$	0.1	150	160	170	180	190	200	140	150	160

Experimental Results

Longitudinal Fatigue

The maximum applied stress versus cycles to failure curves for the negative stress ratio, $R = -1.3$, and the positive stress ratio, $R = 0.1$, tests at the three test temperatures are shown in Figs. 5a and 5b, respectively. The lines shown in the figures are power law curve fits to the data and are represented by the solid, dashed, and dotted lines for the tests at $T = 23$, 163, and 371°C, respectively. All the fatigue data at a given stress level, stress ratio, and temperature were within a factor of 2 from the mean value except for the data obtained at $T = 23$°C, $\sigma = 398$ MPa, and $R = -1.3$. There was one runout test data point that fell outside the factor of 2 scatterband.

Effect of Temperature

The data in Fig. 5a show that the slope of the isothermal S-N curves for the stress ratio, $R = -1.3$, change with an increase in the test temperature. At the lowest stress, 398 MPa, the fatigue life decreased with increasing temperature. The curve fits for Fig. 5a indicate that there was a slight increase in the number of cycles to failure with an increase in temperature at the highest stress, 636 MPa. This behavior has been observed in some low-carbon

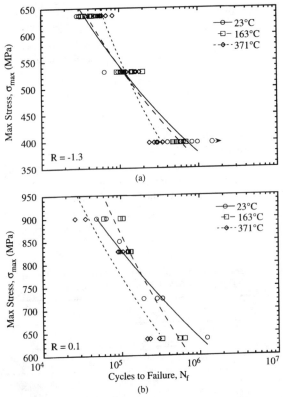

FIG. 5—*Effects of temperature on* S-N *behavior of [0]₁₀ Trimarc-1/Ti-6Al-2Sn-4Zr-2Mo for* (a) R = −1.3 *and* (b) R = 0.1.

steels at moderately elevated temperatures. The increase in fatigue strength for the low-carbon steels at elevated temperatures was attributed to strain aging, which increased the yield strength in load-controlled cyclic testing caused by the reduction in cyclic plasticity [17]. There were not enough data to make any conclusions about the increase in life at the 636-MPa stress level in the TMC without the benefit of matrix material characterization.

The data in Fig. 5b show also that the slopes of the three curve fits for the stress ratio, $R = 0.1$, change with an increase in the test temperature. At the long life, $\sigma_{max} = 636$ MPa, the cycles to failure decrease with an increase in temperature. In contrast, at the shorter lives, $\sigma_{max} = 827$ and 900 MPa, the cycles to failure appears to be independent of the temperature. Interestingly, the 636-MPa data for $R = -1.3$ and 0.1 exhibit opposite trends in the cycles to failure data for increased temperature tests. The $R = -1.3$ data exhibit a slight increase in cycles to failure with an increase in temperature, while the $R = 0.1$ exhibits an obvious decrease in cycles to failure with an increase in temperature. Further conclusions cannot be made without the support of matrix material characterization.

The effects of temperature on the unloading modulus and maximum and minimum strain with applied cycles are shown in Fig. 6a and Fig. 6b, respectively, for $R = -1.3$ at $\sigma_{max} = 398$ MPa. The room temperature test showed no change in either the unloading modulus or the maximum and minimum strains for the first 400 000 cycles. After that point there were indications of damage initiation and accumulation in the unloading modulus and the maxi-

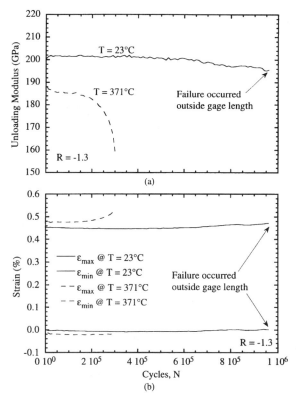

FIG. 6—*Effects of temperature on* (a) *unloading modulus and* (b) *maximum and minimum strain versus applied fatigue cycles for* $[0]_{10}$ *Trimarc-1/Ti-6Al-2Sn-4Zr-2Mo for R = −1.3 and* $\sigma_{max} = 398$ *MPA at T = 23 and 371°C.*

mum strain. The specimen failed after 967 853 cycles outside the gage section, and consequently the data shown were not an accurate representation of the material behavior at those test conditions. The specimen for the elevated temperature test failed inside the gage section after 301 188 cycles. The unloading modulus showed a gradual decline, ≈4.2% over the first ≈200 000 cycles. The remaining life, ≈100 000 cycles, shows a sharp drop in unloading modulus of approximately 12.6%. The maximum and minimum strains also showed acceleration in degradation after ≈200 000 cycles. The minimum strain showed a slight decrease, and the maximum strain showed no change prior to ≈200 000 cycles.

Effect of Stress Ratio

The maximum applied stress versus cycles to failure curve for the three temperatures $T = 23$, 163, and 371°C at the two stress ratios are shown in Figs. 7a, 7b, and 7c, respectively. The lines shown in the figures are power-law curve fits to the data and are represented by the solid and dashed lines for the tests at $R = -1.3$ and 0.1, respectively.

The data in Figs. 7a, 7b, and 7c show that there was a large difference in cycles to failure between the two stress ratios for a given maximum stress. For example, in Figs. 7a and 7b, there was greater than an order-of-magnitude difference in cycles to failure between the two

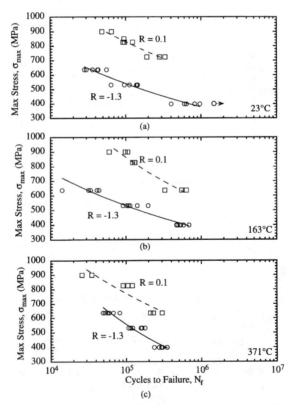

FIG. 7—*Effects of stress ratio on S-N behavior of* $[0]_{10}$ *Trimarc-1/Ti-6Al-2Sn-4Zr-2Mo for* (a) T = 23°C, (b) T = 163°C *and* (c) T = 371°C.

stress ratios at σ_{max} = 636 MPa. In Fig. 7c, at 371°C, the difference was only a factor of five increase between the $R = -1.3$ and $R = 0.1$ curve at σ_{max} = 636 MPa.

The effects of stress ratio on the unloading modulus and the maximum and minimum strain with applied cycles are shown in Fig. 8a and Fig. 8b, respectively, for $T = 371$°C and σ_{max} = 636 MPa. The unloading modulus for the negative stress ratio test shows a slight decreasing trend up to ≈10 000 cycles after which the curve drops sharply until failure. The specimen failed outside the gage section, which was evident from the decrease in the maximum and minimum strain curves in Fig. 8b. The decrease in final unloading modulus was ≈3.7% from the initial measurement. The unloading modulus for the $R = 0.1$ showed a slight increase until ≈10 000 cycles, after which the curve drops sharply until failure with a final decrease of 9.0% from the initial measurement. The maximum and minimum strains provide no indication of damage prior to ≈100 000 cycles.

Comparison with Other TMC Systems

The stress range in the fiber ($\Delta\sigma_f$) versus cycles to failure (N_f) data is plotted in Fig. 9 using the data conducted at 371°C. For a given $\Delta\sigma_f$, $R = -1.3$ yields longer life than $R = 0.1$. This implies that the higher mean stress at $R = 0.1$ results in a decrease of fatigue life.

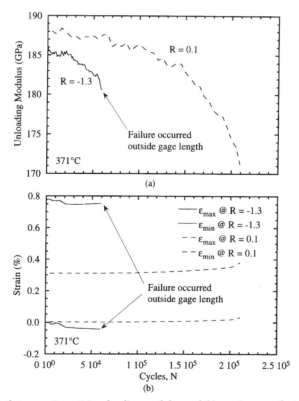

FIG. 8—*Effects of stress ratio on* (a) *unloading modulus and* (b) *maximum and minimum strain versus applied fatigue cycles for* [0]$_{10}$ *Trimarc-1/Ti-6Al-2Sn-4Zr-2Mo for* T = 371° C *and* σ_{max} = 636 MPa.

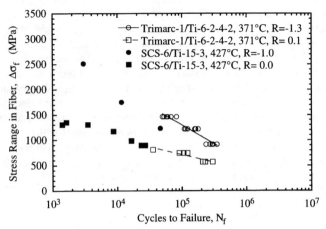

FIG. 9—*Stress range versus cycles to failure for two material systems, Trimarc-1/Ti-6Al-2Sn-4Zr-2Mo and SCS-6/Ti-15-3.*

Similar trends were observed also at temperatures of 23 and 163°C. Based on tests at $R = -1.0$ to 0.7, Lerch and Halford [11] also reported the same behavior for SCS-6/Ti-15-3 tested at 427°C. The data from Lerch and Halford [11] for $R = 0.0$ and -1.0 are also plotted in Fig. 9. These stress ratios are close to the stress ratios tested in this study. Even though SCS-6/Ti-15-3 was tested at a higher temperature and different stress ratios, the data from this composite were consistent with the trend observed for Trimarc-1/Ti-6-2-4-2.

Figure 10 shows the maximum fiber stress versus cycles to failure for $[0]_{10}$ Trimarc-1/Ti-6Al-2Sn-4Zr-2Mo and for other titanium matrices reinforced with SCS-6 fibers. The lines

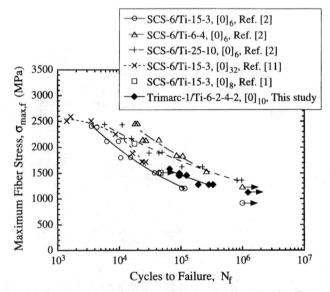

FIG. 10—*Maximum fiber stress in 0° fibers versus cycles to failure for $[0]_{10}$ Trimarc-1/Ti/-6Al-2Sn-4Zr-2Mo and other titanium matrices reinforced with SCS-6 fibers.*

shown in the figure are representative trend lines to the data. The 6-ply layups of SCS-6/Ti-15-3, SCS-6/Ti-6-4, and SCS-6/Ti-25-10 were tested by Jeng et al. [2] at R = 0.1 and 23°C. The [0]$_8$ SCS-6/Ti-15-3 laminates were tested by Johnson [1] at R = 0.1 and 23°C, and the [0]$_{32}$ laminates were tested by Lerch and Halford [11] at R = 0.0 and 427°C. Interestingly, the maximum fiber stress for the composites, at temperatures less than 427°C, follows the same general trend. The contribution of the Trimarc-1™ fibers appears to be similar to that of the SCS-6 fibers, at least at these temperatures. The trends shown in Fig. 10 are also consistent with the fiber-dominated failure observed in layups with 0° fibers [1]. Figure 10 is similar to the plot used by Johnson [1] to compare the 0° fiber stresses in different layups of SCS-6/Ti-15-3. The data shown in Fig. 10 indicate an apparent *in situ* fatigue limit of ≈1200 MPa for the SCS-6 and Trimarc-1 fibers at temperatures < 427°C [1,2].

Correlation of Data Obtained at Two Stress Ratios

We used the Walker-Parameter [18] approach in an attempt to correlate and predict the data obtained during this study. The Walker-Parameter (W) is given by

$$W = \sigma_{max} (1 - R)^m = \Delta\sigma_a^m \, \sigma_{max}^{1-m} \qquad (1)$$

where m is a fit parameter ranging from 0.3 to 0.7 for metals [22]. The Walker-Parameter was related to N_f using the following equation

$$\log(N_f) = A_1 + A_2 \log(W) \qquad (2)$$

Figure 11 shows Eq 2 applied to the data obtained at 23°C for R = −1.3 and 0.1. The constants A_1, A_2, and m, obtained using the R = −1.3, 0.1 data, are reported in Table 2 for the three temperatures. The values of m are 0.47, 0.51, and 0.39 for the data at 23, 163, and 371°C, respectively. These values are within the range reported for metals [22]. In contrast, Lerch and Halford [11] reported a low value of m = 0.12 for SCS-6/Ti-15-3 tested at 427°C.

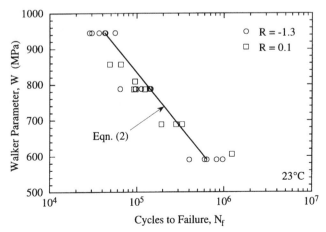

FIG. 11—*Walker-Parameter versus cycles to failure for Trimarc-1/Ti-6Al-2Sn-4Zr-2Mo at* T = 23°C, R = −1.3 and 0.1.

TABLE 2—*Walker-Parameter coefficients for Trimarc-1/Ti-6Al-2Sn-4Zr-2Mo at T = 23, 163, and 371°C at R = −1.3 and 0.1.*

T, °C	A_1	A_2	m
23	23.57	−6.39	0.47
163	21.68	−5.73	0.51
371	15.04	−3.48	0.39

The correlation and predictive capabilities of Eq 2 are shown in Figs. 12a–12c for temperatures 23, 163, and 371°C, respectively. The solid line represents perfect correlation of model with experiment, and the dashed lines represent X2 variation in fatigue life. In Fig. 12b only one data point fell outside a X2 variation from the prediction. In Figs. 12a and 12c all the data points fell inside the X2 variation. The Walker-Parameter constants obtained from the R = −1.3 and 0.1 data were used to predict the cycles to failure for the tests conducted at other stress ratios as shown in Figs. 12b and 12c, respectively. Additional tests are required at different stress ratios to verify the true predictive capabilities of Eq 2 at other temperatures.

Transverse Fatigue

Figure 13 is a plot of maximum applied stress versus cycles to failure for the transverse specimens at the three test temperatures for stress ratio, R = 0.1. There are several instances where there was a factor of three or a larger difference in the cycles to failure for identical test conditions. Often, the low-cycles-to-failure data can be explained by extremely poor fiber spacing as shown in Fig. 2b.

John et al. [19–21] showed that the creep-rupture and fatigue crack growth behavior of unidirectionally reinforced MMC subjected to transverse loading can be predicted using a net-section based model. The model was based on the assumption that the fibers in MMC subjected to transverse loading are essentially ineffective during the majority of the life. A

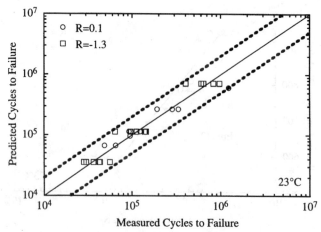

FIG. 12a—*Predicted versus measured cycles to failure using the Walker-Parameter for [0]$_{10}$ Trimarc-1/Ti-6Al-2Sn-4Zr-2Mo at 23°C for R = −1.3 and 0.1.*

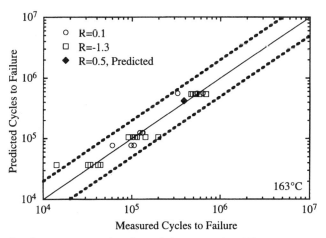

FIG. 12b—*Predicted versus measured cycles to failure using the Walker-Parameter for* $[0]_{10}$ *Trimarc-1/Ti-6Al-2Sn-4Zr-2Mo at 163°C for* R $= -1.3, 0.1,$ *and* $0.5.$

schematic of the transverse model is shown in Fig. 14. The matrix stress range in the ligament between the fibers can be calculated as

$$\Delta\sigma_{lig} = \frac{\Delta\sigma_{90}}{F_n} \qquad (3)$$

where

$$F_n = 1 - \frac{2R_f}{B_p} \qquad (4)$$

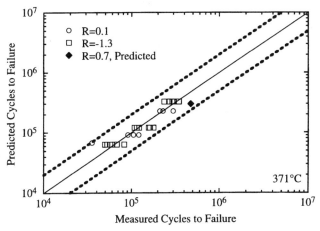

FIG. 12c—*Predicted versus measured cycles to failure using the Walker-Parameter for* $[0]_{10}$ *Trimarc-1/Ti-6Al-2Sn-4Zr-2Mo at 371°C for* R $= -1.3, 0.1,$ *and* $0.7.$

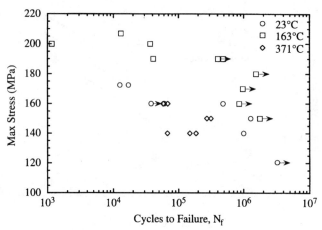

FIG. 13—*Effects of temperature on cycles to failure of* $[90]_{10}$ *Trimarc-1/Ti-6Al-2Sn-4Zr-2Mo for* R = 0.1.

$\Delta\sigma_{90}$ = applied far-field stress range for the [90] composite, $\Delta\sigma_{lig}$ = stress range in the matrix ligament between the fibers, R_f = fiber radius (= 64 µm), B_p = average ply thickness = B/n, B = thickness of composite, and n = number of plies. Using the geometry shown in Fig. 14, Eq 4 can also be written as [20]

$$F_n = 1 - \sqrt{\frac{4 \, V_f \, s}{\pi \, B_p}} \qquad (5)$$

where s = center-center fiber spacing along the length of the [90] composite = 0.195 mm. Equation 5 reduces to the expression derived by Walls et al. [23] for a square array, i.e., when $s = B_p$. Thus, by associating $\Delta\sigma_{lig}$ with the matrix S-N behavior, the composite S-N behavior can be predicted.

The maximum stress, σ_{max} versus N_f data for neat (fiberless) Ti-6-2-4-2 and $[90]_{10}$ Trimarc-1/Ti-6-2-4-2, are shown in Fig. 15 from tests conducted at 163°C with R = 0.1. Note that

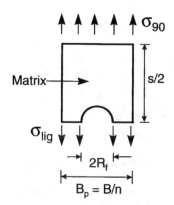

FIG. 14—*Schematic diagram of rectangular net section model used for life prediction of unidirectional [90] orientation composite.*

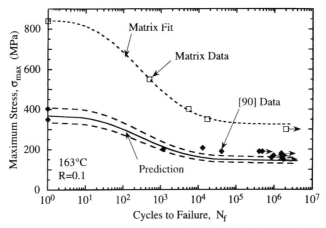

FIG. 15—*Prediction of $[90]_{10}$ S-N behavior from neat Ti-6Al-2Sn-4Zr-2Mo for R = 0.1 at T = 163°C.*

the matrix data were obtained from specimens that were prepared such that the matrix wires were oriented perpendicular to the loading axis. The following expression [21] was used to represent the matrix behavior

$$\sigma_{max} = \sigma_{m,th} + (\sigma_{m,u} + \sigma_{m,th})e^{p(\log N_f)^q} \quad (6)$$

where $\sigma_{m,u}$ = ultimate matrix stress = 840 MPa, $\sigma_{m,th}$ = threshold matrix stress = 325 MPa, and p and q are fit constants. For the fit of the matrix data shown in Fig. 15, $p = -0.054$ and $q = 2.75$. Using Eqs 5 and 6, and assuming $\Delta\sigma_{lig} = \Delta\sigma_m = \sigma_{max*} (1 - R)$, the composite fatigue behavior, i.e., $\Delta\sigma_{90}$ versus N_f, was predicted and plotted as a solid line in Fig. 15. The prediction correlates well with the overall trend of the data. The tensile strength of the composite ($N_f = 1$) and the threshold regime were predicted satisfactorily [21]. The model predicts the near-flat S-N behavior for stresses $\leq$ 200 MPa exhibited by the data. Figure 15 also shows that the fatigue behavior occurs over a small range of stress for the $[90]_{10}$ composite, i.e., the S-N curve is not conducive to a design based on cycle-dependent "design" stress. In contrast, a threshold-type approach, i.e., design stress $< \sigma_{th}$, appears to be more appropriate. The dashed lines close to the prediction correspond to the fatigue behavior predicted for $\pm 10\%$ variation in $\Delta\sigma_{lig}$. These predictions highlight the large variations in predicted N_f due to small changes in $\Delta\sigma_{lig}$ for stress levels $\leq$ 200 MPa.

The model assumes that the fibers are located in a straight row. Hence, the assumed fracture plane is always perpendicular to the applied load and follows the narrowest matrix ligament between the fibers. But, in actual practice, the arrangement of the fibers may be staggered and, consequently, the fracture surface could be tortuous as shown in Figs. 16a–16d.

As seen in Fig. 15, the data for $[90]_{10}$ composite show significant scatter for stresses $\leq$ 200 MPa. This regime coincides with the predicted threshold-type behavior ($N_f > 10^4$) as given by the solid line in Fig. 15. An edge view of the fracture surface of specimens 96-K39 and 96-K43 both tested at $\sigma_{max} = 190$ MPa and $T = 163°C$ is shown in Figs. 16a, b and 16c, d, respectively. Specimen 96-K43 failed at 40 846 cycles, while specimen 96-K39 actually lasted almost ten times longer. Comparing Figs. 16a–d, we see that the fracture surface of 96-K39 is more tortuous than that of 96-K43. Hence, the effective fracture surface

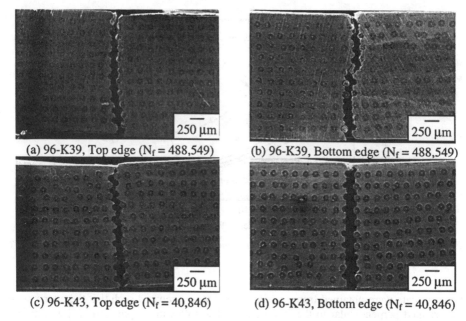

(a) 96-K39, Top edge (N_f = 488,549)   (b) 96-K39, Bottom edge (N_f = 488,549)

(c) 96-K43, Top edge (N_f = 40,846)   (d) 96-K43, Bottom edge (N_f = 40,846)

FIG. 16—*Photomicrographs of fracture planes for two* $[90]_{10}$ *Trimarc-1/Ti-6Al-2Sn-4Zr-2Mo specimens at identical test conditions:* σ_{max} = *190 MPa, R* = *0.1, and* T = *163°C with an order-of-magnitude difference in their cycles to failure.*

area of the matrix ligament between the fibers in 96-K39 could be higher than that of 96-K43. Hence, the effective $\Delta\sigma_{lig}$ for 96-K43 could be lower than that of 96-K39. As shown by the model predictions (dashed lines) in Fig. 15, such small variations in stress can translate easily to large variations in N_f. Hence, establishing a σ_{th} value for design will require numerous tests.

Detailed examinations of the fracture surface and characterization of the variations in fiber spacing, fiber touches, etc. are in progress to determine their effect on the cycles to failure. Preliminary results appear to indicate that an excessive amount of fiber-to-fiber touches does reduce the fatigue life of the transverse specimens.

Conclusions

An experimental investigation of the low-cycle fatigue behavior of the Trimarc-1/Ti-6Al-2Sn-4Zr-2Mo unidirectional composite was conducted at temperatures of 23, 163, 371°C at stress ratios −1.3 and 0.1 for longitudinal tests and 0.1 for transverse tests at a frequency of 3 Hz. The data for different stress ratios were combined using the Walker approach and shown to fall within a X2 variation in fatigue life for the three test temperatures. The results of these tests, along with comparisons with other TMC systems, demonstrate that fatigue data collapse together for a 0° fiber stress range or maximum stress versus cycles-to-failure plot. A net-section model using the matrix *S-N* behavior was used to satisfactorily predict the transverse *S-N* behavior of the composite. The model and the data showed that small variations in the maximum stress can produce large variations in the cycles to failure of the composite under transverse fatigue loading.

Acknowledgments

This research was conducted at the Materials and Manufacturing Directorate (AFRL/ MLLN), Air Force Research Laboratory, Wright-Patterson Air Force Base, OH 45433-7817. D. J. Buchanan, R. John, and K. E. Goecke were supported under on-site contract number F33615-94-C-5200. The authors gratefully acknowledge the efforts of Mr. M. J. Shepard for his many hours on the SEM documenting the fractography of the test specimens.

References

[1] Johnson, W. S., "Fatigue Testing and Damage Development in Continuous Fiber Reinforced Metal Matrix Composites," *Metal Matrix Composites: Testing, Analysis, and Failure Modes, ASTM STP 1032,* W. S. Johnson, Ed., American Society for Testing and Materials, West Conshohocken, PA, 1989, pp. 194–221.

[2] Jeng, S. M., Wang, P. C., and Yang, J.-M., "Fatigue Damage Evolution and Degradation of Mechanical Properties in Silicon-Carbide (SiC) Fiber-Reinforced Titanium Matrix Composites," *Life Prediction Methodology for Titanium Matrix Composites, ASTM STP 1253,* W. S. Johnson, J. M. Larsen, and B. N. Cox, Eds., American Society for Testing and Materials, West Conshohocken, PA, 1996, pp. 377–394.

[3] Bakuckas, J. G., Jr. and Johnson, W. S., "A Methodology to Predict Damage Initiation, Damage Growth, and Residual Strength in Titanium Matrix Composites," *Life Prediction Methodology for Titanium Matrix Composites, ASTM STP 1253,* W. S. Johnson, J. M. Larsen, and B. N. Cox, Eds., American Society for Testing and Materials, West Conshohocken, PA, 1996, pp. 497–519.

[4] Harmon, D. M. and Saff, C. R., "Damage Initiation and Growth in Fiber Reinforced Metal Matrix Composites," *Metal Matrix Composites: Testing, Analysis, and Failure Modes, ASTM STP 1032,* W. S. Johnson, Ed., American Society for Testing and Materials, West Conshohocken, PA, 1989, pp. 237–250.

[5] Wang, P. C., Jeng, S. M., Yang, J.-M., and Russ, S. M., "Fatigue Damage Evolution and Property Degradation of SCS-6/Ti-22Al-23Nb Orthorhombic Titanium Aluminide Composite," *Acta Metallurgica,* Vol. 44, No. 8, 1996, pp. 3141–3156.

[6] Wang, P. C., Jeng, S. M., Yang, J.-M., and Mall, A. K., "Fatigue Life Prediction of Fiber-Reinforced Titanium Matrix Composites," *Acta Metallurgica,* Vol. 44, No. 3 1996, pp. 1097–1108.

[7] Neu, R. W. and Nicholas, T., "Methodologies for Predicting the Thermomechanical Fatigue Life of Unidirectional Metal Matrix Composites," *Advances in Fatigue Lifetime Predictive Techniques: 3rd Volume, ASTM STP 1292,* M. R. Mitchell and R. W. Landgraf, Eds., American Society for Testing and Materials, West Conshohocken, PA, 1996, pp. 1–23.

[8] Nicholas, T., "Fatigue Life Prediction in Titanium Matrix Composites," *Transactions of the ASME,* Vol. 117, 1995, pp. 440–447.

[9] Nicholas, T. and Kroupa, J. L., "Micromechanics Analysis and Life Prediction of Titanium Matrix Composites,"*Journal of Composites Technology & Research,* Vol. 20, No. 2, April 1998.

[10] Blatt, A. P. and Stevens, K. A., "Unpublished Research," Materials and Manufacturing Directorate (WL/MLLN), Air Force Research Laboratory, Wright-Patterson Air Force Base, OH 45433-7817, 1995.

[11] Lerch, B. and Halford, G., "Fatigue Mean Stress Modeling in a [0]$_{32}$ Titanium Matrix Composite," *Proceedings of the 7th Annual HITEMP Review,* Vol. II, Paper No. 21, 1996, pp. 1–10.

[12] Hartman, G. A., III and Russ, S. M., "Techniques for Mechanical and Thermal Testing of Ti$_3$Al/ SCS-6 Metal Matrix Composites," *Metal Matrix Composites: Testing, Analysis, and Failure Modes, ASTM STP 1032,* W. S. Johnson, Ed., American Society for Testing and Materials, West Conshohocken, PA, 1989, pp. 43–53.

[13] Hartman, G. A. and Buchanan, D. J., "Methodologies for Thermal and Mechanical Testing of TMC Materials," *Characterisation of Fibre Reinforced Titanium Matrix Composites,* 77th Meeting of the AGARD Structures and Materials Panel, AGARD Report 796, Bordeaux, France, 27–28 Sept. 1993.

[14] Hartman, G. A., III. "A Thermal Control System for Thermal/Mechanical Cycling," *Journal of Testing and Evaluation,* JTEVA, Vol. 13, No. 5, September 1985, pp. 363–366.

[15] Larsen, J. M., Russ, S. M., and Jones, J. W., "An Evaluation of Fiber-Reinforced Titanium Matrix Composites for Advanced High-Temperature Aerospace," *Metallurgical and Materials Transactions,* Vol. 26A, December 1985, pp. 3211–3223.

[16] Hartman, G. A., Ashbaugh, N. E., and Buchanan, D. J., "A Sampling of Mechanical Test Automation Methodologies Used in a Basic Research Laboratory," *Automation in Fatigue and Fracture:*

Testing and Analysis, ASTM STP 1231, C. Amzallag, Ed., American Society for Testing and Materials, West Conshohocken, PA, 1994, pp. 36–50.

[17] Sandor, B. I. in *Achievement of High Fatigue Resistance in Metals and Alloys, ASTM STP 467*, American Society for Testing and Materials, West Conshohocken, PA, 1970, pp. 254–275.

[18] Walker, K., "The Effect of Stress Ratio During Crack Growth Propagation and Fatigue for 2024-T3 and 7075-T6 Aluminum," *Effects of Environment and Complex Load History on Fatigue Life, ASTM STP 462*, American Society for Testing and Materials, West Conshohocken, PA, 1970, pp. 1–14.

[19] John, R., Lackey, A. F., and Ashbaugh, N. E., "Fatigue Crack Growth Parallel to Fibers in Unidirectionally Reinforced SCS-6/TIMETAL®21S," *Scripta Materialia*, Vol. 35, No. 6, 1996, pp. 711–716.

[20] John, R., Khobaib, M., and Smith, P. R., "Prediction of Creep-Rupture Life of Unidirectional Titanium Matrix Composites Subjected to Transverse Loading," *Metallurgical and Materials Transactions*, Vol. 27A, October 1996, pp. 3074–3080.

[21] John, R., Larsen, J. M., Buchanan, D. J., and Hall, J., "Prediction of Fatigue Strength of Unidirectional Titanium Matrix Composites Subjected to Transverse Loading," to be submitted for publication, 1977.

[22] *Military Standardization Handbook, Metallic Materials and Elements for Aerospace Vehicle Structures*, MIL-HDBK-5E, Vols, 1, 2, 1987.

[23] Walls, D. P., Bao, G., and Zok, F. W., "Mode I Fatigue Cracking in a Fiber Reinforced Metal Matrix Composite," *Acta Metallurgica et Materialia*, Vol. 41, No. 7, 1993, pp. 2061–2071.

Kin Liao,[1] *Carl R. Schultheisz,*[2] *Donald L. Hunston,*[2] *and*
L. Catherine Brinson[3]

Environmental Fatigue of Pultruded Glass-Fiber-Reinforced Composites

REFERENCE: Liao, K., Schultheisz, C. R., Hunston, D. L., and Brinson, L. C., **"Environmental Fatigue of Pultruded Glass-Fiber-Reinforced Composites,"** *Composite Materials: Fatigue and Fracture, Seventh Symposium, ASTM STP 1330,* R. B. Bucinell, Ed., American Society for Testing and Materials, 1998, pp. 217–234.

ABSTRACT: Pultruded glass-fiber-reinforced vinyl ester composite coupons were subjected to four-point-bend environmental fatigue to study their long-term durability for infrastructure applications. Specimens were tested dry and while immersed in water and in solutions of water containing mass fractions of 5 and 10% NaCl salt. Some specimens were also preconditioned by soaking in water or salt solutions for five to six months without loading; the preconditioned specimens showed a fractional decrease of 5 to 13% in flexural strength compared to dry specimens. For specimens cyclically loaded at or above 45% of the average flexural strength of dry coupons, no change in fatigue life was observed for specimens tested while immersed in the fluids (with or without preconditioning) as compared to specimens tested dry. However, water and salt solutions are detrimental to the fatigue life of the material during long-term loading: at a cyclic load of 30% flexural strength, all specimens tested in air survived beyond 10^7 cycles, while all those tested under environmental fatigue did not. It is found that long-term environmental fatigue behavior is not controlled by the quantity of water absorbed; rather, it is governed by a combination of both load and fluid environment. However, a difference in fatigue life in the different fluid environments was not demonstrated. Microscopic examination revealed evidence of degraded fiber/matrix interphase, which is believed to be a controlling factor in the environmental performance of the glass composite.

KEYWORDS: glass-fiber-reinforced composites, four-point-bend fatigue, water, salt solution, fiber/matrix interphase

Glass-fiber-reinforced composites (or glass-fiber-reinforced plastics, GFRP) have been used, but to a limited extent, in the building and construction industry for decades [1–5]. Because of the need to repair and retrofit rapidly deteriorating infrastructure in recent years, the potential for using fiber-reinforced composites for a wider range of applications is now being realized [6–18]. The state of the industry has been described in several review articles and in the proceedings of recent technical conferences pertaining to the subject [6–11].

Fiber-reinforced composites offer excellent resistance to environmental agents and fatigue as well as the advantages of high stiffness-to-weight and strength-to-weight ratios when compared to conventional construction materials. However, one of the obstacles preventing the extensive use of composites has been a lack of long-term durability and performance data. Although there have been numerous studies in fatigue and environmental fatigue of

[1] School of Applied Science, Division of Materials Engineering, Nanyang Technological University, Singapore 639798.
[2] Polymers Division, National Institute of Standards and Technology, Gaithersburg, MD 20899-0001.
[3] Mechanical Engineering Department, Northwestern University, Evanston, IL 60208-3111.

composite materials in the past three or four decades, most of these have not been aimed for applications in infrastructure. Reviews on the fatigue behavior of composite materials can be found in Refs *19–22*.

At this time, the construction industry has focused predominantly on lower-cost glass reinforcement rather than the carbon fiber reinforcement used in aerospace applications. In addition, the needed service life is much longer in infrastructure applications. For instance, some bridges are designed to last for over 50 years. Hence the infrastructure community must be concerned with longer-term behavior as well as different materials and service environments than the aerospace industry. As a result, although data and experience gained from the past may serve as a general guideline, new studies and data pertaining to infrastructure applications are in great demand, especially for composites produced by low-cost, large-volume processing methods such as pultrusion.

Previous studies showed that exposure to water and other corrosive fluids such as acids will degrade the properties and shorten the fatigue life of GFRP. The environmental fatigue performance of GFRP is influenced by all its constituents, that is, the fiber, the matrix, and the fiber/matrix interphase region [23–34]. Some studies focusing on each of these components will be briefly reviewed below.

It is well understood that glass fibers degrade upon exposure to corrosive fluids under stress [35,36]. Metcalfe and Schmitz suggested that the underlying mechanisms of stress corrosion are driven by the exchange of alkali metal ions (Na^+ and K^+) in the glass and hydrogen ions (H^+) of the attacking fluid [37,38]. Schmitz and Metcalfe also proposed that the stress corrosion process consists of two stages: the incubation period, which extends over approximately 95% of the life of the fiber, is controlled by interaction of water with the cations in the flaws to build up hydroxyl concentration (pH) to a critical level; a rapid corrosion period of the silica network follows the incubation period, where flaws in the glass fiber propagate to reach a critical size under corrosion and stress, leading to failure [37,38]. Vauthier et al. have shown that more broken fibers were found in environmentally aged GFRP samples than those without aging, implying degraded fiber strength [39]. Sekine et al. have also shown evidence of fiber degradation during environmental fatigue where elements from glass fibers were found in the fluid [32].

Formation of matrix cracks may have a critical role in environmental fatigue failure. Carswell and Roberts have conducted investigations on the fatigue behavior of a polyester reinforced with a chopped-glass-fiber-strand mat. Their experiments were carried out in air and several liquid media including water and acids [40]. Samples tested in air showed many matrix cracks prior to failure, while only a few cracks were found in samples tested in acid media, which implies that failure followed fairly quickly after the formation of the first matrix crack in the acid media. Hofer et al. also suggested that, in a study of a glass/epoxy composite, moisture entering the matrix crack network affects the strength, elastic modulus, and strain capability of the fiber/matrix interphase region of the plies immediately beyond the last cracked ply, and thus accelerates matrix cracking and shortens the overall life of the composite [41].

The fiber/matrix interphase region has a controlling effect on the environmental fatigue of composites. Fiber/matrix debonding can be found during environmental aging even without externally applied load [24]. When comparing the microscopic features of several types of GFRP fatigued in air and in water under tensile stress, Watanabe had found that the failure surface of the samples in water was uneven, with much debonding [42]. The rate of reduction of fatigue strength in off-axis specimens in water was higher for unidirectional specimens, implying direct impact of water on the interphase region. Fried found significant degradation in shear fatigue strength after seawater exposure of GFRP up to five years, which strongly

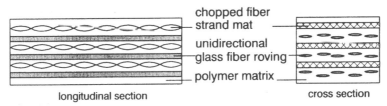

FIG. 1—*Schematic diagram of the structure of a pultruded composite coupon under study.*

points to interface effects [*43*]. It has been shown that surface finish on glass fibers leads to a significant difference in fatigue performance [*44,45*].

Much effort in understanding the fatigue and environmental fatigue behavior of GFRP has been focused on tensile fatigue, with less attention being given to the fatigue behavior under other types of loading [*46–66*]. Long-term data beyond 10^7 cycles is very limited for environmental fatigue of GFRP, especially for the case where the material is undergoing cyclic loading while immersed in fluid. In particular, long-term environmental fatigue data for pultruded GFRP under flexural loading are not well explored. As mentioned earlier, lack of long-term durability data is one of the technical obstacles preventing the extensive use of GFRP in construction. In light of this problem faced by the construction industry, the present work is concerned with generating long-term fatigue and environmental fatigue data for pultruded GFRP, and to develop an understanding of the performance-limiting issues, so that confidence can be gained in using GFRP for applications in large-scale primary and secondary structures.

Material and Method[4]

The material under study is pultruded, E-glass-fiber-reinforced vinyl ester composite (Morrison Molded Fiber Glass[5] (MMFG) EXTREN 625). The as-received materials were 30.5 by 122 by 0.64 cm (12 by 48 by 0.25 in.) plates. Bar specimens of dimensions 12.7 by 1.3 by 0.64 cm (5.0 by 0.5 by 0.25 in.) were cut from the plates according to ASTM Test Methods for Density and Specific Gravity (Relative Density) of Plastics by Displacement (D 790-92) for the four-point-bend test. Structurally, the material consists of unidirectional fiber roving and layers of chopped fiber strand mat embedded in vinyl ester matrix, shown schematically in Fig. 1. The fiber volume fraction is about 60%. As a result of the pultrusion process, the fiber rovings are not uniformly distributed locally. Also, because of the relatively small size of the test specimens (0.83-cm² cross-sectional area), some specimens may contain more fiber roving than others or have the rovings lying in different positions in the specimens.

The variability in the pultruded structure throughout the large plates gives rise to considerable scatter in the data for both the stiffness and strength properties. Variations in strength are to be expected from any material, but the variations in stiffness with this material are much larger than commonly found with typical laminated composites used in aerospace applications.

---

[4] Certain commercial materials and equipment are identified in this paper in order to specify adequately the experimental procedure. In no case does such information imply endorsement by the National Institute of Standards and Technology and Northwestern University, nor does it imply necessarily that the items are the best available for the purpose.
[5] Morrison Molded Fiber Glass have since changed their name to Strongwell-Bristol Division.

Specimens were preconditioned by immersing in three different fluid environments at room temperature:

a. De-ionized water for five months.
b. 5% salt solution for six months (a solution in the proportions of 5 g NaCl in 95 g de-ionized water).
c. 10% salt solution for six months (a solution in the proportions of 10 g NaCl in 90 g de-ionized water).

Water and salt solutions were chosen to simulate rain and salt spray used for de-icing, two commonly encountered outdoor environments. Mass changes reflecting the uptake of fluid for the specimens during preconditioning were recorded at regular time intervals using an electronic balance.

Four-point-bend loading, a very common loading situation in infrastructure, was chosen for this study. For instance, the girder of a bridge sustains four-point-bend loading when vehicles pass over it. Quasistatic flexural strength and flexural modulus for as-received specimens as well as for preconditioned specimens were determined according to ASTM D 790-92 for the four-point-bend test. Specimens were cyclically loaded using a sinusoidal wave function at 10 Hz, with $R = 10$ (where R is the ratio of the maximum to the minimum cyclic load) at room temperature in five different conditions:

A. Air (dry).
B. De-ionized water (specimen without preconditioning).
C. De-ionized water (specimen preconditioned in de-ionized water for five months).
D. 5% salt solution (specimen preconditioned in the 5% salt solution for six months).
E. 10% salt solution (specimen preconditioned in the 10% salt solution for six months).

When testing under cyclic load, specimens were sealed in nylon bags filled with deionized water or salt solution. Cyclic tests, carried out at selected intervals (85, 65, 45, or 30% of the mean quasistatic flexural strength (denoted FS) of dry specimens tested in air at room temperature), were periodically interrupted to measure the flexural modulus by means of a displacement gage (MTS model 632.06H-20) and to record the mass change of the specimen using an electronic balance. Surface-related damage during fatigue was also examined and recorded using a microscope. Specimens from all conditions were tested quasistatically. However, in order to conserve time and because long-term loading is our primary concern, fatigue tests were *not* carried out at all the selected load intervals for each of Conditions A through E. Specimens from Conditions A, B, and C were tested at all levels: 85, 65, 45, and 30% FS. Specimens from Condition D were tested at 65, 45, and 30% FS. Specimens from Condition E were tested only at 30% FS in order to focus on the long-term loading, and because that level of loading shows the effects of fluid sorption most clearly, as well as because the rate of fluid sorption was found to decrease with increasing salt content, suggesting that Condition E may actually be less damaging than Conditions D or C.

Results and Discussion

Quasi-Static Flexural Strength and Modulus

The flexural modulus and quasi-static flexural strength for as-received specimens and those after preconditioning in Conditions a (water), b (5% salt solution), and c (10% salt solution) are shown in Figs. 2 and 3, respectively. Flexural strength, *FS*, is determined as the maximum tensile stress on the lower outer surface:

$$FS = P/bd^2 \qquad (1)$$

where *P* is the failure load, and *L*, *b*, and *d* are the length, width, and thickness of the specimen, respectively. Compared to the as-received specimens, the flexural modulus for those after preconditioning remains essentially the same (Fig. 2). The flexural strength after preconditioning, however, showed some degree of degradation (Fig. 3). Compared to the dry specimens, the mean flexural strength for specimens preconditioned in Conditions a, b, and c showed a 4.8, 12, and 13% decrease, respectively. These results are in general agreement with data from previous studies on environmental aging of GFRP where degradation of tensile and flexural strength of GFRP after immersion in water and salt solution at room temperature was reported [23–34]. A t-test on the data in Fig. 3 using pooled variances indicates that the strengths of specimens immersed in 5 and 10% salt solutions are significantly different from the strength of dry specimens at the 95% confidence level; the strength of specimens immersed in water are significantly different from the strength of dry specimens

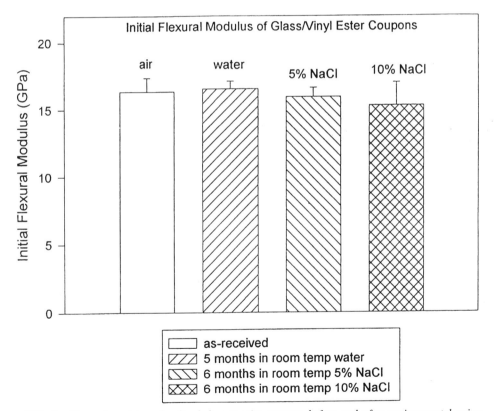

FIG. 2—*Flexural modulus of pultruded composite coupons before and after environmental aging. Error bars represent the standard uncertainty in the experimental data, which is calculated as one sample standard deviation.*

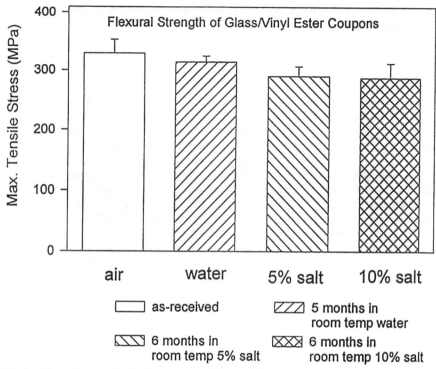

FIG. 3—*Flexural strength of pultruded composite coupons before and after environmental aging. Error bars represent the standard uncertainty in the experimental data.*

at the 90% confidence level. A t-test on the data of Fig. 2 shows no significant difference in the modulus values at the 90% confidence level.

Macroscopic failure modes under monotonic flexural loading for the as-received and preconditioned specimens closely resemble each other. Typical load-displacement curves are shown in Fig. 4. These curves remain quite linear prior to failure. Note that the failure loads for the four specimens shown in Fig. 4 do not represent average values, which were indicated in Figs. 2 and 3. The variability in the material is enough that the deformation behavior of individual specimens may not always be indicative of the averages. The failure process under quasi-static loading begins with the development of parallel cracks in the matrix perpendicular to the longitudinal direction on the tensile surface (the lower surface sustaining tensile stress). The matrix cracking is followed by failure of the glass fibers in the unidirectional roving. The third failure mechanism involves propagation of longitudinal through-the-width cracks on planes of maximum shear between fiber roving and the layers of fiber strand mats, or between the fiber strand mats themselves. These large cracks are referred to as "delaminations" throughout the paper. The strength degradation as a result of preconditioning was attributed to damage of the fiber/matrix interphase region [24,32].

Sorption Behavior

The amount of fluid absorbed by samples during preconditioning and cyclic loading was closely followed. Typical results are presented in Fig. 5, where the sorption data (mass

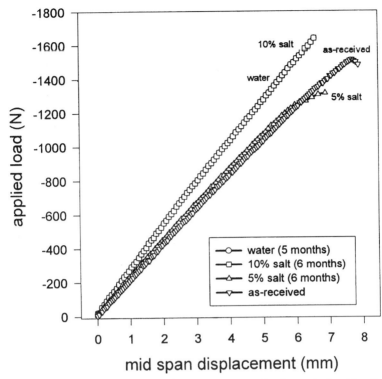

FIG. 4—*Typical load versus mid-span displacement curves under four-point-bend loading. Data for the specimen labeled "water" are hidden behind the data labeled "10% salt." The standard uncertainty in the experimental data is 5 N.*

fraction of water uptake) for eight individual specimens are shown. All specimens shown in the figure had been loaded cyclically at 30% FS. Each line in Fig. 5 represents sorption data of one specimen. Hollow symbols represent sorption data taken from specimens under no externally applied load, while filled symbols represent sorption data taken during cyclic loading (where the ordinate is time under cyclic load). For instance, the two lines connecting hollow circles represent sorption data from two individual specimens preconditioned in water, one for 4000 h, another for 5800 h. These specimens were then cyclically loaded to failure at 30% FS, and the mass of fluid uptake during cyclic loading was represented by filled circles. Similarly, squares and diamonds (hollow or filled) represent sorption data in 5 and 10% salt solution, respectively. Two specimens were cyclically tested at 30% FS without preconditioning, the sorption data during which period are represented by filled triangles.

Sorption behavior in three different fluids (i.e., water, 5% salt solution, and 10% salt solution) without externally applied load all seem to follow Fickian behavior (hollow symbols shown in Fig. 5): a rapid initial mass increase followed by saturation after prolonged immersion. Mass increase in the material depends on the type of media, concentration of the media, as well as the externally applied load. For instance, the average mass fraction increases after 4000 h immersion in water, 5% salt solution, and 10% salt solution are 0.64, 0.52, and 0.45%, respectively. Thus, the rate of water ingress is greater than that for salt solutions, and the higher the salt concentration, the slower the sorption rate. It is also clear that the initial rate of mass increase for specimens tested in water is higher under cyclic

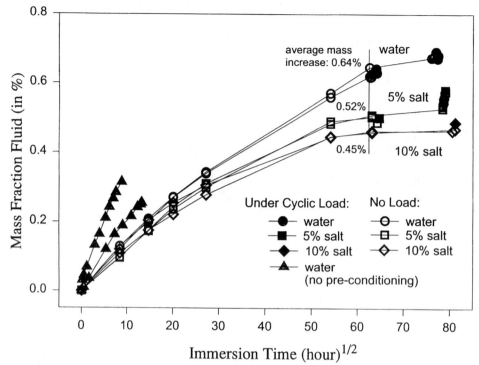

FIG. 5—*Sorption behavior of pultruded composite coupons under various aging conditions. The mass fraction of sorbed fluid is calculated as* $(m - m_0)/m_0$, *where* m_0 *is the initial mass of the sample, and m is the current mass of the sample. The standard uncertainty in the measurement of the mass fraction of sorbed fluid is estimated as 0.01%.*

loading. Compared to fluid mass gain during preconditioning, mass changes during cyclic loading after preconditioning did not increase significantly in most cases. Rapid mass gain during fatigue was seen in one or two cases, which may be attributed to large deformation or development of damage. It should be noted that the time for fluid sorption during cyclic loading at 30% FS was limited to less than 270 h by the life of the sample for all specimens, while the load-free time of immersion for most specimens being preconditioned was more than 5000 h.

Mass increase caused by fluid uptake during cyclic loading depends on load amplitude as well as time under load. For specimens tested under Condition B (specimens were not preconditioned before being cyclically tested in water), the higher the cyclic load, the faster the initial sorption rate. However, the final increase in water content before failure depends on the duration of the cyclic test, as shown in Fig. 6.

How does sorption behavior affect fatigue life? Is fatigue performance affected differently by water and salt solutions, fluid concentration, and quantity of fluid absorbed? These issues will be discussed in the subsequent section.

Fatigue Damage and Failure Mechanisms

Fatigue damage and failure mechanisms are assessed by both quantitative measurements of flexural modulus, and by macroscopic and microscopic observation of specimen surfaces

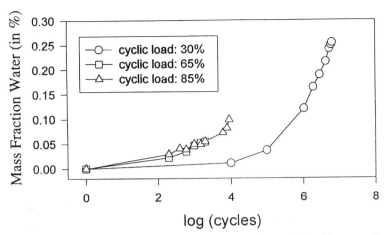

FIG. 6—*Sorption behavior of pultruded composite coupons under cyclic load in water. The standard uncertainty in the measurement of the mass fraction of water is estimated as 0.01%.*

during cyclic loading and after fracture. As has been well documented, change in elastic modulus is a strong indicator of the state of the material [50–56]. Data collected from more than 20 individual specimens tested under Conditions A through E suggest that degradation of flexural modulus is similar in generic pattern, independent of cyclic load levels and environmental conditions. This similarity reflects a resemblance in the essence of the damage development process. Some typical degradation curves (from samples tested under Condition B) are shown in Fig. 7, where flexural modulus and life (cycles to failure) are presented in normalized scale. A flexural modulus degradation curve is typified by a small drop within the first 10% of life, followed by a relatively flat region, indicating a slower rate of damage development, to about 85% of life, then succeeded by a much faster drop during the last stages of life. Such a damage-related pattern of flexural modulus degradation is typical of

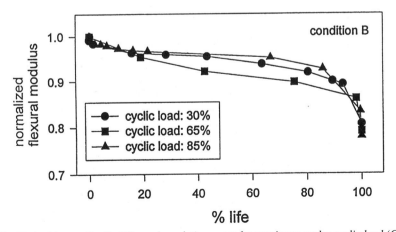

FIG. 7—*Typical (normalized) stiffness degradation cures for specimens under cyclic load (Condition B). The standard uncertainty in the normalized modulus data is estimated as 0.01.*

fiber-reinforced composite materials [53]. Despite the overall similarity in the generic damage process, specific damage events may be load- or environment-dependent.

Transverse matrix cracks usually initiate from the edge of the specimen on the lower (tensile) surface, possibly as a result of microdamage during specimen cutting. For specimens cyclically loaded above 45% FS for all of Conditions A-D, visible transverse surface cracks can be found as early as 10% life, while for those loaded at 30% FS (Conditions A-E), such cracks will not appear until mid- to late-life, and only a few transverse cracks were seen on the latter. Thus, damage initiation is effectively load-dependent but environment-independent above 45% FS. Cyclic loading at 30% FS seems to be a demarcation for the initiation of environment-dependent fatigue damage. Small transverse cracks appear much earlier in Conditions B through E (ranged from 43 to 79% life) than those from Condition A (90% life). Thus, damage initiation is also environment-dependent at 30% FS.

Damage development is largely load-dependent. For specimens loaded at or above 45% FS, more regularly spaced transverse surface cracks appear during mid-life (i.e., between 10 to 85% life). These cracks grow in length with an increase of applied load cycles. They usually accompany the drop in flexural modulus (less than 10%) during early to mid-life. Only a few transverse cracks were found throughout the life of those tested at 30% FS. Debonding between longitudinally oriented fibers and matrix may occur as traversed by advancing transverse matrix cracks (Fig. 8). By about 80 to 85% life, longitudinal cracks between layers of fiber strand mat or between fiber strand mat and unidirectional fiber roving begin to develop near the tensile surface. This "delamination" quickly lengthens (Fig. 9), resulting in a significant drop in flexural modulus and leading to final failure. Delamination leading to final failure also occurs in specimens loaded at 30% FS. Examination by scanning electronic microscope (SEM) reveals that the fiber surface on the delamination is rather clean, suggesting poor adhesion between layers of fiber strand mat and between fiber strand mat and unidirectional roving, which facilitate growth of delamination. Degradation curves for selected specimens tested in Conditions A through E at 30% FS are shown in Fig. 10. Judged by the much faster rate of decrease of flexural modulus for environmentally loaded specimens at 30% FS (Conditions B-E), damage development is environment-dependent at 30% FS.

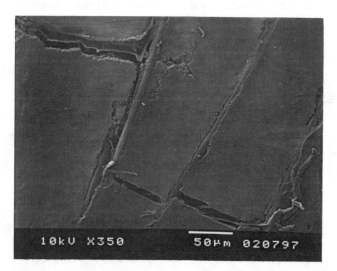

FIG. 8—*Scanning electron micrograph of matrix cracks on the lower specimen surface transverse to the direction of tensile stress; debonding between the matrix and the fibers can also be clearly seen.*

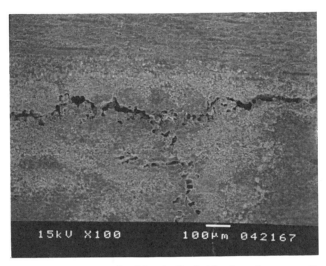

FIG. 9—*Scanning electron micrograph of longitudinal side view of a composite coupon where a horizontal delamination is seen. The delamination (between the fiber strand mat and unidirectional fiber roving) was developed from a vertical matrix crack caused by tensile stress.*

The effect of fluid environment on fatigue life is also apparent in the stress-life (SN) diagram in Fig. 11.

SN Behavior

The SN data are presented in Fig. 11. The abscissa represents number of cycles to failure in the logarithmic scale, while the ordinate represents an estimate of the maximum tensile

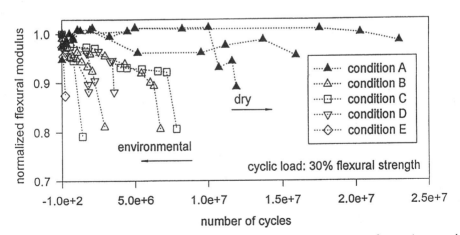

FIG. 10—*Degradation of flexural modulus during fatigue at room temperature for specimens under environmental Conditions A through E. Specimens were cyclically loaded using a sinusoidal wave function at 10 Hz, with R = 10 (where R is the ratio of the maximum to the minimum applied cyclic load). The standard uncertainty in the normalized modulus data is estimated as 0.01.*

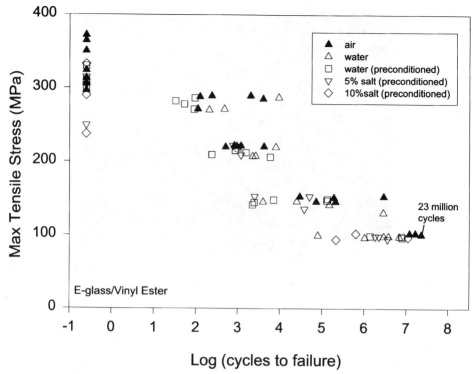

FIG. 11—*Stress-life (SN) data for pultruded composite coupons. The stress is the maximum applied stress calculated from the applied load using Eq 1. Specimens were cyclically loaded using a sinusoidal wave function at 10 Hz, with R = 10 (where R is the ratio of the maximum to the minimum applied cyclic load). The standard uncertainty in the maximum applied stress is estimated as 10 MPa.*

stress on the specimen, σ, calculated as $\sigma = PL/bd^2$; P is the peak of the cyclic load; L, b, and d are the same as those in Eq 1. The life data may be categorized into two regimes, a regime of stress-dependence but environment-independence, and a regime of stress- and environment-dependence.

For specimens cyclically loaded above 45% FS, it seems that the SN data are indistinguishable among Conditions A through D (specimens from Condition E were tested at 30% FS) as data points are overlapping. Although flexural strengths are decreased slightly (see Fig. 3) after preconditioning in water and salt solutions, a difference in fatigue life, if any, did not show up clearly on the SN diagram. This regime of load-dependent and environment-independent fatigue behavior is characterized by high cyclic load levels and low cycles to failure (short loading time).

However, a discrepancy in fatigue life starts to emerge for those specimens tested at lower cyclic load levels. At 45% FS, more dry specimens survived to longer life (number of cycles to failure), despite some overlap that still persists in the data. This level of loading appears to be a transition stage between the two regimes. For specimens cyclically tested at 30% FS, the dry specimens are clearly segregated from those tested under fluid environments. All dry specimens survived beyond 10^7 cycles, while specimens tested under environmental Conditions B through E all failed within 10^7 cycles. Most environmental SN data (i.e., specimens tested under Conditions B, C, and D) tested at 30% FS are clustered within 10^6 to 10^7 cycles.

The flexural moduli all start to show a decreasing trend at around 10^6 cycles. It is clear that water and salt solutions did exert a detrimental effect on the fatigue life during long-term loading. This regime of load and environment dependence is characterized by low cyclic load levels and high cycles to failure (long loading time). A similar stress-life pattern has been reported by Phillips for stress rupture of GFRP, where he has suggested three distinct regimes, namely, environment-independent, environment- and stress-dependent, and stress-independent for the stress versus time-to-failure plot [67]. In our case, the existence of the first two regimes is apparent.

Although it has been shown by previous studies that fluids may exercise their corrosive effect on glass fibers to a different extent depending on their pH value [37,38], this has not been demonstrated by our data obtained so far. However, Metcalfe and Schmitz studied a much broader spectrum of solutions with a range of pH between 2 and 14 [37,38]. Our salt solutions are only slightly acidic, with a pH of approximately 6.5 for the 5% solution and approximately 6.0 for the 10% solution. In fact, because of dissolved gases, the de-ionized water is actually somewhat basic, with a pH of approximately 8.5. Much overlapping of data is seen on the SN plot for specimens tested under environmental Conditions B through E between cyclic loads of 30 to 65% FS (see Fig. 11). It is quite possible that the effect of fluid is overwhelmed by the material variability of the specimens. Comparing those tested under Conditions B (water, no preconditioning) and C (tested in water after preconditioning in water for five months), it seems that the effect of preconditioning in water has no significant influence on the life data, despite the fact that the water content has almost reached saturation for specimens from Condition C. In fact, Bonniau and Bunsell have indicated that damage is not related to the quantity of water absorbed, but to the time of exposure and temperature after the water concentration limit was passed [25]. Comparing those specimens tested in Conditions D (5% salt) and E (10% salt) at the 30% FS level, the effect of salt concentration does not seem to have a significant role in fatigue life. In addition, a difference in life data between those tested in salt solutions and those tested in water is not found.

Environmental Fatigue Degradation Mechanism

It has been suggested by a number of previous studies that the fiber/matrix interphase has a controlling effect in the environmental fatigue performance of fiber-reinforced composites [41–44]. If the fiber/matrix interphase region is damaged or destroyed during fluid ingression, matrix cracks may propagate more easily because fibers have lost their reinforcing function. This assumption matches well with the fact that small matrix cracks appear earlier on the edge of specimens under environmental fatigue than those tested in air. It seems that the competing effect of matrix plasticization as a result of fluid uptake (which increases the failure strain) is overwhelmed by the environmental damage to the interphase region. Examination of the failure surfaces under SEM revealed a morphological difference on the fiber surface. Typically, more matrix residue adhered to the fiber surface of dry specimens (Fig. 12), while "cleaner" fiber surfaces with much less matrix adhering to the fibers are seen from specimens failed under environmental fatigue (Fig. 13). The morphological difference on the fracture surface indeed suggested that fluid action degrades the adhesion between the fibers and the matrix. The same observation is also reported by Sekine et al. [32].

The process of environmental fatigue, then, involves coupled interaction between interphase degradation and matrix cracking: once cracks in the matrix are developed, ingress of corrosive fluid into the material is at a much faster rate, which further accelerates the degradation process of the interphase region. Degradation of fiber strength cannot be deduced from data in this paper. However, strong evidence of fiber strength degradation was found in previous studies [32,39].

FIG. 12—*Scanning electron micrograph of fiber surfaces of a specimen cyclically loaded in air.*

Concluding Remarks

This study considered fatigue loading of pultruded E-glass fiber-reinforced vinyl ester composite plates in a four-point-bending geometry. The specimens were tested under five different environmental conditions:

A. Air (dry).
B. De-ionized water (specimens without preconditioning).

FIG. 13—*Scanning electron micrograph of fiber surfaces of a specimen failed in environmental fatigue in water.*

C. De-ionized water (specimens preconditioned in de-ionized water for five months).
D. 5% salt solution (a solution in the proportions of 5 g NaCl in 95 g de-ionized water, specimens preconditioned in the 5% salt solution for 6 months).
E. 10% salt solution (a solution in the proportions of 10 g NaCl in 90 g de-ionized water, specimens preconditioned in the 10% salt solution for six months).

Based on the findings in this study, the following can be concluded:

- Immersion in water and salt solutions results in degraded flexural strength of pultruded GFRP. A t-test on the data in Fig. 3 indicates that the strengths of specimens immersed in 5 and 10% salt solutions are significantly different from the strength of dry specimens at the 95% confidence level; the strength of specimens immersed in water are significantly different from the strength of dry specimens at the 90% confidence level. A t-test on the data of Fig. 2 shows no significant difference in the modulus values at the 90% confidence level.
- Above a cyclic load of 45% FS, specimens show little difference in fatigue life, whether tested in air or in the other four environments. Fatigue life in this regime is stress-dependent but environment-independent. This regime is characterized by fatigue life within 10^6 cycles.
- At a cyclic load of 30% FS, however, results indicate that water and salt water have a significant detrimental effect on the life of the coupon, a regime of stress- and environment-dependence. This regime is characterized with fatigue life beyond 10^6 cycles.
- Fatigue life of specimens preconditioned in water and salt solution for about five months did not appear to differ from those without preconditioning from data collected so far.
- Effect of fiber/matrix interfacial damage on fatigue performance is evident.

A fatigue limit has not been demonstrated from the data collected thus far, for the fatigue behavior is still environment- and load-dependent at the 30% FS load level. A regime of environment- and load-independence, if it exists, must be below 30% FS. The concept that the matrix cracking stress in air may be taken as an allowable design stress for environmental fatigue (as matrix cracking allows faster fluid ingression) [40] is inconsistent with the data presented in this paper, because matrix cracking is load-and environment-dependent.

Acknowledgment

The authors thank Daniel Witcher of Morrison Molder Fiberglass (now Strongwell-Bristol Division) for the donation of pultruded GFRP plates.

References

[1] Chambers, R. E., "Structural Fiber Glass-Reinforced Plastics for Building Applications," *Plastics in Buildings*, I. Skeist Ed., Reinhold Publishing Co., New York, 1965, pp. 72–118.
[2] Hollaway, L., *Glass Reinforced Plastics in Construction*, John Wiley & Sons, New York, 1978.
[3] Makowski, Z. S., "Symbiosis of Architecture and Engineering in the Development of Structure Uses of Plastics," in *Plastics in Material and Structural Engineering*, Elsevier Scientific Publishing Co., New York, 1982, pp. 59–72.
[4] Starr, T. F., "Structural Applications for Pultruded Profiles," *Composite Structures*, Vol. 22, I. H. Marshall Ed., Applied Science Publishers, New York, 1983, pp. 192–211.
[5] Green, A., "Glass-Fiber-Reinforced Composites in Building Construction," *Transportation Research Record 1118*, Transportation Research Board, Washington, DC, 1987, pp. 73–76.
[6] Barbero, E. and GangaRao, H. V. S., "Structural Applications of Composites in Infrastructure, Part I," *SAMPE Journal*, Vol. 27, No. 6, November/December 1991, pp. 9–16.

[7] Barbero, E. and GangaRao, H. V. S., "Structural Applications of Composites in Infrastructure, Part II," *SAMPE Journal*, Vol. 28, No. 1, January/February 1992, pp. 9–16.

[8] GangaRao, H. V. S. and Barbero, E., "Construction, Structural Applications," in *International Encyclopedia of Composites*, Vol. 6, VCH, New York, 1991, pp. 173–176.

[9] *Advanced Composite Materials in Civil Engineering Structures*, S. L. Iyer and R. Sen, Eds., American Society of Civil Engineers, New York, 1991.

[10] *Infrastructure: New Materials and Methods of Repair, Proceedings of the Third Materials Engineering Conference*, K. D. Basham Ed., San Diego, California, 13–16 Nov. 1994.

[11] *Fiber Composites in Infrastructure, Proceedings of the First International Conference on Composites in Infrastructure (ICCI '96)*, H. Saadatmanesh and M. R. Ehsani, Eds., Tucson, Arizona, 5–7 Jan. 1996.

[12] Sprecher, N., "Composites and Buildings: New Applications," *Composites, Plastiques Reinforces, Fibres De Verre Textile*, Vol. 32, No. 3, 1992, pp. 250–263.

[13] Green, A., Bisarnsin, T., and Love, E. A., "Pultruded Reinforced Plastics for Civil Engineering Structural Applications," *Proceedings, 47th Annual Conference, Composites Institute*, The Society of the Plastics Industry, Inc., Cincinnati, OH: February 1992, Section 15-B.

[14] Tsuji, Y., Kanda, M., and Tamura, T., "Applications of FRP Materials to Prestressed Concrete Bridges and Other Structures in Japan," *PCI Journal*, July-August 1993, pp. 50–58.

[15] Salama, M. M., "Advanced Composites for the Offshore Industry: Applications and Challenges," *Revue De L'Institut Francais Du Petrole*, Vol. 50, No. 1, January-February 1995, pp. 19–26.

[16] Tarricone, P. "Composite Sketch," *Civil Engineering*, May 1995, pp. 52–55.

[17] "*High-Performance Construction Materials and Systems: An Essential Program for America and Its Infrastructure*," Executive Report 93-5011.E, The Civil Engineering Research Foundations, April 1993.

[18] Loud, S., "Three Steps Toward a Composites Revolution in Construction," *SAMPE Journal*, Vol. 32, No. 1, January/February, 1996, pp. 30–35.

[19] Harris, B., "Fatigue and Accumulation of Damage in Reinforced Plastics," *Composites*, October 1977, pp. 214–220.

[20] Stinchcomb, W. W. and Reifsnider, K. L., "Fatigue Damage Mechanisms in Composite Materials: A Review," *Fatigue Mechanisms, ASTM STP 675*, American Society for Testing and Materials, West Conshohocken, PA, 1979, pp. 762–787.

[21] Agarwal, B. D. and I. J. Broutman, in *Analysis and Performance of Fiber Composites*, John Wiley & Sons, New York, 1980.

[22] Konur, O. and Matthews, F. L., "Effect of the Properties of the Constituents on the Fatigue Performance of Composites: a Review," *Composites*, Vol. 20, No. 4, July 1989, pp. 317–328.

[23] Romanenkov, I. G., "Dependence of the Mechanical Properties of GRPs on their Water Absorption," *Soviet Plastics*, Vol. 2, 1967, pp. 74–75.

[24] Pritchard, G. and Taneja, N., "Water Damage in Polyester/Glass Laminates, Part II: Microscopic Evidence," *Composites*, Vol. 4, No. 4, 1973.

[25] Bonniau, P. and Bunsell, A. R., "Water Absorption by Glass Fibre Reinforced Epoxy Resin," *Composite Structures*, I. H. Marshall, Ed., 1981, pp. 92–105.

[26] Martin, J. R. and Gardneer, R. J., "Effect of Long Term Humid Aging on Plastics," *Polymer Engineering and Science*, Vol. 21, No. 9, June 1981, pp. 557–565.

[27] Apicella, A., Migliaresi, C., Nicodemo, L., Nicolais, L., Iaccarino, L., and Roccotelli, S., "Water Sorption and Mechanical Properties of a Glass-Reinforced Polyester Resin," *Composites*, October 1982, pp. 406–410.

[28] Garg, A. C. and Pawar, S. K., "Environmental Effect on Fracture of Glass Fibre Reinforced Polyester," *Fibre Science and Technology*, Vol. 17, 1982, pp. 133–139.

[29] Garg, A. C. and Paliwal, V., "Effect of Water on the Fracture Behavior of Glass Fibre Reinforced Polyester," *Fibre Science and Technology*, Vol. 17, 1982, pp. 63–69.

[30] Rege, S. K. and Lakkad, S. C., "Effect of Salt Water on Mechanical Properties of Fibre Reinforced Plastics," *Fibre Science and Technology*, Vol. 19, 1983, pp. 31–324.

[31] Komai, K., Minoshima, K., and Shiroshita, S., "Hygrothermal Degradation and Fracture Process of Advanced Fibre-Reinforced Plastics," *Materials Science and Engineering*, A143, 1991, pp. 155–166.

[32] Sekine, H., Shimomura, K., and Hamana, N., "Strength Deterioration and Degradation Mechanism of Glass Chopped Reinforced Plastics in Water Environment," *ISME International Journal*, Series I, Vol. 31, No. 3, 1988, pp. 619–626.

[33] Bradley, W., Chiou, P. B., and Grant, T., "The Effect of Seawater on Polymeric Composite Materials," *Composite Materials for Offshore Operations: Proceedings of the First International Work-*

shop, NIST Special Publication 887, S. S. Wang and D. W. Fitting, Eds., National Institute of Standards and Technology, August 1995, pp. 193–202.

[34] Grami, M., Moilanen, M. J., and Rosenow, M. W. K., "Environmental Degradation of Glass Fibers in Polyurethane Matrix Composites," *Proceedings of ICCM-10,* Vol. VI, Whistler, B.C., Canada, August 1995.

[35] Charles, R. J., "Static Fatigue of Glass. I," *Journal of Applied Physics,* Vol. 29, No. 11, November 1958, pp. 1549–1553.

[36] Charles, R. J., "Static Fatigue of Glass. II," *Journal of Applied Physics,* Vol. 29, No. 11, November 1958, pp. 1554–1560.

[37] Schmitz, G. K. and Metcalfe, A. G., "Stress Corrosion of E-Glass Fibers," *I&EC Product Research and Development,* Vol. 5, No. 1, March 1966, pp. 1–8.

[38] Metcalfe, A. G. and Schmitz, G. K., "Mechanism of Stress Corrosion in E-Glass Filaments," *Glass Technology,* Vol. 13, No. 1, February 1972, pp. 5–16.

[39] Vauthier, E., Chateauminois, A., and Bailliez, T., "Hygrothermal Aging and Durability of Unidirectional Glass-Epoxy Composites," *Proceedings, 10th International Conference of Composite Materials,* Vol. VI, August 1995, pp. 185–192.

[40] Carswell, W. S. and Roberts, R. C., "Environmental Fatigue Stress Failure Mechanism for Glass Fibre Mat Reinforced Polyester," *Composites,* April 1980, pp. 95–99.

[41] Hofer, K. E., Skaper, G. N., and Bennett, L. C., "Effect of Moisture on Fatigue and Residual Strength Losses for Various Composites," *Journal of Reinforced Plastics and Composites,* Vol. 6, January 1987, pp. 53–65.

[42] Watanabe, M., "Effect of Water Environment on Fatigue Behavior of Fiberglass Reinforced Plastics," *Composite Materials: Testing and Design (Fifth Conference), ASTM STP 674,* S. W. Tsai, Ed., American Society for Testing and Materials, West Conshohocken, PA, 1979, pp. 345–367.

[43] Fried, N., "Degradation of Composite Materials: The Effect of Water on Glass-Reinforced Plastics," *Mechanics of Composite Materials, Proceedings of the Fifth Symposium on Naval Structural Mechanics,* F. W. Wendt, H. Liebowitz, and N. Perrone, Eds., Pergamon Press, New York, 1967, pp. 813–837.

[44] Hofer, K. E., Jr., Benett, L. C., and Stander, M., "Effect of Various Fiber Surface Treatments on the Fatigue Behavior of Glass Fabric Composites in High Humidity Environment," *Proceedings, 31st Annual Technical Conference of SPI,* Washington, DC, February 1976, Section 6-A.

[45] Shih, G. C. and Ebert, L. J., "The Effect of the Fiber/Matrix Interface on the Flexural Fatigue Performance of Unidirectional Fiberglass Composites," *Composites Science and Technology,* Vol. 28, 1987, pp. 147–161.

[46] Boller, K. H., "Fatigue Characteristics of RP Laminates Subjected to Axial Loading," *Modern Plastics,* June 1964, pp. 145–188.

[47] Dally, J. W. and Carrillo, D. H., "Fatigue Behavior of Glass-Fiber Fortified Thermoplastics," *Polymer Engineering and Science,* November 1969, Vol. 9, No. 6, pp. 434–444.

[48] Hashin, Z. and Rotem, A., "A Fatigue Failure Criterion for Fiber Reinforced Materials," *Journal of Composite Materials,* Vol. 7, 1973, pp. 448–464.

[49] Amijima, S. and Tanimoto, T., "The Effect of Glass Content and Environment Temperature on the Fatigue Properties of Laminated Glass Fiber Composite Materials," *Mechanical Behavior of Materials, Proceedings of the 1971 International Conference on Mechanical Behavior of Materials,* Vol. V, The Society of Materials Science, Kyoto, Japan, 1972, pp. 269–278.

[50] Owens, M. J., "Fatigue Damage in Glass-Fiber-Reinforced Plastics," in *Composite Materials: Fracture and Fatigue,* Vol. 5, L. Broutman, Ed., Academic Press, New York, 1974, pp. 313–341.

[51] Dharan, C. K. H., "Fatigue Failure Mechanisms in a Unidirectionally Reinforced Composite Material," *Fatigue of Composite Materials, ASTM STP 569,* American Society for Testing and Materials, West Conshohocken, PA, 1975, pp. 171–188.

[52] Davis, J. W. and Sundsrud, G. J., "Fatigue Data on a Variety of Nonwoven Glass Composites for Helicopter Rotor Blades," *Composite Materials: Testing and Design (Fifth Conference), ASTM STP 674,* S. W. Tsai, Ed., American Society for Testing and Materials, West Conshohocken, PA, 1979, pp. 137–148.

[53] Mandell, J. F., Huang, D. D., and McGarry, F. J., "Tensile Fatigue Performance of Glass Fiber Dominated Composites," *Composites Technology Review,* Vol. 3, No. 3, Fall 1981, pp. 96–102.

[54] Reifsnider, K. L., Schulte, K., and Duke, J. C., Jr., "Long-Term Fatigue Behavior of Composite Materials," *Long-Term Behavior of Composites, ASTM STP 813,* T. K. O'Brien, Ed., American Society for Testing and Materials, West Conshohocken, PA, 1983, pp. 136–159.

[55] Jessen, S. M. and Plumtree, A., "Fatigue Damage Accumulation in Pultruded Glass/Polyester Rod," *Composites,* Vol. 20, No. 6, November 1989, pp. 559–567.

[56] Xiao, J. and Bathias, C., "Fatigue Behavior of Unnotched and Notched Woven Glass/Epoxy Laminates," *Composites Science and Technology,* Vol. 50, 1994, pp. 141–148.

[57] Echtermeyer, A. T., Engh, B., and Buene, L., "Lifetime and Young's Modulus Changes of Glass/Phenolic and Glass/Polyester Composites Under Fatigue," *Composites,* Vol. 26, No. 1, 1995.

[58] Donaldson, S. L. and Kim, R. Y., "Life Prediction of Glass/Vinylester and Glass/Polyester Composites Under Fatigue Loading," *Proceedings of ICCM-10, Vol. I,* Whistler, B. C., Canada, August 1995, pp. 577–584.

[59] Jones, C. J. Dickson, R. F., Adam, T., Reiter, H., and Harris, B., "Environmental Fatigue of Reinforced Plastics," *Composites,* Vol. 14, No. 3, July 1983, pp. 288–293.

[60] Fesko, D. G., "Flexural Fatigue of Unidirectional Fiberglass-Reinforced Polyester," *Polymer Engineering and Science,* Vol. 17, No. 4, April 1977, pp. 242–245.

[61] Romans, J. B., Sands, A..G., and Cowling, J. E., "Fatigue Behavior of Glass-Filament-Wound Epoxy Composites in Water," *Naval Research Laboratory Report* 7246, April 1971.

[62] Smith, E. W. and Pascoe, K. J., "The Role of Shear Deformation in the Fatigue Failure of Glass Fibre-Reinforced Composites," *Composites,* October 1977, pp. 237–243.

[63] Bevan, L. G., "Axial and Short Beam Shear Fatigue Properties of CFRP Laminates," *Composites,* October 1977, pp. 227–232.

[64] Sims, D. F. and Brogdon, V. H., "Fatigue Behavior of Composites Under Different Loading Modes," *Fatigue of Filamentary Composite Materials, ASTM STP 636,* K. L. Reifsnider and K. N. Lauraitis, Eds., American Society for Testing and Materials, West Conshohocken, PA, 1977, pp. 185–205.

[65] Agarwal, B. D. and Joneja, S. K., "Strain-Controlled Flexural Fatigue of Unidirectional Composites," *Composites Technology Review,* Vol. 4, No. 1, Spring 1982, pp. 6–13.

[66] Newaz, G. M., "Fatigue Damage Growth Rate in Unidirectional Composites in Flexural Loading," *Journal of Materials Science Letters,* Vol. 4, 1985, pp. 197–199.

[67] Phillips, D. C. and Scott, J. M., "The Shear Fatigue of Unidirectional Fibre Composites," *Composites,* October 1977, pp. 233–236.

G. Zaffaroni,[1] *C. Cappelletti,*[1] *M. Rigamonti,*[1] *L. Fambri,*[2] *and A. Pegoretti*[2]

Comparison of Two Accelerated Hot-Wet Aging Conditions of a Glass-Reinforced Epoxy Resin

REFERENCE: Zaffaroni, G., Cappelletti, C., Rigamonti, M., Fambri, L., and Pegoretti, A., "Comparison of Two Accelerated Hot-Wet Aging Conditions of a Glass-Reinforced Epoxy Resin," *Composite Materials: Fatigue and Fracture, Seventh Volume, ASTM STP 1330,* R. B. Bucinell, Ed., American Society for Testing and Materials, 1998, pp. 235–245.

ABSTRACT: Two accelerated hot-wet aging tests of glass-reinforced epoxy resin were performed at 45 and 70°C at the same level of relative humidity (RH = 84%). Mechanical and physical properties of "dry" and differently saturated composites are compared. It has been found that the higher the conditioning temperature, the higher the equilibrium moisture content. The glass transition temperature decreases from 138°C (dry) to 108 to 111°C (for both moisture-saturated cases). Moreover, it has been found that moisture absorption reduces the static properties while not modifying the endurance in fatigue tests.

KEYWORDS: polymeric composites, hygrothermal effects, accelerated test

Nomenclature

log   Decimal logarithm
ln   Natural logarithm
σ Stress
ε Strain
ω Angular frequency
W_f Work of fracture
T_g Glass transition temperature
$\tan(\delta)$ Tangent of loss angle
M_∞ Moisture content at saturation (weight percent)
E_r Relative loss of rigidity
N Number of fatigue cycles
E_{11} Elastic modulus
σ_R Strength
ε_R Elongation at failure

In the past 20 years, much effort has been devoted to the study of moisture in composites both from industrial and academic research. It has been well documented [1–3] that environmental aging lowers the performances of fiber-reinforced epoxy resins. The effect is

[1] Technical Development Laboratory, AGUSTA, via G. Agusta, C. Costa di Samarate, Italy.
[2] Department of Materials Engineering, University of Trento, via Mesiano, Trento, Italy.

especially great when polymeric composite tests are carried out at high temperatures. In particular, the modulus and the glass transition temperature of the resin decrease, determining that the resin behavior dominates some mechanical properties such as compression and shear.

Since natural aging times are too long compared with the industrial time scale, it is usual to submit structural elements to artificially accelerated hygrothermal aging before testing. In fact, an increase of the conditioning temperature reduces aging times because the moisture transport into the composite is accelerated. In a similar way this is true for every low-molecular-weight penetrant. However, a too high temperature could damage the composite matrix in an overloaded way (see, for example, the Hopfemberg-Frisch chart [4]). This paper presents a comparison of the effects generated by two different kinds of hot-wet aging on the performance of glass-reinforced epoxy resin quasi-isotropic laminates with respect to the "dry" condition. The same relative humidity of 84% and temperatures of 45 and 70°C have been used for conditioning. We call these "accelerated" for the higher temperature and "standard" for the lower temperature.

Background

For the interpretation of some experimental results it is useful to highlight the various factors that were involved in the testing of a polymeric composite when submitted to a dynamic load. To do this a quasi-thermodynamic treatment that follows the work of Mr. Burns [5] is presented. The approach is not presumed to be exact, but is only a rough estimation of correlation among cause, viscoelastic characteristics, and effect.

The mechanical work, dW, expended to increase the crack of dA in a specimen submitted to a stress σ and a deformation $d\varepsilon$ can be written as:

$$dW = \sigma d\varepsilon - J dA \qquad (1)$$

where J is the crack-driving force, and A is the crack area. It is straightforward to find that

$$J = \int_0^\varepsilon - (\partial\sigma/\partial A)_\varepsilon d\varepsilon \qquad (2)$$

For a viscoelastic material submitted to a sinusoidal stress:

$$\sigma = \sigma_0 \cdot \sin(\omega t) \qquad (3)$$

where t is the time, and ω is the angular frequency. The resulting strain can be written as $\varepsilon = \varepsilon_0 \cdot \sin(\omega t - \delta)$, where δ is the loss factor or

$$\sigma = \varepsilon_0 E' \sin(\omega t) + \varepsilon_0 E'' \cos(\omega t) \qquad (4)$$

where E' and E'' are the storage and loss moduli, respectively. Moreover, it is

$$\sigma = E^* \cdot \varepsilon \qquad (5)$$

where $E^* = E' + iE''$ is the complex modulus.

For a first approximation we may suppose that the out-of-phase modulus E'' (which is strongly related to the polymer matrix viscoelastic behavior) is almost constant during the fatigue experiment and therefore:

$$(\partial\sigma/\partial A)_\varepsilon = \varepsilon \cdot (dE'/dA) \tag{6}$$

and combining Eqs 1 to 6 we obtain

$$J = -(\varepsilon^2/2)\, dE'/dA = -\tfrac{1}{2}\, \sigma \cdot \varepsilon\, d(\ln E^*)/dA \tag{7}$$

It is intuitive to understand that E^* depends on the number of cycles N with a decreasing trend. For the sake of simplicity we suppose that $\ln(E^*)$ is varying linearly with the number of cycles or, in other words, $\ln(E^*)$ is proportional to ωt. Then

$$d(\ln(E^*)) \approx k\, d(\omega t) \tag{8}$$

where k is a constant.

During a fatigue loading-unloading cycle where the crack area changes from A_c to $A_c + \Delta A_c$, the work of fracture W_f is

$$W_f = \int_{A_c}^{A_c+\Delta A_c} J\, dA = \int_0^{2\pi} -\,\sigma \cdot \varepsilon/2\; d(\ln E^*) \tag{9}$$

Introducing Eqs 3 and 4

$$W_f = -\tfrac{1}{2}\, \sigma_0\, \varepsilon_0\, k\, \pi\, \cos(\delta) \tag{10}$$

Therefore an increase of the energy dissipation characteristics of a polymer (equivalent to an increase of δ) corresponds to a decrease of the work of fracture. In other words, the crack propagates less with respect to the perfectly elastic case ($\delta = 0$). As expected, an increase in the applied maximum load increased W_f.

Experimental Methodology

The composite used in the present work was manufactured by laying up and curing the 3M prepreg SP250-S29. This composite system is an S2 glass fiber-reinforced epoxy resin and it used widely in helicopter components. The composite plates were cured as per the manufacturer's recommendations.

The layup of specimens was: $+45°/-45°/90°/0°/0°/90°/-45°/+45°$. Cured specimens were divided into three groups. The first one was immediately tested. The second set was put in an environmental chamber set up at 45°C/84% relative humidity until saturation and then tested. The specimens of the third group were submitted to accelerated aging at 70°C/84% RH until saturation and then tested. By saturation we mean the equilibrium situation. The following measurements have been carried out on the above specimens. At least three specimens were tested for each experimental condition.

Dynamic Mechanical Thermal Analysis (DMTA)

To determine the dynamic storage and loss moduli (and then the tan (δ) curve) we carried out DMTA tests on specimens machined from the same plate from which the mechanical samples were obtained. The lateral dimensions were 5 by 40 mm. The experimental conditions were: heating rate, 3°C/min; load mode, 3-point bending (sinusoidal); frequency, 1 Hz; displacement, 64 μm; temperature range, from 23 to 250°C.

Static Tests

Specimens have been tested in tension by applying a quasi-static increasing load with a constant crosshead speed of 0.5 mm/min. Specimen dimensions were 300 by 38 by 1.5 mm.

Fatigue Tests

Each specimen was submitted to tension-tension sinusoidal load with R = (minimum load)/(maximum load) = 0.1. Tests were carried out at a frequency of 5 Hz with different maximum loads in order to generate a stress versus number-of-cycles-to-failure plot. On each side of the specimen a linear variable differential transducer was bonded in order to monitor the rigidity of the specimen at different levels of damage. Specimens that did not break after 10^6 cycles were considered to have reached their infinite life, and a static test was carried out to determine the "residual static strength." The load at which the specimens do not break after 10^6 cycles is indicated as the fatigue endurance. Specimen dimensions were 300 by 38 by 1.5 mm.

Moisture Absorption Kinetics

Twelve specimens, called travelers, were cut from the same laminate that was used to manufacture the mechanical sample. The lateral dimensions of the sample were 50 by 50 mm. By regularly weighing three travelers in each hygrothermal condition, the moisture absorption was monitored. From the initial slope of the weight gain versus the square root time plot an apparent diffusion coefficient was found. To frame the above coefficient, additional determinations at 26°C/84% RH and at 33°C/84% RH were carried out. The evolution of the damage in static and fatigue tests was assessed by visual observation. In fact, by using a lamp situated behind the specimen it was easy, with this kind of specimen, to see the ply damage (white bands developing) and delamination growth (with the aid of a liquid penetrant).

Experimental Results

The resin content of the specimens was found to be 32% (weight) and the void content less than 1% (the determination was carried out by acid digestion using an Agusta method).

Moisture Absorption

Figure 1 shows the absorption curves for the two types of conditioning considered here. It is clear that the aging at 70°C/84% RH generates a moisture saturation level (indicated with M_∞) 10% higher than the 45°C/84% RH, reaching the approximate equilibrium values of 0.98 and 0.86%, respectively.

Figure 2 shows an Arrhenius plot for the diffusion coefficient. The solid line in the figure is the best fit of the three first points, corresponding to 26°, 33°, and 45°C. It is clear that the diffusion coefficient determined at 70°C/84% RH (the point at 2.90E-3 in Fig. 2) does not lie over the regression curve. This suggests that the absorption mechanism could be different from the mechanism that takes place at lower temperatures. In addition, from Fig. 3 (see next paragraph) it could be found that 70°C is near the T_g of the material. This observation can be supported by the tan(δ) value at 70°C in Table 1. Therefore it can be concluded that the 70°C/84% RH is an anomalous absorption condition. In fact, it is known

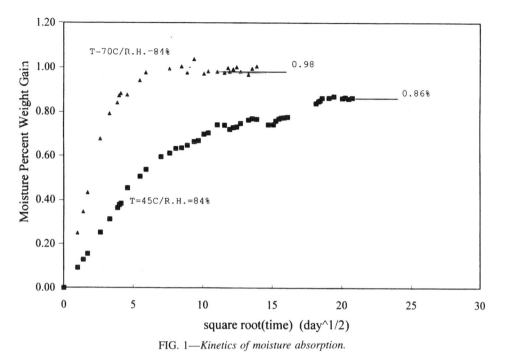

FIG. 1—*Kinetics of moisture absorption.*

that the higher the environmental temperature, the higher the probability to have non-Fickian transport.

Dynamic Mechanical Thermal Analysis

Figure 3 shows the tanδ-DMTA curves for the three kinds of specimens under examination. It is easy to find that the tan(δ) curves for the two wet cases are shifted upward with respect to the dry case.

Note that the two curves for the moisture-saturated material are positioned at the same temperature but have very different intensities. These observations could be correlated with the supposed increased plasticization of the matrix.

Relevant results of DMTA tests are summarized in Table 1. A small decrease of the T_g going from the 45°C/84% RH to the 70°C/84% RH aging situation could be seen if the T_g is taken as the maximum of the tan(δ) peak. In any case, the decrease in T_g is not proportional to the increase of the moisture level (10% higher). In fact, for the present composite from Ref 6 with the saturation level shown in the preceding paragraph, $M_\infty = 0.86\%$ ($T = 45°C$) and $M_\infty = 0.98\%$ ($T = 70°C$)—a T_g decrease of approximately 5°C is expected.

Static Tests

The evolution of damage may be summarized in four steps:

1. 90° ply damage.

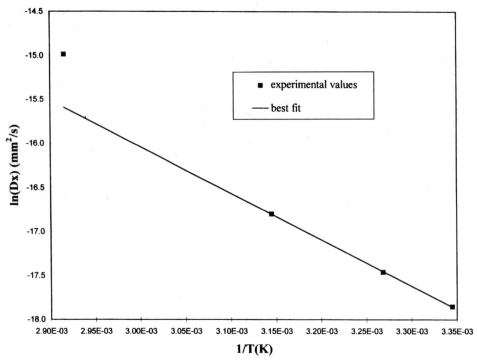

FIG. 2—*Arrhenius plot for diffusion coefficients.*

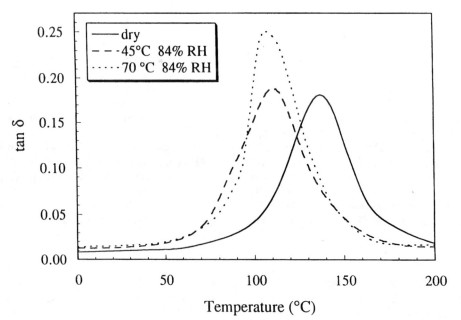

FIG. 3—*DMT Analysis: tan(δ) curves.*

TABLE 1—*DMTA results.*

Condition	$\tan(\delta)$ $(T = 70°C)$	T_g^* (°C)	$T_g^°$ (°C)
Dry	0.020	113	138
Saturated 45°C/84% RH	0.035	80	111
Saturated 70°C/84% RH	0.039	81	108

* T_g taken as the onset of the E' inflection curve.
° T_g taken as the maximum of the $\tan(\delta)$ peak.

2. 45° ply damage and edge delaminations.
3. Growth of the edge delaminations until:
4. Failure of the specimen with 0° ply "burst."

Figure 4 shows an example of the stress-strain curve. Let $\varepsilon(1)$ and $\sigma(1)$ be the strain and the stress at which the first step of the damage sequences takes place. Analogously it could be defined as $\varepsilon(2)$ and $\sigma(2)$ for the second step. Let E_{11} be the initial tangent modulus, ϵ_R the elongation at failure, and σ_R the corresponding strength. The results for tests carried out at RT are summarized in Table 2.

For specimens aged at 70°C/84% RH, it is worth mentioning that ply damage is not well displayed and the ±45° delamination does not propagate.

Fatigue Tests

Fatigue loading provokes *intra*laminar and *inter*laminar fracture that decreases the rigidity of the laminate. During the fatigue tests, the damage evolution follows the same sequence of the static test, that is:

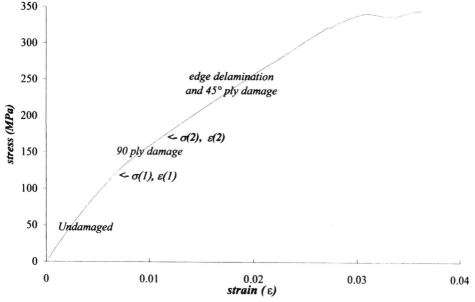

FIG. 4—*Typical stress-strain static test curve.*

TABLE 2—*Synthesis of static tests results (shown in parentheses is the percent standard deviation). The values are the means of three test results (8 for E_{11} and σ_R of the dry case).*

Condition	E_{11}(GPa)	$\epsilon(1)$	$\sigma(1)$ (MPa)	$\epsilon(2)$	$\sigma(2)$ (MPa)	ϵR	σR (MPa)
Dry	19.8	0.0065	130	0.0111	160	0.0310	340
	(3.7%)	(2.6%)	(3.1%)	(4.0%)	(5.0%)	(7.0%)	(2.4)
Saturated at	18.6	0.0170	224	0.0215	258	0.0250	296
45°C/84%	(1.0%)	(8.5%)	(4%)	(2%)	(3%)	(4.4%)	(2.8%)
RH							
Saturated at	18.7	0.0100	152	0.0150	210	0.0240	285
70°C/84%	(2.4%)	(5%)	(3%)	(3%)	(3%)	(2.2%)	(1.0%)
RH							

1. 90° ply damage.
2. 45° ply damage and edge delaminations.
3. Increasing of edge delaminations until:
4. Failure of the specimen.

It is noteworthy that the volume of damaged composites seems to be higher in the two wet cases than in dry specimens. As usual the number-of-cycles-to-failure depends on the applied load. Figure 5 shows the maximum sigma versus log(number of cycles) curve for the three cases under examination. Fatigue lives higher than 10^6 cycles are considered as infinite lives, and after this the residual static strength and the moduli were assessed. Test results are reported in Table 3. From this table it seems that the 45°C/84% RH aging has

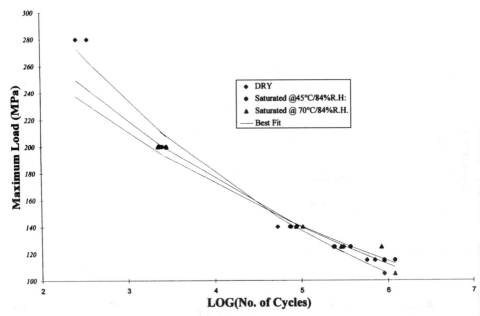

FIG. 5—*Maximum load versus logarithm of (number of cycles to failure) curves for fatigue tests.*

TABLE 3—*Residual static strength after 10^6 fatigue cycles: mean of three values. Shown in parentheses is the percent standard deviation.*

Aging Condition	Residual Static Strength (MPa)	E_{11} (GPa)	Deformation at Failure
Dry	255 (2.5%)	17.4 (2.0%)	0.0310
Saturated 45°C/84% RH	184 (5.0%)	12.9 (1.8%)	0.0250
Saturated 70°C/84% RH	278 (2.0%)	15.9 (1.7%)	0.0240

more marked effects than the 70°C/84% RH aging. It is remarkable that even if the moisture contents are different, the fatigue behavior of the two hot-wet cases is very similar. Moreover, the absorbed moisture increases slightly or, at least, does not modify (the data are not enough to verify which is the case) the behavior at low loads, which, for convenience, shall be called the fatigue endurance.

Figure 6 shows the loss of rigidity (E) during the fatigue test. Experimental points are well fitted by the linear expression: E versus the logarithm of the number of cycles to failure (N).

For comparative purposes it is worth considering the relative loss of rigidity, defined as: $E_r = E(N)/E(N = 0)$. The data and the best-fit lines are shown in Fig. 7. The regression model used is: $E_r = a_0 + a_1 \cdot \ln(N)$. The three best-fit lines shown in Fig. 7, which correspond to the three aging conditions, are approximately parallel. Moreover, it is clear that the E_r trend for the two "wet" cases is very similar: the two sets of points seem to lie over the same line. In addition, as expected, from Fig. 7 it can be seen that the relative loss of rigidity is higher in the dry case with respect to the wet cases.

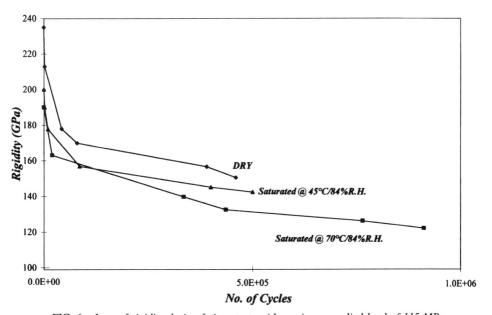

FIG. 6—*Loss of rigidity during fatigue tests with maximum applied load of 115 MPa.*

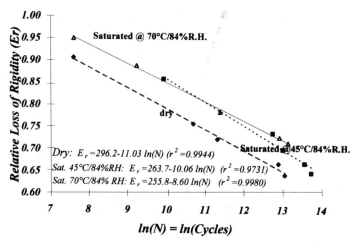

FIG. 7—*Relative loss of rigidity versus logarithm of (number of cycles to failure) lines.*

Discussion

From the point of view of the static properties, it is clear that the moisture uptake has a great effect on both rigidity and strength. Moreover, notwithstanding the higher saturation level, the more drastic conditioning does not have marked effects on the static performances compared with standard aging. On the other hand, the hot/wet (both types) aging seems to improve the fatigue life at low load or, at least, the behavior is not modified (we have not enough data to distinguish which is the case). Then, in this case, the absorbed moisture seems to have an opposite effect with respect to the static results. At high load the same trend as static tests is followed.

These facts could be explained by supposing that:

1. The moisture reduces the performances of the resin and/or resin-fiber interface in terms of strength and of elastic properties. Moreover, it is known that the moisture decreases the strength of the glass fiber [7].
2. In the case of accelerated aging (70%C/84% RH), part of the water molecules or molecular clusters in excess with respect to standard aging are probably localized in microcavities and are less active than the remaining molecules.

The reduction of static strength generated by aging could be explained by the first hypothesis. The modulus reduction can be understood on this basis also.

The observation that $\tan(\delta)$ increases with humidity absorption could explain the augmented "plasticity" of the resin itself, which results in a different static failure sequence (for example, ply damage is not well displayed). The similarity of the behavior of the wet cases in all the experimental determinations (static properties: i.e., E_{11}, σ_R) can be explained on the basis of the second assumption.

As far as the endurance of the wet specimens submitted to fatigue tests is concerned, it could be stated that part of the transferred energy during a fatigue loading-unloading cycle is irreversibly dissipated in heat instead of increasing the microcracks. This was clearly shown by Eq. 10 in the "Background" section. Moreover, it is easy to understand that this effect is more important at high N values (low loads) because the total dissipated energy is higher. In the other case (high loads) it could be masked by the increased moisture damage (similar to the static case). In addition, from Fig. 3 it can be seen that tan(δ) at room temperature (the temperature at which the fatigue tests were carried out) is very similar for both the humidity saturated situations, but is higher in the dry case. This could explain the similarity in the dynamic behavior. It is easy to understand the fact that, on the same basis, the relative loss of rigidity is higher in the dry case with respect to the wet cases.

Conclusions

In this paper the effects of two kinds of hot-wet accelerated aging are compared. It has been found that the more drastic the conditioning, the higher the moisture saturation level. As expected, the moisture absorption reduces the performances of the material (elastic properties and T_g).

The different water content induces only a slight variation in the glass transition temperatures. Only the static strength seems to be a little lower after conditioning at the higher temperature. The two kinds of aging seem to have similar effects on the fatigue properties. In addition, neither of the hot-wet aging processes decreases the endurance with respect to the dry situation.

Acknowledgments

The authors wish to thank Mr. S. Risetti for performing much of the experimental work, Mr. V. Pagani for his support, and Mr. C. Zanotti for his encouragement and helpful comments.

References

[1] Maymon, G., Briley, R. P., and Rehfield, L. W., Influence of Moisture Absorption and Elevated Temperature on the Dynamic Behavior of Resin Matrix Composites: Preliminary Results, ASTM STP 658, American Society for Testing and Materials, 1978, p. 221.
[2] Hahn, H. T. and Kim, R. Y., Swelling of Composite Laminates, ASTM STP 658, American Society for Testing and Materials, 1978, p. 98.
[3] Springer, G. S., Ed., Environmental Effects on Composite Materials, Vols. 1 and 2, Technomic Publications, Westport, VA, 1981.
[4] Hopfemberg, H. B. and Frisch, H. L., Journal of Polymer Science, Part B7, 1969, p. 405.
[5] Burns, S. J., "Application of Thermodynamic to Fracture," Environmental Degradation of Engineering Materials in Aggressive Environment, Proceedings of the Second International Conference on Environmental Degradation of Engineering Materials, Blacksburg VA, September 1981.
[6] Cappelletti, C., Rivolta, A., and Zaffaroni, G., "Environmental Effects on Mechanical Properties of Thick Composite Structural Elements," Journal of Composites Technology & Research, Vol. 17, No. 2, April 1995, p. 107.
[7] Antoon, M. K. and Koenig, J. L., "The Structure and Moisture Stability of the Matrix Phase in Glass-Reinforced Epoxy Composites," Journal of Macromolecular Science-Review Macromolecular Chemistry, Vol. C19, No. 1, 1980, pp. 135–173.

Impact

Alan T. Nettles[1]

The Effects of Tensile Preloads on the Impact Response of Carbon/Epoxy Laminates

REFERENCE: Nettles, A. T., **"The Effects of Tensile Preloads on the Impact Response of Carbon/Epoxy Laminates,"** *Composite Materials: Fatigue and Fracture, Seventh Volume, ASTM STP 1330,* R. B. Bucinell, Ed., American Society for Testing and Materials, 1998, pp. 249–262.

ABSTRACT: Low-velocity instrumented dropweight impact tests were performed on carbon/epoxy laminates. The composite plates were 8-ply $(+45, 0, -45, 90)_s$ laminates supported in a clamped-clamped/free-free configuration with varying amounts of in-plane load, N_x, applied. The amount of damage induced into the specimen was evaluated using the instrumented impact data and X-ray inspection. Results showed that for a given impact energy level, more damage was induced into the specimen as the external in-plane load, N_x, was increased. The majority of damage observed consisted of backface splitting of the matrix parallel to the fibers in that ply, associated with delaminations emanating from these splits. A free-edge delamination model was used to explain the type and extent of the major delaminations caused by the preload/impact combinations.

KEYWORDS: composites, damage resistance, preload, instrumented impact, toughened epoxy, nondestructive evaluation, delamination

Foreign-object impact damage to laminated fiber-reinforced plastics is a loading condition in which many complex failure modes are present. Many studies, both analytical and experimental, have been performed in the past few decades. In most of these studies, the composite laminates were free from external stresses and deformations in the initial state before the impact event. In practice, a component is frequently under some form of external load (a preload or prestrain) just before a foreign-object impact event occurs. For example, a structure such as a pressure vessel may be under a tensile load (hoop stresses) during an impact event, which may change the resulting behavioral response of the composite laminate that makes up the vessel. A composite laminate may be under a state of compressive, shear, or tensile preload (or any combination of the three) that may alter the laminate's response to a transverse impact event. In order to maintain a well-defined scope, this paper addresses only the case of tensile preloads. The objective of this study is to examine the effects of a tensile preload on the impact response of carbon/epoxy laminates. The responses examined are maximum plate deflection, maximum load of impact, absorbed energy, total time of the impact event, and amount of damage formed as determined by radiography. Tensile prestrains between 0 and 8000 μstrain were applied to rectangular 8-ply laminates in a clamped-clamped/free-free support configuration, and three levels of impact energy were used to evaluate the changes in the measured responses as a result of the changing prestrains. This

[1] Materials research engineer, NASA Langley Research Center, Hampton, VA 23681.

is of interest because a tensile preload can cause a stiffening effect on the laminate being impacted and thus change its mechanical response.

A review of past investigations on this subject is given followed by a description of the material and testing techniques used in this study. An energy balance model of the problem is presented, and then experimental data on the maximum plate deflection, maximum load of impact, absorbed energy, total time of the impact event, and amount of damage formed are given followed by some concluding remarks.

Review of Past Studies

Olster and Roy [1] performed ballistic penetration tests on tensile preloaded graphite/ epoxy plates and found that the combination of impact and preload could cause a 45% reduction in ultimate tensile strength, whereas penetrated plates without a preload would fail at approximately a 38% reduction in ultimate tensile strength. This suggests a slight inter- action between the preload stresses and the stresses set up by the impact event.

For nonpenetrating impacts on preloaded specimens, Avva [2–5] has produced the most experimental data. Typically, a "failure threshold curve" is obtained by plotting specimen preload (tension or compression) versus impactor kinetic energy and drawing a faired curve between the data in which the specimens broke upon impact (a "catastrophic" break) and those which survive the impact. For the data presented on tensile loading, a faired curve could easily be drawn between regions of catastrophic and noncatastrophic failures as seen in Fig. 1. For the specimens that did not experience catastrophic failures, the residual tensile strengths were much above the catastrophic failure curve. Specimens that were impacted while under no preload also demonstrated a residual strength versus impact energy curve higher than the catastrophic failure curve. The residual strength data, from the preloaded specimens that did not catastrophically fail, fall close to the no preload residual strength curve. An additional set of tests was conducted using the catastrophic failure curve as a guide for setting preloads at three different impact energies. All of these specimens were preloaded in the region of the catastrophic failure curve. Most of the specimens failed cat- astrophically upon impact, and those that did not fail had very little residual strength re- maining. These results are sketched out graphically in Fig. 1 for clarity and comparison. Only pertinent data points are included to reduce clutter in the plots. Thus from this study it appears that a tensile preload can reduce the residual strength-carrying capability of the impacted laminate, but only when preloaded near the "catastrophic failure" curve.

A C-scan analysis of impacted beams under a tensile preload was conducted by Sankar and Sun [6]. Although only one initial preload value was used (about one third of ultimate load), the results did show differences in those beams impacted with an initial tensile stress and those impacted with no initial stress. The beams were 20-ply graphite/epoxy with a stacking sequence of $[0_2,90_2,0_2,90_2,0_2]_S$ supported as clamped-clamped/free-free. A 1.27-cm (0.5-in.)-diameter steel ball was used as a projectile at speeds of 10 to 40 m/s (33 to 131 ft/s). The span of the impacted beams was set at 18 cm (7 in.). The width of the beams was 3.8 cm (1.5 in.). The relatively high aspect ratio for these plates could possibly allow more global deflection than the specimens impacted in the previously cited studies even though ballistic projectiles were used (which tends to minimize global deflections). Results of note are that larger zones of delamination were found to exist for those specimens impacted while preloaded. The only exception to this trend was a reduction in delamination area near the back plies on preloaded specimens for larger energy impacts. Residual tensile strength was also determined for these specimens. It is evident that the tensile preload causes a reduction in residual tensile strength by approximately 10% for the impact velocities used in this study.

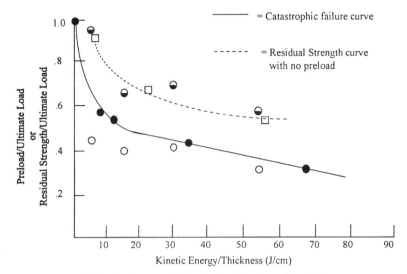

● — Catastrophic Failure
○ — Survived Impact
◐ — Residual Strength
□ — Residual Strength with no Preload

——————— = Catastrophic failure curve

- - - - - - = Residual Strength curve
with no preload

FIG. 1—*Representative tension data from Avva* [3].

Analytical treatments of this problem have been undertaken in a few studies [7–10]. In all of these studies it was shown that a tensile prestress will decrease the maximum deflection due to impact and raise the maximum load of impact when no damage is present. Compressive preloads will do the opposite.

Experimental Details

Material and Specimens

The fiber/resin system used for this study was Hercules' IM7™ fiber in Fiberite's 977-2™ toughened epoxy. The lay-up configuration chosen for the impact tests was quasi-isotropic [+45,0,−45,90]$_S$. This configuration is representative of the basic lay-up unit being used for many composite structures. The laminates were processed from unidirectional pre-preg tape as flat panels using a hot press. The prepreg tape was of a 60% fiber volume fraction no-bleed variety. The resulting panels had a nominal ply thickness of approximately 0.12 mm (0.0047 in.), giving an overall laminate thickness of 0.0955 mm (0.0376 in.). Specimens 9.5 cm (3.75 in.) wide and 29.8 cm (11.75 in.) long were cut from these panels. The instrumentation to be used on the specimens to assure a uniform preload across their width was determined from a series of preliminary tests on specimens that contained numerous strain gages on their surfaces. From these tests it was found that bending was never introduced into the specimen, so gages need only be placed on the same surface. In addition it was found that the strain remained constant along the longitudinal axis of the specimen and the only variations in strain were across the width of the specimen. As long as the outermost gages (1.3 cm (0.5 in.) from each edge) had the same strain, the ones between them also had this same strain; thus, it was determined that only the two gages nearest the edges and on the same side need to be used to assure that a uniform strain field was being set up in the specimen.

Pre-Tensioning Device

The device to apply tensile preloads to the specimens basically consisted of two wedge grips connected to a load cell and a hydraulic piston; this is schematically presented in Fig. 2. As the piston was pumped with hydraulic fluid, it would cause the grips to move apart, applying tensile load to the composite specimen.

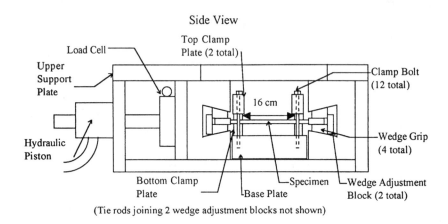

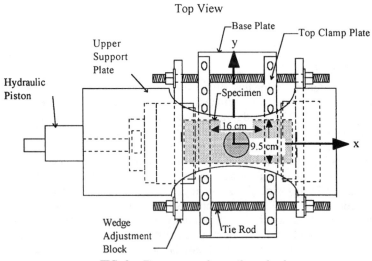

FIG. 2—*Fixture to apply tensile preloads.*

The wedge adjustment blocks were first tightened together to lodge the specimen in the grips. These blocks also prevented the wedges from moving apart (and thus stop gripping) during the application of the preload. As the preload was applied, the strain gages were monitored. When one gage deviated from the other, that side reading the lower strain could have its adjustment blocks loosened, allowing the specimen to expand slightly along that side and thus causing the strain to rise on that side. These adjustments were made until the final level of strain was reached and the adjustment blocks could be "fine tuned" to give a nearly uniform prestrain across the width of the specimen. Thus a fixed displacement was achieved at the ends of the specimen, resulting in the preload and prestrain.

Once the level of prestrain was achieved, the top clamp plates were tightened down with the twelve bolts (see Fig. 2), each at a torque of 28 N-m (250 in.-lb). This clamping gives the clamped-clamped/free-free boundary condition to the specimens with a span of 16 cm (6.4 in.). Measurements of the strain before and after impact indicated that the specimen did not move significantly in the clamps for high levels of prestrain. At low levels of prestrain, the overall strain was seen to drop after the impact event, thus indicating some lateral movement of the plate in the clamps. The high torque on the clamp-down bolts assured that no bending moments could occur under the clamps.

Impact Testing

Once the desired level of prestrain was achieved, the preloading device was positioned under a Dynatup 8200 drop tower. The height of the crosshead was adjusted to give the desired impact energy. In this study, the mass of the crosshead with tup was 2.4 kg (5.22 lb). Once a desired drop height was decided upon, all impact tests at that corresponding impact energy level were conducted before the crosshead was moved again. This assured uniformity in the impact energy levels used. A total of three impact energies were used in this study: 3.4, 4.5, and 6.0 J (2.5, 3.3, and 4.4 ft-lb). These levels induced damage from nonvisible to clearly visible fiber breakage. Data acquisition was performed using Dynatup software and a personal computer.

Post-Impact Testing

In order to assess the effects of tensile preloads on the damage resistance of the specimens tested, two forms of post-impact inspection were used. The first type of inspection was a visual one to see if any damage could be detected from either surface of the plate. This was followed by a nondestructive X-ray inspection.

After each specimen was impacted, it was carefully removed from the preload device and inspected for detectable damage. No magnifying devices were used; however, if damage could be felt and not seen, this was recorded as detectable damage since no special tools were needed to identify this damage. The extent and type of detectable damage was recorded for each impacted specimen. The type of damage that could be detected consisted of backface (nonimpacted side) matrix splitting, front-face (impacted side) indentation damage, and fiber fracture on the front and back faces after the more severe impacts.

For specimens that did contain detectable damage, a circular dam of plumber's putty was placed around the damage on one of the faces to hold in a zinc iodide (ZnI) dye penetrant solution that is opaque to X-rays. For specimens that did not contain observable damage, a small hole of ~0.5 mm (0.02 in.) diameter was drilled through the specimen at the impact site to allow the dye penetrant to be exposed to any through-the-thickness damage that may have occurred within the laminate. The specimens were wiped clean of the plumber's putty and excess penetrant after at least a 24-h soak and exposed to X-rays. The recording film was placed directly under the specimen to obtain a permanent copy of the X-ray signature of each specimen. The dye penetrant was observed to seep into all matrix delaminations and cracks and was seen easily on the developed film.

Results and Discussion

In this section the experimental results obtained for effects of a tensile preload on the visual damage, maximum plate deflection, maximum load of impact, absorbed energy, total time of the impact event, and amount of damage formed are presented. Load-deflection curves from the instrumented impact data and X-ray detection of damage due to the impact event will be discussed.

Visual Examination

In general, as the preload increased, the amount of visual damage also increased. The first sign of visible damage was always splitting of the matrix between fibers on the back 45° face. A summary of the detectable damage is given in Table 1. Note that contact stresses produced visible damage at the impact site on specimens impacted at 6.0 J (4.4 ft-lb) and

TABLE 1—*Detectable damage to specimens tested.*

Impact Energy, J	Prestrain, $\mu\varepsilon$	Detectable Damage
3.4	400	None
3.4	987	None
3.4	2045	None
3.4	3387	None
3.4	4334	19-mm split on back face
3.4	5012	26-mm split on back face
3.4	6060	24-mm split on back face
3.4	6972	27-mm split on back face
4.5	452	18-mm split on back face
4.5	833	17-mm split on back face
4.5	1960	18-mm split on back face
4.5	2430	17-mm split on back face
4.5	3229	None
4.5	4205	None
4.5	4268	30-mm split on back face
4.5	5182	22-mm split on back face
4.5	5564	47-mm split on back face
4.5	6698	60-mm split on back face. Dent with small split on front
4.5	7399	116-mm split on back face. Dent with short split on front
4.5	7970	65-mm split on back face. Dent with small split on front
6.0	498	21-mm split on back face
6.0	634	19-mm split on back face
6.0	1594	20-mm split on back face
6.0	2132	18-mm split on back face
6.0	3053	53-mm split on back face
6.0	4156	45-mm split on back face. Dent with small split on front
6.0	5384	Split across width on back face. Dent with small split on front
6.0	7173	Split across width on back face. Dent with splits on front

at a preload above 3000 με and on the specimens impacted at 4.5 J (3.3 ft-lb) above 6600 με.

Maximum Deflection

The maximum deflection of impact decreases with increasing tensile prestrain. Plots of the measured data are shown in Fig. 3*a*. These results appear to be unaltered even when damage is present. From the experimental data, all of the specimens impacted at 6.0 J (4.4 ft-lb) sustained damage with front-face contact force damage (dents) occurring in the highly preloaded specimens, yet the maximum deflection versus prestrain plot decreases at a fairly linear rate for all of the values of prestrain tested.

Maximum Load of Impact

The maximum load of impact results for each of the three impact levels used are shown in Fig. 3*b*. In general, as the prestrain increases, the maximum load of impact increases. However, at higher levels of prestrain (~3000 με for specimens impacted at 6.0 J and ~6000 με for specimens impacted at 4.5 J), the maximum load of impact tends to drop or remain about the same with increasing prestrain since the larger prestrain causes significant damage

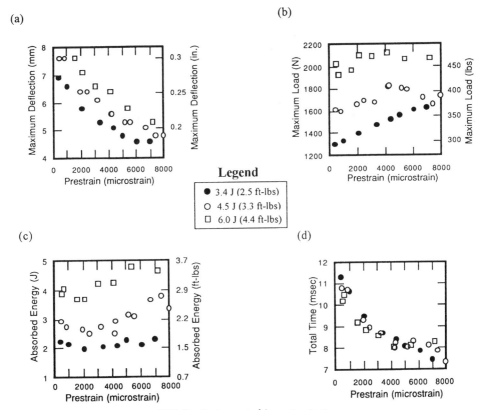

FIG. 3—*Instrumented impact output.*

to form during the impact event. This causes a "softening" effect on the laminate, which induces smaller maximum transverse loads.

Absorbed Energy of Impact

The amount of energy lost during the impact event is termed "absorbed energy." The absorbed energy versus prestrain plots are given in Fig. 3c. For low prestrains the absorbed energy is seen to decrease with increasing prestrain up to a critical level (~2000 $\mu\varepsilon$ for all levels of impact), at which point it begins to rise. This is thought to be a result of compliance in the preloading device. Since a true clamped condition was not fully achieved at low prestrains, the specimen could displace a given amount in the x-direction in the fixture, thus losing a given amount of energy. As the prestrain increased, the grips tended to help hold the specimen in a true clamped configuration and thus less energy was lost due to this movement of the laminate being tested. For prestrain above ~2000 $\mu\varepsilon$, it can be seen that for the two highest impact energies, as the preload increased, the absorbed energy increased. This is due to more damage forming in the specimen at the higher prestrains. It should be noted that "absorbed energy" is not equal to "energy due to damage" since damping of the specimen and vibrations in the impact apparatus are sources of "lost" energy. This can be seen from the results in Fig. 3c, where a significant amount of absorbed energy is recorded for the 3.4-J (2.5-ft-lb) energy level, yet little damage is inflicted upon the specimen, especially at the low prestrains.

Duration of Impact

Figure 3d shows that as the tensile prestrain increased, the duration of the impact event decreased. Since the maximum deflection decreases with increasing prestrain, it is expected that the duration of impact should follow the same trend.

Assuming a simple spring-mass model of the impact event where the transverse force is proportional to the transverse displacement (i.e., $P \propto w$), the resulting equation of motion in the transverse direction is

$$w(t) = w_{max} \sin\left[\left(\sqrt{\frac{P_{max}}{w_{max}\, m}}\right)t\right] \qquad (1)$$

where

$w(t)$ = transverse displacement at time t,
w_{max} = maximum transverse displacement,
P_{max} = maximum load of impact, and
m = mass of impactor.

This implies that the time to maximum deflection (t_{max}) can be obtained by

$$\left(\sqrt{\frac{P_{max}}{w_{max}\, m}}\right)t_{max} = \frac{\pi}{2} \qquad (2)$$

and hence

$$t_{max} = \frac{\pi}{2}\sqrt{\frac{w_{max}\, m}{P_{max}}} \qquad (3)$$

From the spring-mass-model, the total energy put into the system is

$$E_{tot} = \frac{1}{2} P_{max}\, w_{max} + E_{dam} = \text{impact energy} \qquad (4)$$

Putting Eq 4 into Eq 3 gives

$$t_{max} = \frac{\pi}{2}\sqrt{\frac{(E_{tot} - E_{dam})m}{(P_{max})^2}} \qquad (5)$$

Since E_{dam} is small compared with E_{tot}, from this equation it can be seen that as the impact energy (E_{tot}) increases and as the maximum load of impact (P_{max}) also increases (as it does with increasing impact energy) that the total time of impact will not vary much with impact energy. This is evident from Fig. 3d, in which the data are fairly independent of impact energy. As damage is induced into the specimen, P_{max} becomes smaller, thus causing the total time of impact to not decrease as rapidly. This can be seen in Fig. 3d, where the data begin to increase for the highest impact energy at high preloads where significant damage is formed.

Load-Deflection Curves

For all of the specimens tested, the load-deflection plots show that a significant amount of energy is lost during the impact event, even at the lowest-impact energy level used. Typical load/deflection curves are shown in Fig. 4 for the lowest level impact at the smallest prestrain and for the highest level impact at the highest prestrain. These two plots represent the two extreme cases of impact tested in this study. Between 60 to 80% of the incident impact energy was measured as "absorbed" for all of the specimens tested, even though some of these specimens had very little or no detectable damage from X-ray examination. It must be

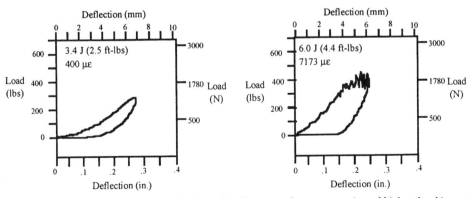

FIG. 4—*Load/displacement curves for lowest level impact at lowest prestrain and highest level impact at highest prestrain.*

concluded that most of the energy is lost to mechanisms other than damage. This has been noted elsewhere [*11*]. Thus the absorbed energy data can only be used for qualitative and not quantitative purposes.

X-Ray Inspection

As expected, the specimens that contained more visible damage also contained more internal damage, mostly in the form of matrix splitting and delaminations. The specimens that had no externally visible damage did demonstrate that some delaminations occurred (the small drilled hole allowed the penetrant to reach these areas). These delamination zones were relatively small compared to the specimens that had visual damage associated with them. For the specimens that did contain externally visible damage, the matrix crack on the back 45° face was clearly visible on the X-rays and always had delaminations emanating from the crack. These delaminations tended to form a lobe shape.

The effects of the preload on the resulting impact damage can be seen in Figs. 5–7. These comparisons show X-ray signatures of specimens impacted at equal incident energy, but high and low preloads. The higher prestrains clearly induced more damage into the specimens, especially in respect to the length of the crack on the back 45° face. This crack was consistently seen to grow by a factor of 4 or more from the lowest prestrain to the highest. This is not to be totally unexpected if the stress from the preload is superimposed upon the impact stresses.

The characteristic shape of the induced damage consistently was of the form shown in Fig. 8*a*, 8*b*. The bottom ply split was the major feature of the damage, with delamination "lobes" emanating from this major crack. The lobes were generally symmetric about the 45° crack, with the lobe on a given side of the split always appearing furthest away from the loading direction. Figure 9 is a free-body diagram of the bottom ply of the plate that contains the matrix crack. Since the externally applied and impact induced x-direction stresses dominate, only these are considered (no y-direction stresses are considered). During the impact event, the x-direction stresses become larger nearer the x-location of the impact (i.e., in the center of the plate). Before the matrix crack forms, the σ_x stresses shown in Fig. 9 are balanced out by similar σ_x stresses acting on the opposite face. However, as the crack forms, the ply can no longer hold the σ_x stresses through its thickness since the surface formed must be traction free. Thus the moment about the line A-B in Fig. 9 must be balanced by

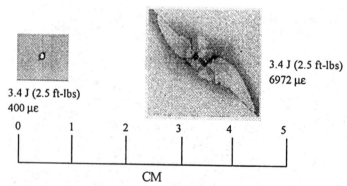

3.4 J (2.5 ft-lbs)
400 µε

3.4 J (2.5 ft-lbs)
6972 µε

0   1   2   3   4   5

CM

FIG. 5—*X-ray signatures of specimens impacted at 3.4 J with small and large prestrains.*

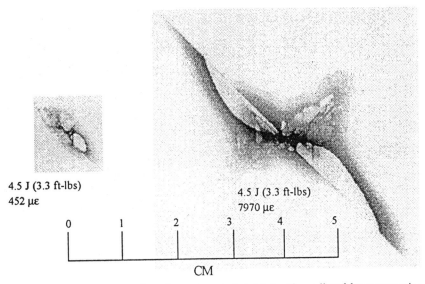

4.5 J (3.3 ft-lbs)
452 με

4.5 J (3.3 ft-lbs)
7970 με

0   1   2   3   4   5

CM

FIG. 6—*X-ray signatures of specimens impacted at 4.5 J with small and large prestrains.*

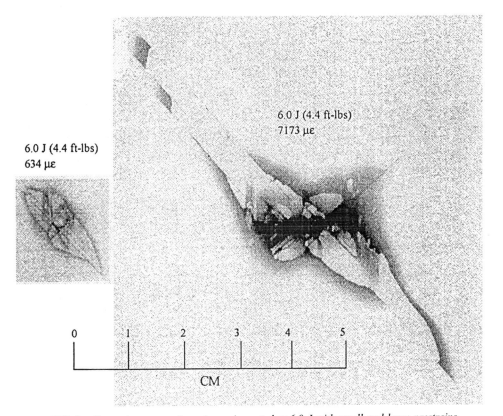

6.0 J (4.4 ft-lbs)
634 με

6.0 J (4.4 ft-lbs)
7173 με

0   1   2   3   4   5

CM

FIG. 7—*X-ray signatures of specimens impacted at 6.0 J with small and large prestrains.*

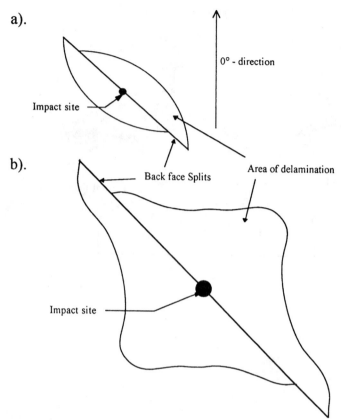

FIG. 8—*Schematic of bottom ply splitting with associated delaminations as determined by radiography: (a) low preload, (b) high preload.*

normal σ_z stresses on AC. This is analogous to the "free edge effect" outlined by Pipes and Pagano [12].

At the tip of the crack (Point C in Fig. 9) the balance in moments about the line A-B can come from the shear stress τ_{xy} along the face containing B-C since this face is still attached to material and can thus carry tractions. Moving along the crack from Point C towards Point A, the z-direction stresses rapidly become significant and large enough to peel (delaminate) the bottom ply away from the ply adjacent to it. This area of peeling is represented by the dashed arc in Fig. 9. As the center of the plate is approached, the z-direction stresses rise rapidly since the x-direction stresses rise towards the center of the plate and the moment arm decreases in this direction. As the preload increases, the x-direction stresses become quite a bit larger, causing the z-direction stresses to increase very rapidly to produce an area of peeling (delamination) such as that shown in Fig. 8b. At some point, the moments about A-B become balanced and the z-direction stresses decrease as Point A is approached. It should be noted that at a distance away from the crack, compressive z-direction stresses are set up to balance the tensile z-direction stresses that caused the balance of moments about A-B. The tensile/compressive distribution of z-direction stresses is similar to that at a free edge due to in-plane tensile loads [12].

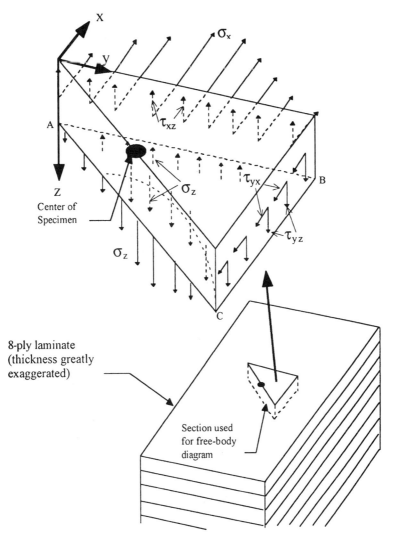

FIG. 9—*Free-body diagram of bottom ply of the plate with a 45° matrix crack.*

Concluding Remarks

While most studies of the impact response of composite laminates deal with unstressed specimens, this study examined the impact response of stressed plates. More specifically, the effects of a tensile preload on the maximum plate deflection, maximum load of impact, absorbed energy, total time of the impact event, and amount of damage formed were investigated.

For other impact parameters, a tensile prestress tends to increase the maximum force of impact until damage forms. The superposition of stresses causes damage to be formed, mostly in the form of a longer damage zone. As the preload becomes larger, a decrease is seen in the duration of impact and in the maximum transverse deflection due to impact. The dominant type of damage formed is matrix cracking (splitting) along the back ply with delaminations due to normal tensile stresses σ_z that are set up along the edge of the matrix cracking. The induced midplane stretching from the large deflections dominates the load/displacement response of the plate.

Acknowledgments

This work was supported by a grant from the Marshall Space Flight Center Director's Discretionary Fund. The help of Dr. Ronald B. Bucinell of Union College is greatly appreciated.

References

[1] Olster, E. F. and Roy, P. A., "Tolerance of Advanced Composites to Ballistic Damage," *Composite Materials: Testing and Design (Third Conference)*, ASTM STP 546, American Society for Testing and Materials, 1974, pp. 583–603.

[2] Sharma, A. V., "Low-Velocity Impact Tests on Fibrous Composite Sandwich Structures," *Test Methods and Design Allowables for Fibrous Composites*, ASTM STP 734, American Society for Testing and Materials, 1981, pp. 54–70.

[3] Avva, V. S., "Impact Initiated Damage in Laminated Composites," Air Force Office of Scientific Research Technical Report 82-1038, 1982.

[4] Avva, V. S., "Fatigue/Impact Studies on Laminated Composites," Air Force Wright Aeronautical Laboratories Technical Report 83-3060, May 1983.

[5] Avva, V. S., Vala, J. R., and Jeyaseelan, M., "Effect of Impact and Fatigue Loads on the Strength of Graphite/Epoxy Composites," *Composite Materials: Testing and Design (Seventh Conference)*, ASTM STP 893, American Society for Testing and Materials, 1986, pp. 196–206.

[6] Sankar, B. V. and Sun, C. T., "Low-Velocity Impact Damage in Graphite-Epoxy Laminates Subjected to Tensile Initial Stresses," *AIAA Journal*, Vol. 24, No. 3, 1985, pp. 470–471.

[7] Sun, C. T. and Chen, J. K., "On the Impact of Initially Stressed Composite Laminates," *Journal of Composite Materials*, Vol. 19, November, 1985, pp. 490–504.

[8] Chen, J. K. and Sun, C. T., "Dynamic Large Deflection Response of Composite Laminates Subjected to Impact," *Composite Structures*, Vol. 4, 1985, pp. 59–73.

[9] Sun, C. T. and Chattopadhyay, S., "Dynamic Response of Anisotropic Laminated Plates Under Initial Stress to Impact of a Mass," *Journal of Applied Mechanics*, September 1975, pp. 693–698.

[10] Bert, C. W. and Birman, V., "Response of Prestressed Cylindrically Curved Composite Structures Subjected to Low-Velocity Impact," *Proceedings of the Fourth Japan-U.S. Conference on Composite Materials*, American Society for Composites, 1988, pp. 43–52.

[11] Nettles, A. T., "Instrumented Impact and Residual Tensile Strength Testing of Eight-Ply Carbon/Epoxy Specimens," NASA Technical Paper 2981, National Aeronautics and Space Administration, 1990.

[12] Pipes, R. B. and Pagano, N. J., "Interlaminar Stresses in Composite Laminates Under Uniform Axial Extension," *Journal of Composite Materials*, Vol. 4, October 1970, pp. 538–548.

Joel E. Patterson[1]

Differences in the Impact Response Mechanisms of Graphite/Epoxy Composite Cylinders

REFERENCE: Patterson, J. E., **"Differences in the Impact Response Mechanisms of Graphite/Epoxy Composite Cylinders,"** *Composite Materials: Fatigue and Fracture, Seventh Volume, ASTM STP 1330,* R. B. Bucinell, Ed., American Society for Testing and Materials, 1998, pp. 263–272.

ABSTRACT: Impact damage to composites has been identified as a major limiting factor in the wider use of composite materials as structural components. Graphite fiber is often selected as a structural fiber due to its low density and high strength, both desirable properties in rocket motorcase and launch tube design. An understanding of the damage mechanisms involved is critical in the determination of composite structural integrity and for the development of damage models and design codes. To this end, a test program was conducted to characterize the impact response of graphite/epoxy structures. Overall design of the test article utilized for this program was directed towards a generic thin-walled structure applicable for use as either a rocket motorcase or launch tube. Low-energy impacts between 1.25 and 7.75 J (0.9 and 5.6 ft-lb) were imparted to empty cylinders and to cylinders whose casewalls were strengthened to simulate launch tube and rocket motorcase configurations. Significant differences between the test configurations were identified with regard to visual damage, impact loads, absorbed energy, and casewall deflection. However, with all the readily apparent differences noted, the post-impact residual strength capability was quite similar. The data indicate that significant strength losses can be expected at impact energies as low as 1.3 J (0.94 ft-lb).

KEYWORDS: composites, low-energy impact, graphite/epoxy, residual strength, filament wound, rocket motor case, launch tube, impact damage

Graphite/epoxy materials have become the composites of choice in rocket motorcase and launch tube design because of their high strength-to-weight ratio. Lighter weight means improved missile performance with regard to increased flight range and/or payload capacity. However, low-energy impact has been identified as a major threat to the structural integrity of composite systems [1–5]. Tactical situations demand impact-tolerant structures not only to ensure operational capability but also to prevent possible harm to the missile platform or operator. An understanding of the damage mechanisms involved is critical in the determination of composite structural integrity and for the development of damage models and design codes. Development of impact-tolerant structures is complicated by the very nature of composites themselves. A multitude of different fiber types can be utilized with a myriad assortment of resin matrix systems. The structure, with alternating lay-ups and plies, further complicates the effort to develop impact models and codes. One often-overlooked aspect of the process lies in the end-use application of the structure. A critical impact for a motorcase

---

[1] Aerospace engineer, U.S. Army Aviation and Missile Command, AMSMI-RD-PS-CM, Redstone Arsenal, AL 35898.

TABLE 1—*Test vehicle geometry.*

Parameter	Value
Inner diameter	12.7 cm
Total length	31.75cm
Test length	25.4 cm
Volume	3212 cm^3

(case with propellant backing) may not be the same for a launch tube and vice versa. In the study presented here, a baseline graphite/epoxy cylindrical structure was designed to simulate a generic rocket motorcase or launch tube. Controlled levels of impact were induced to the baseline empty cylinder, to cylinders filled with an inert propellant grain, and to cylinders whose casewalls were stiffened by interior rings. While some studies [6,7] have investigated the effect of casewall backing on the residual strength of impacted composites, casewall thickness and associated impact energies involved were measurably greater than what is presented in this report.

Program Specifics

The Test Article

The baseline test vehicle was a filament-wound graphite/epoxy open-ended cylinder. Its nominal dimensions were 12.7-cm inside diameter, a length of 31.75 cm, wall thickness of 1.4 mm, and a weight of 276 g. The wind pattern consisted of two 30° helical layers (where the winding advances along a helical path) and two hoop layers (where the fibers run parallel at near 90° to the component centerline). The sequence was XOXO, where X symbolizes a helical layer and O represents a hoop layer. The cylinder was designed for a minimum burst strength of approximately 30.1 MPa based on a maximum expected operating pressure (MEOP) of 20.6 MPa and a 1.5 safety factor. Tables 1 and 2 present design criteria and parameters.

Test Configuration

As stated previously, three specimen configurations were utilized in this study, the first being the baseline test vehicle described above. This first configuration, identified as baseline, was impacted and tested as an empty cylinder. the other configuration types employed in-

TABLE 2—*Test vehicle design.*

Design Parameter	Criteria
Fiber/matrix	Hercules IM-6 Graphite/Epon 826
Fabrication method	Filament wound
Fiber lay-up	XOXO^a (30,90,30,90)
Wall thickness	1.39 mm
Design MEOP	20.6 MPa
Design burst	31.1 MPa

^{a} NOTE—X denotes helical wrap, O denotes hoop wrap.

ternal devices to increase stiffness along the cylinder length. In order to simulate a motorcase configuration, an inert propellant grain was cast to serve as a casewall backing medium. As it was not possible to cast more than one grain, it was necessary that the grain be inserted into a cylinder prior to impact and then removed before hydrotesting. While this did not allow for a bonded casewall/grain interface, the fact that the grain was cast into an identical case section provided a "fit" between the case and grain that was quite close. Launch tube designs can incorporate interior runners or rails that act not only to provide for a safe exit of the munition but also provide some additional stiffness to the tube. For this study, aluminum rings (approximately 12.7-cm outer diameter, 2.54 cm wide) were inserted approximately 5 cm into each end of the cylinder. This provided for an unsupported span length of 20 cm.

Test Procedure

Impact Test—Impact testing was conducted using a commercially available drop weight impact station. Hammer weight of the system was 5.15 kg. The impact tup was spherically shaped with a diameter of 1.27 cm. The instrumented impact tup used by the system not only imparts the impact energy to the target but also acts as a loadcell to record target resistance. Measured data included impact energy, impact load, and event time. The system utilizes the measured load-time data and instantaneous tup velocity at the moment of impact with standard equations of motion to calculate cylinder deflection and the amount of absorbed/rebounded energy.

The cylinders were placed in a V-block test fixture for test purposes. No restraints were applied to the specimens prior to testing, which allowed the cylinders free movement in reaction to the impact. Drop heights were selected to simulate a tool drop scenario. The resulting levels of impact energies utilized in this program ranged from approximately 1.27 to 7.75 J. Impacts were targeted to be mid-cylinder strikes. Multiple impacts to the cylinders were prevented by use of a rebound brake.

Hydroburst Evaluation—After impact, the residual strength of the cylinders was determined by pressurization testing of the cylinders. The cylinders were placed in an end-plate /tie rod test fixture (Fig. 1). This configuration allowed the tube to "float" at either end,

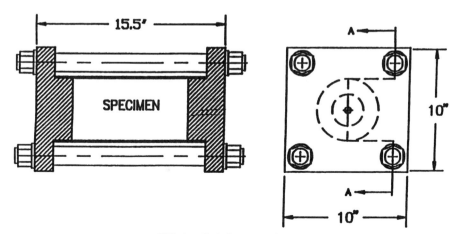

FIG. 1—*Hydroburst test fixture.*

which resulted in the absence of any axial loads such that the loading in the tubes was completely radial in nature. An interior bladder was used to contain the pressurization fluid. Pressurization rate was 500 psi/s.

Test Results

Impact Testing—(a) <u>Visual Damage.</u> Some evidence of exterior fiber cracking was seen for each of the three <u>configuration types.</u> Of these, measurable cracking was most severe for the baseline specimens where maximum crack length observed was found to be 4.9 cm at an impact energy of 7.67 J. An interesting note for the baseline cylinders was the observation that crack length appeared to follow a linear relationship with impact energy (Fig. 2). The equation of fit has a correlation coefficient (R^2) of 0.9808. Largest crack lengths observed for the motorcase and launch tube simulants were measured at 1.27 and 2.54 cm, respectively. No obvious relationship was noted between impact energy and crack length for these configurations. In all cylinders, the cracks were observed running parallel to the cylinder's longitudinal axis across the outer hoop bands of fiber. Interior damage was limited to slight fiber tow separation from the wall surface.

(b) <u>Composite Structural Response.</u> Table 3 presents results associated with impact testing of the <u>three configuration types.</u> A review of the data indicates several differences in the structures' response to impact. The driver of these differences is most probably the effect of casewall deflection on case response. As expected, restricted casewall deflections were observed for the specimens employing internal support as compared to the baseline "empty" cylinders (Fig. 3). Impact load on the structure is a measure of the structure's inherent resistance to deformation. The minimized deflection seen for the motorcase and launch tube simulants is reflected in the increased loads observed for these two structural configuration types (Figs. 4 and 5). Differences are also noted in the recorded event times, i.e., time to

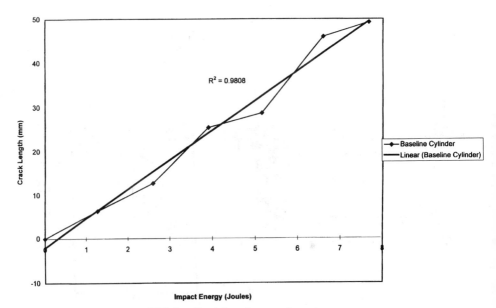

FIG. 2—*Crack length versus impact energy.*

TABLE 3—*Impact test results.*

Specimen Type	Energy, J [ft-lb]					Time, ms		
Motorcase Simulant	Impact	Absorbed	Rebound	Load, kg lb	Max. Load	Total	Deflection, mm [in.]
023	1.27 [0.94]*	...a	...a	119.0 [261.9]	3.95	10.40	...a
028	2.66 [1.96]	1.94 [1.43]	0.72 [0.53]	170.0 [373.9]	4.17	10.20	3.0 [0.12]
047	3.73 [2.75]	2.96 [2.18]	0.77 [0.57]	210.5 [463.1]	3.80	10.05	3.6 [0.14]
041	4.80 [3.54]	4.08 [3.01]	0.72 [0.52]	240.8 [529.7]	3.67	10.27	3.8 [0.15]
035	6.28 [4.63]	5.02 [3.70]	1.26 [0.93]	287.5 [632.6]	3.77	9.85	4.3 [0.17]
037	6.80 [5.01]	6.05 [4.46]	0.75 [0.55]	317.5 [698.6]	3.20	9.55	4.3 [0.17]
Launch Tube							
13-1	1.27 [0.94]*	...a	...a	80.4 [177.0]	5.97	14.1	...a
11-3	2.66 [1.96]	1.56 [1.15]	1.10 [0.81]	105.0 [231.0]	6.97	14.55	4.3 [0.17]
11-1	3.89 [2.87]	2.30 [1.70]	1.59 [1.17]	128.0 [281.7]	6.65	14.68	5.3 [0.21]
14-2	4.94 [3.64]	3.52 [2.60]	1.42 [1.04]	131.0 [288.2]	6.45	16.08	6.6 [0.26]
10-3	6.37 [4.70]	4.42 [3.26]	1.95 [2.14]	136.4 [300.0]	7.17	17.08	7.6 [0.30]
14-3	7.77 [5.73]	4.89 [3.59]	2.90 [2.14]	149.7 [329.4]	7.80	17.40	8.6 [0.34]
Baseline Cylinder							
026	1.27 [0.94]*	...a	...a	75.8 [166.8]	7.88	16.60	...a
017	2.60 [1.92]	0.87 [0.64]	1.75 [1.29]	87.2 [191.9]	8.70	18.90	5.6 [0.22]
045	3.90 [2.88]	1.41 [1.04]	2.49 [1.84]	105.5 [232.2]	8.25	18.55	6.9 [0.27]
021	5.15 [3.80]	1.90 [1.40]	3.25 [2.40]	119.1 [262.0]	8.75	18.88	8.1 [0.32]
020	6.60 [4.87]	2.44 [1.80]	4.16 [3.07]	130.5 [287.2]	9.18	19.25	9.1 [0.36]
055	7.67 [5.66]	2.96 [2.18]	4.71 [3.48]	144.6 [318.1]	8.77	18.48	9.4 [0.37]

a NOTE—Data not available.
* Theoretical value.

maximum load and total time. It is believed that these differences are also linked to the restricted casewall deflections.

If the specimens had been perfectly elastic, absorbed energy and rebound energy should be equal. A review of Table 3 indicates that the composite systems tested here are not perfectly elastic. While impact energies are seen to be comparable, energy absorbed by both the motorcase and launch tube simulants was found to be significantly greater than for the baseline structure (Fig. 6).

Hydroburst Evaluation

As stated previously, all cylinders were pressurized until casewall failure. For all test configurations, failure appeared to emanate from the locus of the impact site. Failure modes for the baseline and motorcase simulant were identical, that is, total failure of the cylinder at the lower impact energy levels and localized wall failure at the higher energy levels. Failure of the launch tube simulant was characterized as localized wall failure regardless of the level of impact energy. Recorded burst pressures are presented in Table 4 and plotted in Fig. 7. From the plot, the similarity of the operational response of the specimen types is easily discernible. Each of the configurations exhibits significant strength losses in burst

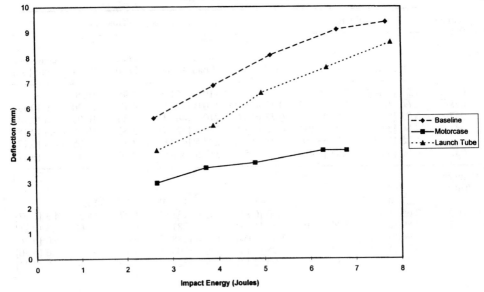

FIG. 3—*Casewell deflection versus impact energy.*

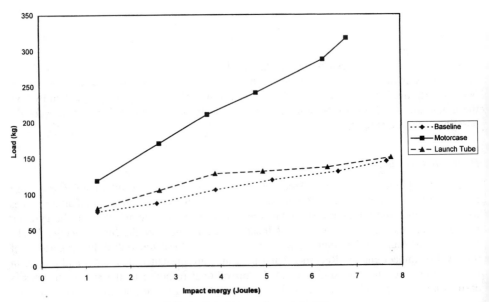

FIG. 4—*Load versus impact energy.*

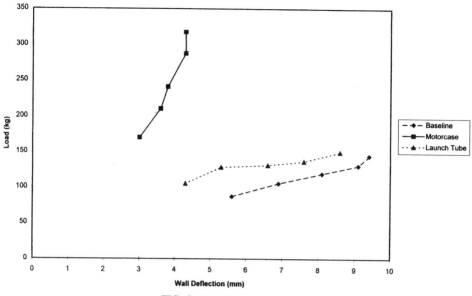

FIG. 5—*Load versus deflection.*

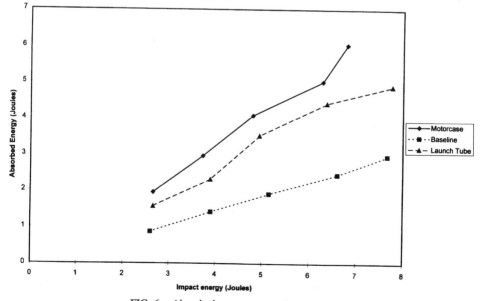

FIG. 6—*Absorbed energy versus impact energy.*

TABLE 4—*Hydroburst test results.*

Specimen Type	Impact Energy, J	Burst Pressure, MPa
Baseline		
000	0.0	30.1
026	1.27	20.2
017	2.60	16.2
045	3.90	15.8
021	5.15	13.0
020	6.60	15.1
055	7.67	10.8
Motorcase		
001	0.0	30.0
023	1.27	19.6
028	2.66	18.6
047	3.73	...[a]
041	4.80	16.0
035	6.28	13.6
037	6.8	14.0
Launch tube		
003	0.0	30.0
13-1	1.27	14.9
11-3	2.66	18.6
11-1	3.89	...[a]
14-2	5.15	...[a]
10-3	6.37	15.2
14-3	7.77	12.0

[a] NOTE—Data not available.

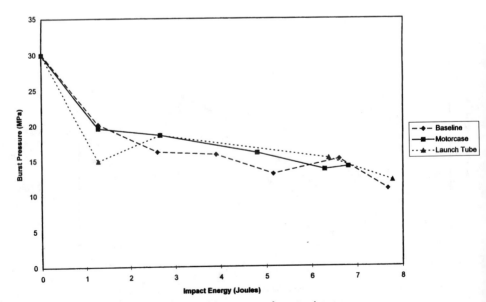

FIG. 7—*Residual burst strength versus input energy.*

strength at impact energies at just over 1.0 J. It should be remembered that the test articles were designed for a 31-MPa burst strength based on a maximum expected operating pressure of 20.6 MPa. The data shown here indicate that operational limits are lost at very low impact levels. Increasing levels of impact are seen to further degrade structural performance.

Discussion of Results

Significant structural differences were identified between the test configurations with regard to (1) induced load, (2) energy (both that absorbed by the test article or rebounded away), (3) casewall deflections, (4) visible exterior damage, and (5) overall event times. It appears that the internal support provided by the inert grain and rings served not only to restrict casewall deflection but also had the additional effect of increasing impact load and total energy absorbed by the structure. Still, with all the readily apparent differences noted, post-impact residual strength capability was quite similar.

It is difficult to attempt to determine the definitive cause of case structural degradation. However, in each case where the cylinders suffered some form of exterior fiber damage, i.e., cracking, the failure appeared to initiate at that location. This observation when combined with the data in Table 4 can offer a possible explanation of post-impact behavior. It is the opinion of the author that damage inflicted by low-energy impact is bounded by a series of thresholds that in turn are imposed by the design of the structure. For example, in the case studied here the cylinder was designed for a minimum burst strength of approximately 30 MPa. As previously mentioned, the fixture used for the burst tests loads the cylinder radially, which essentially means that the hoop layers carry the majority of the load (tensile). The data in Table 4 show a major loss in burst strength occurring at just over 1.0 J followed by a leveling off or plateau region. This plateau approximates a burst strength that is equal to almost half of the design strength. It would appear that a range of impact energies will cause sufficient damage to a layer, in this case the outer hoop layer, so that it fails at a very low stress level, thereby reducing the overall strength of the structure. In fact, if fiber cracking is determined to be the limiting factor in case structural performance, the data imply that even slight fiber damage should be cause for concern. Indications are that crack depth is more critical than crack length is to overall strength reduction. Obviously in a thin-walled structure such as the one tested here, the ramifications of a damaged ply are much greater than for a heavy wall structure.

Conclusions

The purpose of this program was to evaluate the structural and operational response of thin-walled graphite/epoxy structures that had been subjected to controlled levels of impact energy. Three specimen configurations were studied. A baseline structure was fabricated and tested and then adapted to simulate a typical rocket motorcase or launch tube configuration. An inert solid propellant grain was inserted into the baseline cylinder to duplicate, as much as possible, the actual conditions of a live rocket motor. The launch tube simulant employed internal rings to provide additional stiffness along the length of the cylinder. Comparisons were then made with the previously generated damage effects data obtained from the baseline specimens.

The results of this program reaffirm the potential for severe damage to occur in thin-walled graphite/epoxy structures as a result of what may be considered insignificant impact events. The data indicate that significant strength losses can be expected at impact energies as low as 1.0 J. As yet, the predominant damage-induced failure mode has not been identified. The inherent complexity of composite structures may require that impact damage programs be performed on a case-by-case basis to identify potential problem areas. These programs should include experimental testing, finite element analysis, and nondestructive evaluation techniques.

References

[1] Lloyd, B. A. and Knight, G. K., "Impact Damage Sensitivity of Filament Wound Composite Vessels," Morton Thiokol, Inc., *Proceedings,* 1986 JANNAF Propulsion Meeting, Chemical Propulsion Information Agency.

[2] Patterson, J. E., "Graphite Composite Case Assessment Strain-Rate Behavior and Impact Damage Thresholds," Technical Report TR-RD-PR-88-4, U.S. Army Missile Command, Redstone Arsenal, AL, Feb. 1988.

[3] Adler, W. F., Carlyle, and Dorsey, J. J., "Damage Tolerance Assessment of Six-inch Composite Pressure Vessels," Technical Discussion, CR-84-1263, *General Research,* Vol. 1, April 1984.

[4] Patterson, J. E. and McGuire, K. B., "Initial Comparison of Impact Tolerance of Composite Fibers," Technical Report TR-RD-PR-94-11, U.S. Army Missile Command, Redstone Arsenal, AL, July 1994.

[5] Patterson, J. E., Jaklitsch, D. J., and Evans, R. N., "Study of Low Energy Impact Damage to JAVELIN Launch Tube Assembly," Technical Report TR-RD-ST-95-5, U.S. Army Missile Command, Redstone Arsenal, AL, Dec. 1994.

[6] Dombrowski, D. C. and Knight, G. K., "Residual Strength of Carbon/Epoxy Pressure Vessels Subjected to Low Velocity Impact," Morton Thiokol, Inc., *Proceedings,* 1988 JANNAF Composite Motor Case Subcommittee Meeting, Chemical Propulsion Information Agency, Columbia, MD.

[7] Tapphorn, R. M., Pelligrino, P. M., and Beeson, H. D., "Nondestructive Evaluation and Health Monitoring of Graphite/Epoxy Composite Overwrapped Pressure Vessels," NASA Johnson Space Center, White Sands Test Facility, NM.

Dahsin Liu[1] and Xinglai Dang[1]

Testing and Simulation of Laminated Composites Subjected to Impact Loading

REFERENCE: Liu, D. and Dang, X., "**Testing and Simulation of Laminated Composites Subjected to Impact Loading,**" *Composite Materials: Fatigue and Fracture, Seventh Volume, ASTM STP 1330,* R. B. Bucinell, Ed., American Society for Testing and Materials, 1998, pp. 273–284.

ABSTRACT: Because composite laminates are very susceptible to impact loading even at low velocity, low-velocity impact is an important subject in laminated composite analysis. The impact-induced damage is usually invisible to the naked eye and can cause serious structural degradation. Many low-velocity impact tests were performed in previous studies; however, most were phenomenological analysis. In an effort to further understand the responses of composite laminates under low-velocity impact and to develop an accurate and efficient quantitative simulation in the future, this study was aimed at performing instrumented impact tests and computer simulations. A commonly used computer code—LS-DYNA3D—was evaluated in this study, and the results are valuable for future development of a new computer code. In the study, a low-velocity impact event investigated by Sun and Chen with an indentation law and verified by experiments was used to justify the finite element model and contact parameters. Once the computational scheme was established, it was used for a broader investigation consisting of composite laminates with various thicknesses, fiber angles, and impact velocities. Computational results revealed that the peak contact force and maximum deflection were strongly affected by the thickness of composite laminates, while the fiber angles investigated seemed to play a less significant role. In addition, it was concluded that because delamination modeling was not included in the LS-DYNA3D, the computer code needed to be modified if it was to be used for accurate impact simulations.

KEYWORDS: impact, laminated composites, contact force, maximum deflection, energy absorption

Laminated composites are very susceptible to impact loading. Composite damage such as fiber breakage, matrix cracking, fiber-matrix debonding, and delamination can take place in composite laminates even when they are subjected to impact forces at low velocity. These damage modes usually cannot be detected by the naked eye. However, their effects on composite structural degradation are always very significant.

Many impact tests were performed and can be found in the review articles by Abrate [1,2]. In some previous studies [3,4], Liu et al. also concluded that matrix cracking and delamination were the major damage modes in composite laminates subjected to low-velocity impact. A correlation between bending stiffness mismatch and delamination size was established. The relation was successfully used for phenomenological explanations of delamination size, location, and orientation in impacted composite laminates. In an effort to further understand the response of composite laminates under low-velocity impact and to establish an accurate and efficient quantitative simulation, the objectives of the present study are to

[1] Professor and graduate research assistant, respectively, Department of Materials Science and Mechanics, Michigan State University, East Lansing, MI 48824.

perform instrumented impact tests and to simulate the impact responses with a commonly used computer code—LS-DYNA3D. The results from this study are believed to be very valuable for future improvement and/or development of computer modeling and simulation for impact analysis.

In modeling the composite response under low-velocity impact, Tan and Sun [5] verified that an indentation law based on a quasi-static test could be used to investigate low-velocity impact. For a set of composite material and impactor, an indentation test was required to identify the corresponding indentation parameters. Once the indentation law was established, it could be integrated into a computational scheme for various studies [6,7]. In their study [8], Wu and Yen also investigated the relationship between impact force and laminate indention. The numerical method they used was derived from three-dimensional anisotropic elasticity. Effects of material and geometrical parameters on force-indentation relation were also examined in their study. Their results seemed to agree well with Sun and Chen [7].

Because of its strong dependence on parameters obtained from experiments, the approach based on the indentation law usually gave good prediction of composite responses under low-velocity impact. However, it should be pointed out that a new characterization for indentation parameters was needed each time the material or geometry of the composite or impactor was altered. As a consequence, for investigating the effects of material type, lamination, fiber orientation, and thickness on impact response of composite laminates, a computational technique free of experiment-dependent parameters might prove to be more efficient. In search of a more efficient technique for studying impact responses of various composite laminates and impactors, the present study examined an existing computational scheme, namely LS-DYNA3D. The effects of laminate thickness, fiber angle, and impact velocity on composite response were of primary concern along with the feasibility of using LS-DYNA3D for impact analysis.

Instrumented Impact Testing

The impact tests in this study were performed on a DYNATUP impact testing machine. The impactor was a steel rod that had a diameter of 12.5 mm and was attached to the dropping head as shown in Fig. 1. A force transducer having a maximum force limit of 13 350 N was mounted at the tip of the rod and encapsulated by a spherical head. Specimens with dimensions of 125 by 100 mm were clamped by two steel holders from both sides with a rectangular opening of 100 by 75 mm in the center. The specimens and the specimen holders were then C-clamped to the bottom of the impact testing frame, which was nearly fixed to the ground with many dead weights.

As the impactor dropped and approached a specimen, it triggered the timer and started the recording clock. The force history detected by the force transducer was then recorded in a computer. The corresponding displacement of the transducer or the deformation of the specimen could be calculated through double integrations. The force history, the deformation history, and the energy history were found to be the important characteristics of individual composite laminates under impact. In addition, the maximum peak force, the maximum deflection, the contact duration, and the energy absorption capability were also found to be important parameters for impact analysis.

Computational Technique

A dynamic finite element code named LS-DYNA3D was used in this study for impact simulations. In performing the finite element analysis, finite element models for both com-

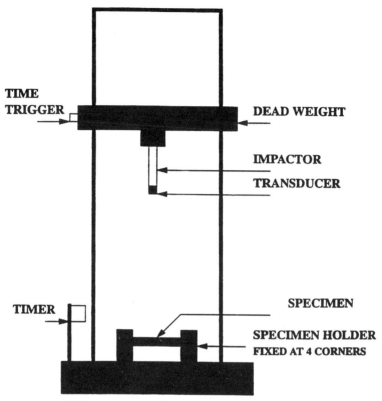

FIG. 1—*Schematic diagram of impact testing setup.*

posite laminates and a semi-spherical impactor were built. The composite material was AS4/3501-6 and had the following elastic constants:

$$E_{11} = 120 \text{ GPa}, E_{22} = E_{33} = 7.9 \text{ GPa}, G_{12} = G_{23} = G_{31} = 5.5 \text{ GPa}, v_{12} = v_{13} = v_{23} = 0.3$$

The composite laminates had dimensions of 100 by 75 mm. The thickness of individual ply was considered to be 0.25 mm. The solid element was used in the finite element models because it gave better results than the shell elements. Full models were built for all specimens, resulting in a mesh of 40 by 40 by 15 for a 15-ply laminate. The laminate models were then fixed around four edges and subjected to a spherical impactor with a diameter of 12.5 mm and a mass of 24 g at the center of the laminates. The impact velocity ranged from 1 to 5 m/s. These finite element models were used throughout the investigations of this study unless otherwise mentioned.

Instead of an indentation law to identify the contact force and indentation, the LS-DYNA3D used a contact algorithm to calculate the contact-impact response. Accordingly, besides an initial impact velocity, a contact parameter called penalty coefficient needed to be determined through comparisons with experimental results and then assigned in the finite element simulations. The LS-DYNA3D gave results of force history and deformation history. Other information such as peak force, contact duration, maximum deflection, and energy absorption engaged in the impact events could be drawn from the histories. Since this study was focused on the comparison between the composite performance and computer simulation under low-velocity impact, the failure process available in the LY-DYNA3D was also imposed in the feasibility studies.

Experimental Results

The following three types of studies were performed: thickness effect, fiber angle effect, and velocity effect. Table 1 shows the details of the stacking sequence and the impact parameters of primary concern, while Figs 2, 3, and 4 are the force-displacement relations for the thickness effect, fiber angle effect, and velocity effect, respectively. The following statements can be summarized from these results:

A. *Thickness effect*—The laminate stiffness and strength increase as the laminate thickness increases, resulting in the increase of peak contact force and decreases of contact duration, maximum deflection, and energy absorption.
B. *Fiber angle effect*—Due to the alternative lay-up and near-square geometry, fiber angle does not play a significant role in impact response for composite laminates with the same thickness.
C. *Velocity effect*—For thin composite laminates under subperforation impact, contact duration, peak contact force, maximum deflection, and energy absorption all increase as the impact velocity increases.
D. *Velocity effect*—Once penetration takes place, both contact duration and peak contact force decrease while maximum deflection remains at about the same level as the impact velocity increases.
E. *Velocity effect*—There is a significant jump in energy absorption due to penetration. The energy absorption remains at about the same level even when the impact velocity increases.

TABLE 1—*Experimental results of impact parameters.*

Types of Effect Stacking Sequences	Impact Velocity m/s	Contact Duration, ms	Peak Force, kN	Maximum Deflection, mm	Energy Absorption, J
Thickness:					
$[0_3/90_3/0_3/90_3/0_3]$	0.98	4.65	3.34	1.47	0.89
$[0_3/90_3/0_3/90_3/0_3/90_3/0_3]$	0.98	4.00	4.41	1.19	0.84
Fiber Angle:					
$[0_3/90_3/0_3/90_3/0_3]$	0.98	4.65	3.34	1.47	0.89
$[30_3/-30_3/30_3/-30_3/30_3]$	0.97	4.55	3.33	1.52	0.81
$[45_3/-45_3/45_3/-45_3/45_3]$	0.97	4.30	3.50	1.27	0.81
$[60_3/-60_3/60_3/-60_3/60_3]$	0.97	4.35	3.46	1.52	0.81
Velocity:					
$[0_3/90_3/0_3]$	0.99	8.15	1.78	2.79	1.04
$[0_3/90_3/0_3]$	1.59	8.43	2.95	4.32	2.59
$[0_3/90_3/0_3]$	2.77	10.05	3.61	8.38	14.25
$[0_3/90_3/0_3]$	3.40	6.25	3.70	13.46	25.76
$[0_3/90_3/0_3]$	3.95	4.70	3.34	12.70	23.15
$[0_3/90_3/0_3]$	4.49	3.20	3.32	11.94	24.81

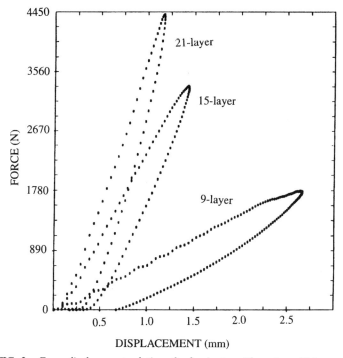

FIG. 2—*Force-displacement relations for laminates with various thicknesses.*

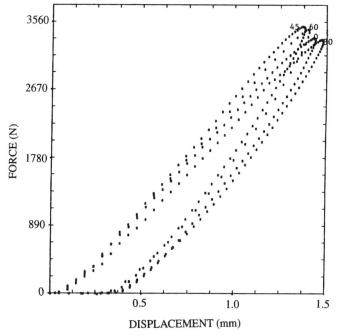

FIG. 3—*Force-displacement relations for laminates with various fiber angles.*

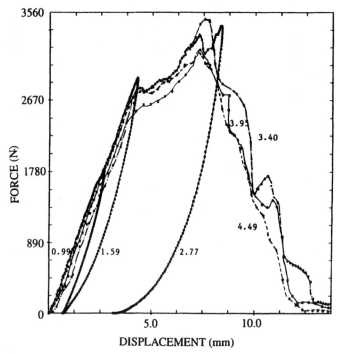

FIG. 4—*Force-displacement relations for laminates subjected to various impact velocities.*

Computational Results

Justification

Before investigating the effects of material properties and geometrical parameters on the impact response of composite laminates, the finite element models and the associated contact parameters were justified by comparing the LS-DYNA3D results with those obtained from the approach based on the indentation law by Sun and Chen [7]. In the justification, the stacking sequence and the impactor of the finite element simulations were identical to those defined in Ref 7, but different from those used in later investigations. It was found that a 40 by 40 mesh was able to give good agreement with Sun and Chen's. Figures 5 and 6 show the force history and deflection history, respectively. The solid line represents the results from LS-DYNA3D, while the enhanced dots are those from Ref 1.

Summary

Similar studies regarding the thickness effect, fiber angle effect, and velocity effect were also performed by the LS-DYNA3D simulation. Figures 7, 8, and 9 show the comparisons between the experimental results and the finite element simulations for the 15-layer laminate, i.e. $[0_3/90_3/0_3/90_3/0_3]$, under 1 m/s impact. The experimental results are designated by the asterisk symbol, while the finite element simulations are the solid lines. Notations A and B for the solid lines represent the results for the impactor and the specimen, respectively. In addition to the impact velocity of 1 m/s, impact velocities of 2.8 and 3.5 m/s were also performed and the results based on various failure criteria and associated progressive damage

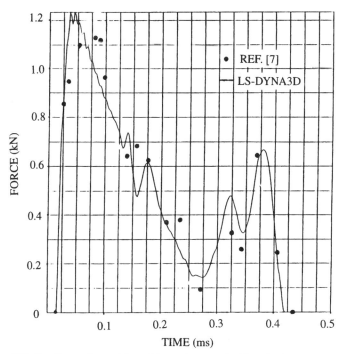

FIG. 5—*Comparison of force history between Ref 7 and LS-DYNA3D.*

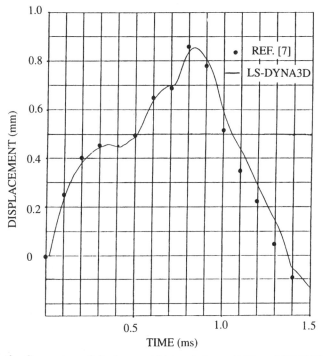

FIG. 6—*Comparison of displacement history between Ref 7 and LS-DYNA3D.*

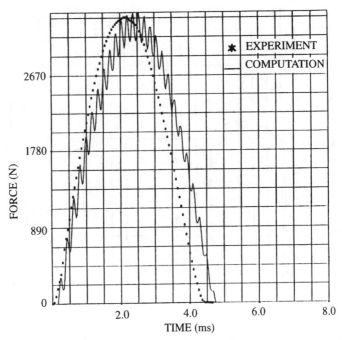

FIG. 7—*Force history of 15-layer laminate with fixed boundary condition and subjected to 1 m/s impact.*

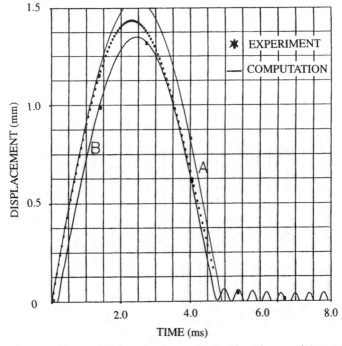

FIG. 8—*Displacement history of 15-layer laminate with fixed boundary condition and subjected to 1 m/s impact.*

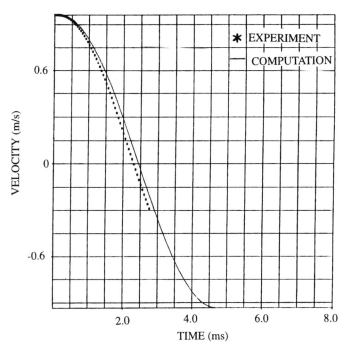

FIG. 9—*Velocity history of 15-layer laminate with fixed boundary condition and subjected to 1 m/s impact.*

models, such as the Tsai-Wu failure criterion [9], the Chang-Chang failure criterion [10], MTL54 (partially Chang-Chang failure criterion) [11], and pure elastic assumption, given in Figs. 10, 11, and 12.

Based on the computational results, the following statements can be summarized:

A. *Thickness effect*—Based on comparisons of force history, deflection history, and velocity history, the computational results for 9-layer laminates agree well with the experimental counterparts if fixed boundary conditions are imposed in the simulations. However, simply supported boundary conditions seem to give better results for 15- and 21-layer simulations.

B. *Fiber angle effect*—Simply supported boundary conditions also give good results for angle-ply laminates that consist of 15 layers.

C. *Velocity effect*—At subperforation impact, the lower the velocity, the better the simulation.

D. *Velocity effect*—Both Tsai-Wu and MTL54 (partially Chang-Chang failure criterion) seem to give reasonable simulation before penetration takes place.

Conclusions

The LS-DYNA3D gives accurate predictions for the response of impacted composites before delamination takes place. In order to closely simulate the progress of impact-induced damage, a new type of finite element and a new failure criterion that account for interlaminar stresses are required.

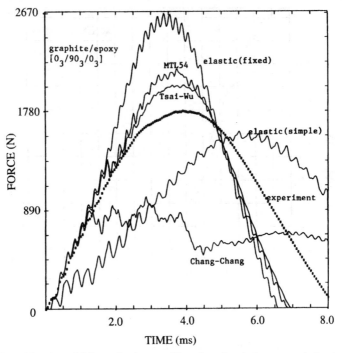

FIG. 10—*Force histories of 9-layer laminate subjected to 1 m/s impact and simulated by various failure criteria.*

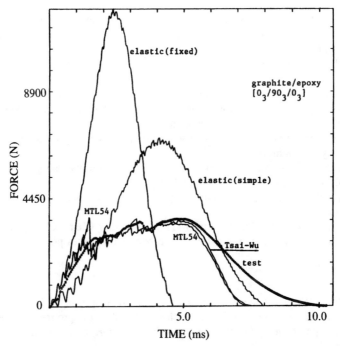

FIG. 11—*Force histories of 9-layer laminate subjected to 2.8 m/s impact and simulated by various failure criteria.*

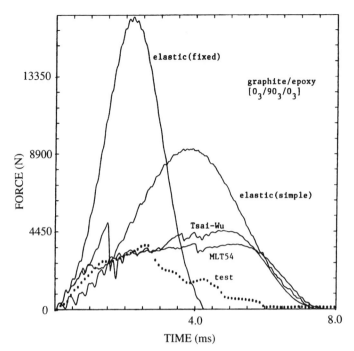

FIG. 12—*Force histories of 9-layer laminate subjected to 3.5 m/s impact and simulated by various failure criteria.*

Acknowledgments

The authors wish to express their sincere thanks to Ford Motors Company for financial support through the University Research Initiation Program. Partial support from the U.S. Army TARDEC (Warren, Michigan) is also greatly appreciated.

References

[1] Abrate, S., "Impact on Laminated Composite Materials," *Applied Mechanics Review,* Vol. 44, No. 4, 1991, pp. 155–190.
[2] Abrate, S., "Impact on Laminated Composites: Recent Advances," *Applied Mechanics Review,* Vol. 47, No. 11, 1994, pp. 517–544.
[3] Liu, D. and Malvern, L. E., "Matrix Cracking in Impacted Glass/Epoxy Plates," *Journal of Composite Materials,* Vol. 21, No. 7, 1987, pp. 594–609.
[4] Liu, D., "Impact-Induced Delamination—A View of Material Property Mismatching," *Journal of Composite Materials,* Vol. 22, No. 7, 1988, pp. 674–691.
[5] Tan, T. M. and Sun, C. T., "Use of Statistical Indentation Laws in the Impact Analysis of Laminated Composite Plates," *Journal of Applied Mechanics,* Vol. 52, 1985, pp. 6–12.
[6] Chen, J. K. and Sun, C. T., "Dynamic Large Deflection Response of Composite Laminates Subjected to Impact," *Composite Structures,* Vol. 4, 1985, pp. 59–73.
[7] Sun C. T. and Chen, J. K., "On the Impact of Initially Stressed Composite Laminates," *Journal of Composite Materials,* Vol. 19, No. 6, 1985, pp. 490–504.
[8] Wu, E. and Yen, C., "The Contact Behavior Plates and Rigid Spheres," *Journal of Applied Mechanics,* Vol. 61, 1994, pp. 60–66.

[9] Tsai, S. W. and Wu, E. M., "A General Theory of Strength for Anisotropic Materials," *Journal of Composite Materials,* Vol. 5, No. 1, 1971, pp. 58–80.

[10] Chang, F.-K. and Chang, K.-Y., "Post-Failure Analysis of Bolted Composite Joints in Tension or Shear-Out Mode Failure," *Journal of Composite Materials,* Vol. 21, No. 5, 1987, pp. 809–833.

[11] LS-DYNA3D User's Manual.

Matthew D. Lansing,[1] James L. Walker,[2] and Samuel S. Russell[3]

Residual Strength Prediction of Impact-Damaged Composite Structures by Optical and Acoustical Computer Sensing with Neural Network Techniques

REFERENCE: Lansing, M. D., Walker, J. L., and Russell, S. S., "**Residual Strength Prediction of Impact-Damaged Composite Structures by Optical and Acoustical Computer Sensing with Neural Network Techniques,**" *Composite Materials: Fatigue and Fracture, Seventh Volume, ASTM STP 1330,* R. B. Bucinell, Ed., American Society for Testing and Materials, 1998, pp. 285–297.

ABSTRACT: The proliferation of composites as a structural material has increased the necessity for detection and characterization of impact damage. Impact damage, due either to mishandling or in-service environment, in the form of delaminations, matrix cracking, and fiber chopping, may be sufficient to reduce the load-bearing ability of the structure below safety limits and may still not be visible to the unaided observer. Such damage may exist on the surface of an internal cavity, which may prevent it from being viewed, or it may exist within the thickness of the structure and not be visible from either side of the laminate. Impact damage that is partially visible often extends further along and through the material than that which can be seen. Computerized methods for damage detection offer greater assessment capabilities than visual inspections and are capable of detecting subsurface or internal flaws. For these reasons, as well as the declining cost of more powerful hardware, computer sensing methods for nondestructive evaluation (NDE) have grown increasingly popular in recent years.

A disadvantage of some computerized NDE methods is that vast amounts of data are generally obtained from any given test, and often it is up to the inspection conductor to interpret these results subjectively based upon prior experience. An alternative is to train the computer sensing system to analyze the huge data arrays, which it can digest more easily than can its human counterpart, and to provide objective interpretations. This paper presents the results of an experimental study in which two computer sensing techniques were used to monitor filament-wound pressure vessels during pressurization. Acoustic emission registers the sound generated by microscopic damage propagation. Video image correlation is a noncontact computer vision technique that simultaneously measures full-field in-plane surface displacements and strains, both linear and angular, with subpixel accuracy. Neural networks were used to predict the burst pressures of impacted pressure vessels based upon data obtained at less than approximately one third of the expected burst pressure for an undamaged specimen.

KEYWORDS: composites, failure, impact, damage, prediction, neural networks, acoustic emission, video image correlation, strain, strain measurement, nondestructive testing (NDT), nondestructive evaluation (NDE), pressure vessels, rocket motors

[1] Senior research associate, University of Alabama in Huntsville-Research Institute, RI-E47, Huntsville, AL 35899.
[2] Senior research associate, University of Alabama in Huntsville-Center for Automation and Robotics, RI-A6, Huntsville, AL 35899.
[3] Mechanics of materials engineer, NASA Marshall Space Flight Center, MSFC, AL 35812.

This study is part of a larger damage assessment program concerning impact-damaged filament-wound pressure vessel (FWPV) test articles. When filled with a simulated inert propellant, these "bottles" may be used as subscale models of solid rocket motor casings. Unfilled bottles may be used to model liquid propellant storage vessels. Additionally, FWPV of this design may be filled with "live" solid propellant and used as igniters for larger solid rocket motors. The test articles were fabricated from graphite or aramide fibers at the NASA Marshall Space Flight Center Productivity Enhancement Complex. The undamaged burst pressure was approximately 20.7 MPa for both types of specimens.

The 14.6-cm-diameter graphite/epoxy and aramide/epoxy vessels fabricated for this study conform to ASTM Standard D 2585-68 (1985), as shown in Fig. 1. The graphite vessels were all tumble wound and rotisserie cured using a graphite fiber preimpregnated with one of three epoxy resins. The aramide/epoxy vessels were tumble wound "wet" using an aramide fiber and a single epoxy resin over a rubber bladder and rigid mandrel.

A calibrated dead weight drop fixture produced impact damage in the mid-hoop (cylindrical) region of each vessel ranging from that which was barely visible to obvious fiber breakage and delamination. A wooden cradle with rubber padding was used to support and position the specimens during impact. One vessel from each resin class was used as a control sample and left undamaged. Impact damage was then produced by dropping a weighted hammer onto the center of the hoop regions utilizing either a blunt or sharp hemispherical impactor tip. Electronic shearography (ES) and video image correlation (VIC) techniques showed that the blunt-tipped impactors generally produced a wide damaged zone with some localized delaminations, while the sharp tip tended to break fibers at the impact point [1]. The damage in the aramide vessels differed from the graphite vessels in that the damage was more extensive in the form of delaminations, rather than fiber breakage, due to the toughness of the aramide fiber.

Acoustical Method

Acoustic emission is a method that monitors microscopic damage propagation over continuous load differentials. An increase in the load to which a damaged composite structure is subjected may result in delamination growth, matrix cracking, or fiber breakage. If the loading rate and maximum load are sufficiently low, the propagation of damage does not extend to macroscopic levels and does not appreciably weaken the structure compared to the initial damage state. The damage propagation results in the release of mechanical energy in

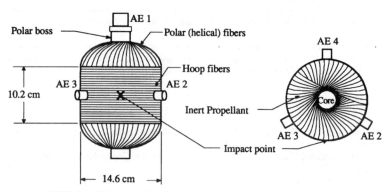

FIG. 1—*Filament-wound pressure vessel test article geometry.*

the form of acoustical waves. It may be argued that different failure mechanisms result in acoustical waves with different characteristics. For example, the acoustical energy released by delamination growth may have a longer duration and lower amplitude than that released by the sudden breakage of a fiber under tension. These waves initiate at the sight of damage propagation and travel through the material in all directions, in accordance with the material's symmetry of stiffness, until the signal is reflected by a boundary or dies out due to dampening.

Piezoelectric crystals may be coupled to the surface of the test article and thus pick up acoustical energy caused by damage propagation and transmitted across the coupling boundary. The acoustical energy causes the piezoelectric crystal to resonate at a known frequency, producing an electrical signal proportional in amplitude to the absorbed energy. A high-speed analog-to-digital converter (ADC) samples the amplified voltage from the piezoelectric transducer with typical sampling frequencies less than 32 MHz. A voltage spike above some predetermined amplitude is referred to as a hit, and these are counted by the acoustic emission software. A hit lock-out time, typically in microseconds, is set such that each hit is not accidentally counted more than once. The result is a histogram of the number of hits that fall into certain amplitude categories.

Qualitatively, it could be argued that high-amplitude fiber breakage is a more severe form of damage propagation than is low-amplitude delamination in a filament wound pressure vessel loaded internally. If this is the case, then an AE amplitude distribution that tends toward higher amplitudes may indicate a test article with a lower burst pressure than one that tends toward lower amplitudes. In theory then, a neural network could be trained to formulate an empirical relationship between the AE amplitude distribution and the burst pressure for impact-damaged FWPV.

Apparatus

The acoustic activity of the test articles during internal pressurization using an air-driven water pump was collected by a Physical Acoustics Corporation (PAC) SPARTAN acoustic emission (AE) system. A PAC R15I transducer, which resonates at 150 kHz and has a 40-dB integral preamplifier and 100 to 300-kHz bandpass filter, was bonded with hot melt glue onto the pipe plug, sealing the upper polar boss (Fig. 1). The signals from this transducer, labeled AE1 in Fig. 1, are used in developing the neural network burst pressure prediction models. The pipe plug serves as a wave guide or collector, providing an efficient and repeatable path for transferring waterborne acoustic signals to the AE sensor. Signals produced at any other point on the vessel can be received by this central transducer with minimal relative distortion, which could otherwise result from the proximity of the sensors to the damage site. In other words, sensor orientation and distance from the damage site influence the signals received from the polar boss mounted sensor less than they would the signals received from a sensor mounted to the composite shell itself. Adverse variations in signal characteristics between vessels due to variation in sensor positioning are thus minimized.

Three PAC R15 (150-kHz) transducers, labeled AE2, AE3, and AE4 in Fig. 1, were bonded symmetrically around the mid-hoop line to provide source location capabilities. These sensors were connected to external PAC 1220A preamplifiers, with 40-dB gain and a 100 to 300-kHz bandpass filter. A 20-dB internal gain and 60-dB signal threshold were used to establish the system's sensitivity. The AE system's timing parameters defined the acoustic hits with a 30-μs peak detection time, 80-μs hit detection time, and a 300-μs hit lock-out-time. With these settings, 0.5-mm 2H lead breaks performed approximately 2 in. from each sensor produced signal amplitudes in the ranges of 80 dB for the graphite vessels and 70 dB for the aramide vessels, verifying adequate sensor coupling.

Procedures

AE data for subsequent burst pressure prediction was acquired during an initial proof pressurization ramp to 5.172 MPa at a rate of 69 kPa/s. The pressure was released after a 2-min hold and then ramped back to a level of 6.875 MPa, with a 5-min hold for VIC imaging after each 1.724-MPa increase. The pressure was again reduced to zero before application of the final pressure ramp to failure. The AE amplitude distribution data collected through the first hold at 5.172 MPa, along with the known burst pressures, were then tabulated for use in the neural network analysis.

Analysis

A back propagation neural network (BPNN) was developed to model the effects of the impact damage on burst pressure using NeuralWorks Professional II/PLUS software, by NeuralWare, Inc. Back-propagation neural networks are composed of processing elements (PE) that are used in a manner analogous to biological neurons, creating the architecture necessary to provide the basis for learning [2]. Weighting functions control the contribution of the input signals that will be passed on to each processing element. Each PE then performs a simple summation of the weighted input values and produces a single output response based upon a continuously differentiable transfer function, in this case a hyperbolic tangent.

The PE in a BPNN (Fig. 2) are arranged into an input layer, an output layer, and at least one middle layer. Data are introduced to the neural network through the input layer. Here

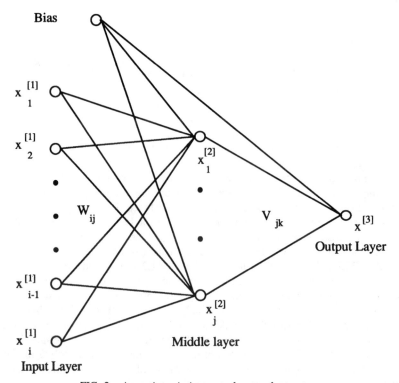

FIG. 2—*Acoustic emission neural network structure.*

the discrete values of the amplitude distribution histogram are entered as an input vector. Each input processing element is fully connected by a series of weighting factors to the middle layer, and these in turn are fully connected by another series of weighting factors to the output layer. If more than one middle layer is used, their PEs are also fully connected. The middle layers serve to map nonlinear variations in the dataset. A bias processing element may also be weight connected to the PE of the middle and output layers to serve as an offset value in the network. Ultimately, the weighting factors serve as the memory of the trained network by providing a multiplier between an initial PE's output value and a subsequent PE's input value, as determined by previous training.

The learning process begins by assigning initially randomized weights to the interconnections of the network and calculating an output value in response to an input vector. A global error results from the sum of the differences between the desired output and the actual output over a specified number of input vectors, referred to as the epoch size. A measure of the local error "e" at the jth processing element in layer "s" is given by the rate of change of the global error with respect to the summed input to the processing element. Only a portion of the actual local error is passed to each weighted connection, multiplying by a constant known as the "learning coefficient." The learning coefficient is kept as small as realistically possible to allow the network to iteratively converge on the absolute error minimum. Care must be taken when training a network not to falsely converge to a local minima [2] since these often lead to poor repeatability in the projected output. To overcome this problem, a momentum term is added to the change in each weight factor to keep the network moving towards the absolute minimum. A portion of the previous weight factor change is added back to the new delta weight.

The data from Sensors AE2, AE3, and AE4 in Fig. 1 were used to locate regions of active flaw growth during the pressurization cycles. The proximity of the damage state to the individual hoop sensors may have a biasing effect on the subsequent amplitude distribution [3], and since in non-laboratory situations the location of the damage may not be known, inclusion of these data for burst pressure prediction was not necessarily valid. For this reason, only the data from sensor AE1 are used for burst pressure prediction.

The amplitude distribution data from sensor AE1, over the range from 60 and 100 dB, in categories of 1 dB, were introduced to the network through a 41-PE input layer. The input layer was fully connected by a series of weighting factors to the first of two 13-PE middle layers. The two middle layers were also connected by weighting factors. Finally, the second middle layer was fully weight connected to a single output PE. The values from the output PE were interpreted as burst pressure values. A learning coefficient of 0.001 and a momentum of 0.1 were used during the computation of the update delta weights for each training cycle. The epoch size was set at 3, matching the number of training set input vectors. Each input vector corresponds to the AE amplitude distribution from a single FWPV during proof pressurization. The hyperbolic tangent transfer function applied progressively smaller step sizes to the change in weights as the normalized training error decreased. Several network architectures were attempted before this final version was determined. Networks with single and dual middle layers consisting of as few as 3 PEs to as many as 21 PEs in each middle layer were attempted. The best training results were obtained with the 41-13-13-1 network.

Results

The three resin systems behaved very different from one another acoustically. For example, the amount of AE activity recorded on Sensor AE1 through the end of the first hold at 5.172 MPa varied from an average of 517 hits for Resin A to 118 hits for Resin B and only 11 hits for Resin C. These results were expected since Resins B and C were formulated to be tougher than the brittle Resin A, and Resin C was formulated for more toughening than Resin B. Matrix toughening resulted in a structure that could better redistribute stresses around stress concentrations, increasing the resistance to failure.

Figure 3 illustrates the relationship between burst pressure and impact energy for the 17 vessels in this study. Overall, Resin B produced the highest burst pressures, implying the least sensitivity to impact damage. As expected, the burst pressures decreased with increasing sharp tip impact energy as this results in more fiber breakage. The blunt tip impacted vessels, on the other hand, showed an increase in burst pressure with larger impact energies. The delaminations generated during these impacts appear to provide stress relief between individual plies. This produces a more uniform overall stress state that results in higher net burst pressures. However, the burst pressure can be expected to drop again at higher impact energies.

A summary of the burst pressure values and the neural network training and prediction results are provided in Tables 1 and 2. By training on the amplitude distribution data collected from three vessels for each resin class, during the initial stages of loading, burst predictions were made with an average overall error (both training and testing) of only 2.8%. The load during data acquisition was less than one quarter of the expected burst pressure for an unimpacted test article and thus resulted in negligible weakening of the structures beyond that due to initial impact damage.

Optical Method

Video image correlation (VIC) is a noncontact computer vision technique that provides full-field in-plane deformation and strain data over a load differential [4]. This method was used to determine hoop and axial displacements, hoop and axial linear strains, and the in-plane shear strains and rotations in the regions surrounding impact sites on the FWPV test articles during internal pressurization subsequent to the AE data acquisition cycle. Determining the remaining life of the pressure vessels based upon these deformation measurement

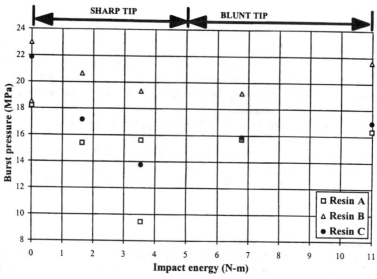

FIG. 3—*Burst pressures reduced by impact damage.*

TABLE 1—*Acoustic emission neural network training results.*

Resin Type	Bottle I.D.	Impact Status	Actual Burst, MPa	Predicted Burst, MPa	% Error
A	A003-004	None	18.19	17.93	−1.5
	C077-078	BT-10.98 N · m	16.36	16.42	0.4
	A017-018*	ST-3.53 N · m	9.453	9.832	4.0
B	C115-116	None	22.99	22.81	−0.8
	C141-142	BT-6.78 N · m	19.21	19.20	−0.00
	C131-132	ST-1.63 N · m	20.66	20.74	0.04
C	A025-026	None	21.86	21.53	−1.5
	A047-048	BT-10.98 N · m	16.98	17.01	−0.1
	C093-094	ST-3.53 N · m	13.76	14.04	2.1
				Abs(Average)	1.2

* Dome Failure

values requires a complex theoretical model or numerical simulation due to the complex behavior of the composite material. Both analytical and numerical techniques are time consuming and complicated. It is difficult to account for variations between coupons used to determine material properties, typically flat uniaxial tension specimens, and the actual material properties of the FWPV. These variations are due to different manufacturing methods. Previous results using neural network methods had been successful in predicting the burst pressure for graphite/epoxy pressure vessels based upon acoustic emission (AE) measurements in similar tests. The neural network associates the character of the AE amplitude distribution, which depends upon the extent of impact damage, with the burst pressure. Similarly, higher amounts of impact damage are theorized to cause a higher amount of strain concentration in the damage-affected zone at a given pressure and result in lower burst pressures. This relationship suggests that a neural network might be able to find an empirical relationship between the VIC strain field data and the burst pressure, analogous to the AE method, with greater speed and simplicity than theoretical or finite element modeling.

Apparatus

Figure 4 illustrates the VIC data acquisition hardware. Approximately 6.5 cm² around the impact zone on each impacted bottle, or at a random location on each unimpacted bottle,

TABLE 2—*Acoustic emission neural network testing results.*

Resin Type	Bottle I.D.	Impact Status	Actual Burst, MPa	Predicted Burst, MPa	% Error
A	C069-070	BT-6.78 N · m	15.71	15.35	−2.3
	A013-014	ST-1.63 N · m	15.39	16.24	5.6
	A023-024	ST-3.53 N · m	15.62	18.70	19.7
B	C139-140	None	18.49	19.25	4.1
	C117-118	BT-10.98 N · m	21.60	21.46	−0.6
	C155-156	ST-3.53 N · m	19.33	20.24	4.7
C	A029-030	BT-6.78 N · m	15.87	15.74	−0.8
	C087-088	ST-1.63 N · m	17.16	17.59	2.5
				Average Magnitude	5.0(2.9)*

* Average error excluding outlier.

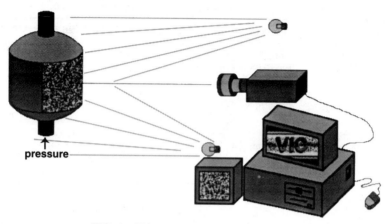

FIG. 4—*Video image correlation apparatus.*

were viewed by a CCD camera with illumination by two 500-W halogen/quartz shop lamps. A random black and white speckle pattern was applied to the region of interest on each bottle by overspray from ordinary flat or low-gloss spray paint to assist in the image correlation process. A PC-based image digitization board was used to acquire images during holds in the pressurization cycle. These images correspond to approximately 0, 8, 17, 25, and 33% of the undamaged burst pressure for an unimpacted bottle. To minimize vibration effects, five frames were averaged for each image.

Procedures

The NASA Sub-Pixel Digital Video Image Correlation (SDVIC) image processing software [5] was used to correlate each nonzero image with the image acquired at 0 pressure. This software, which is available from NASA COSMIC, utilizes a pattern recognition algorithm to determine with subpixel resolution the relative position, and thus deformation, of small image subsets between two images. An automated routine repeats this process for a grid of subsets covering the entire region of interest, resulting in tabulated and false color plotted full-field displacement and strain data. Figure 5 is a false color plot of the hoop strains observed for an impact-damaged graphite/epoxy pressure vessel at an internal pressure of 6.875 MPa. At the load level for which VIC data are presented, visible damage did not exceed the 2-mm dimple caused by the impact tip. However, a vertically oriented linear strain concentrated region, approximately 20 mm long, is evident in the VIC hoop strain field. At higher pressures, the outer hoop plies ripped vertically at the same location, in the same orientation, and initially at approximately the same size as the strain concentration indication. The initial hoop ply failure eventually propagated along the cylindrical region of the bottle, causing the bottle to burst completely, as shown in the post-test photograph of Fig. 6.

Data Reduction

In the case of the AE neural network analysis, the amplitude distribution is divided into a series of discrete categories, with the population of each supplied as the input to the corresponding input layer PE. The six deformation fields (in-plane displacement and the in-

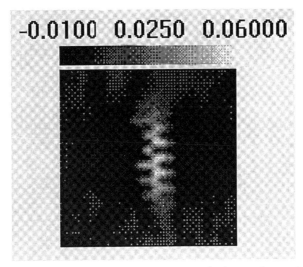

FIG. 5—*Video image correlation hoop strain measurement for impact-damaged graphite/epoxy test article.*

plane strains and rotations) generated by VIC processing are each composed from a grid of 134 × 134 measurements, for a total of 17 956 data points. It is not practical to attempt processing with this many input PEs. Therefore, the tabulated VIC deformation field data are summarized by a group of distributions analogous to the AE amplitude distribution. With some qualitative foresight as to the failure characteristics of these specimens, the hoop strain was chosen as the single parameter that most strongly represents the damage and is the only input to the neural net.

It is theorized that an increase in impact damage severity, corresponding to a decrease in burst pressure, will cause more of the hoop strain field to contain higher strain values due to the strain-concentrating effect of that damage. That is, at a given internal pressure, an unimpacted bottle may be expected to have a narrow distribution of hoop strain values about some average. An impacted bottle with the pressure, field of view, and all image correlation parameters repeated should have a lower and wider distribution due to a shift toward higher strain values. Figure 7 illustrates that a bottle with a burst pressure of 17.9 MPa has a taller, narrower distribution than one with a burst pressure of only 11.7 MPa.

The minimum and maximum strain varies from specimen to specimen, and the neural network approach requires the same number of inputs for each specimen. Thus, each individual strain field was converted to a strain distribution with the same number of categories, but not necessarily the same category ranges. For example, if there are 20 strain categories and input PEs, then the first input PE always receives the number of data points that fall in the lowest 20th (or 5%) of the strain distribution.

The hoop strain distributions used in this test are shown in Fig. 8 in order of increasing burst pressure from front to back. This set of nine aramide/epoxy specimens represent a range of impact damage levels from none to that which reduced the actual burst pressure by approximately one quarter of the undamaged value. Seventeen strain categories and input PEs were used. The data used here were obtained during a 6.89-MPa hold in the proof cycle, or at approximately one third of the undamaged burst pressure. This level of pressurization should cause no damage to an unimpacted specimen.

FIG. 6—*Failed graphite/epoxy pressure vessel.*

Analysis

A software program called VICNet was written to convert the tabulated SDVIC strain field data into strain distribution tables for the entire population of specimens at once. The program then input the strain distributions and actual burst pressures for three of the nine bottles, which had been designated as the training dataset. The VICNet BPNN routine analyzed the training set to adjust internal weights and biases such that the neural network output was within an average of 5% deviation from each actual burst pressure. In addition to the 17 input PEs and single output PE, a middle or hidden layer of three intermediate PEs was used. The smaller number of middle layer neurons, compared to the AE BPNN, indicates a higher degree of linearity. Qualitatively, the BPNN may be interpreted as a relationship between strains and stresses, and the linearity of the network may thus be likened to Hooke's law. The result of this training, which required only 85 iterations and less than 10 s, are shown in Table 3.

The VICNet BPNN routine then input the strain distributions for the remaining specimens, which were designated as the test dataset. These distributions were processed by the neural network in a single pass using the internal weights and biases determined from training. The actual burst pressures were in no way accessible to the software algorithm. The test results

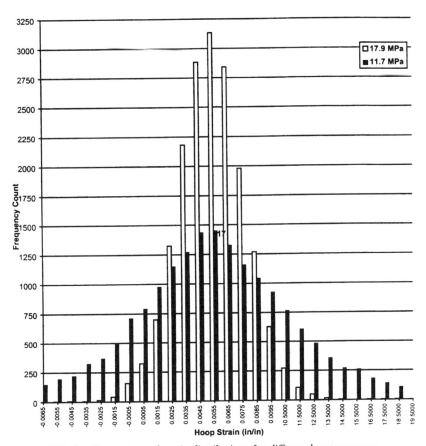

FIG. 7—*Comparison of strain distributions for different burst pressures.*

are shown in Table 4. Multiple independent repetitions of this training and testing have yielded similar average uncertainties.

Conclusions

It has been shown that back-propagation neural network routines can, with some degree of accuracy, be trained to predict the burst pressure of impact-damaged structures based upon acoustic emission or video image correlation strain field data collected from other similar structures proof tested to less than one third of the expected burst pressure. In the case of the filament wound pressure vessels studied here, an average testing error of about 5% for acoustic emission and about 6% for video image correlation have been demonstrated.

To be completely thorough, future testing will be conducted in which these methods are used to predict the burst pressure of the specimens prior to actual burst testing. It will also be determined whether similar analysis at lower proof loads will provide similar levels of uncertainty. Further research will also be conducted to determine the extent to which a neural network that has been trained on AE or VIC data from one size of specimen may be used to predict failure of another size of similar specimen. If this is successful, then a network

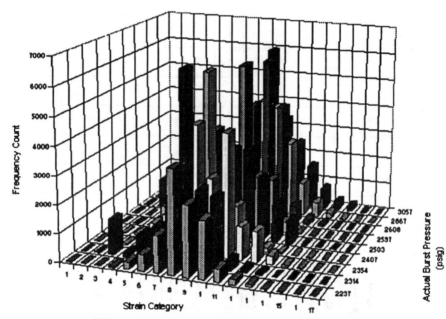

FIG. 8—*VICNet hoop strain distributions for BPNN input.*

TABLE 3—*VICNet training results.*

Actual Burst Pressure, MPa	VICNet Output	Error, %
15.4	15.9	3.4
18.0	18.7	4.3
21.1	19.5	7.3
Average Magnitude, %		5.0

TABLE 4—*VICNet testing results.*

Actual Burst Pressure, MPa	VICNet Output	Error, %
15.9	15.1	−5.2
16.2	16.5	2.0
17.2	18.0	4.5
17.9	19.2	7.2
19.7	18.2	−8.0
16.5	15.2	−8.2
Average Magnitude, %		5.9

that has been trained on an appropriated set of smaller, less expensive specimens may be used to predict the failure of a larger, more expensive service article that may have sustained some form of damage.

Acknowledgments

Funding for this research effort was made available through the NASA Marshall Space Flight Center (MSFC) in cooperation with the University of Alabama in Huntsville (UAH).

References

[1] Walker, J. L., Lansing, M. L., Russell, S. S., Workman, G. L., and Nettles, A., "Materials Characterization of Damage in Filament Wound Composite Pressure Vessels," American Society for Non-destructive Testing, 1995 Spring Conference National Conference, 20–24 March 1995.
[2] Caudill, M. and Butler C., *Understanding Neural Networks, Volume I: Basic Networks,* Massachusetts Institute of Technology, Cambridge, MA, 1992.
[3] Ely, T. E. and Hill, E. v. K., "Longitudinal Splitting and Fiber Breakage Characterization in Graphite Epoxy Using Acoustic Emission Data," *Materials Evaluation,* 1995, pp. 288–294.
[4] McNeill, S. R., Russell, S. S., and Lansing, M. D., "Computing Displacements and Strains from Video Images," *NASA TechBriefs,* March 1996.
[5] "Sub-pixel Digital Video Image Correlation," *Software Catalog,* NASA Computer Software Management Information Center, University of Georgia.

Perspective

James K. Reilly[1]

Fatigue and Fracture of Art Made from Composite Materials

REFERENCE: Reilly, J. K., **"Fatigue and Fracture of Art Made from Composite Materials,"** *Composite Materials: Fatigue and Fracture, Seventh Volume, ASTM STP 1330,* R. B. Bucinell, Ed., American Society for Testing and Materials, 1998, pp. 301–306.

ABSTRACT: Art, in the form of sculpture and masonry murals, has been and is often made from composite materials. Natural fiber materials are used often for sculpture and murals. Examples of art are discussed in relation to most of the important topics of the symposium: fracture by impact; thermal fatigue and fracture; multiaxial loading failure; new composite materials for art; and the monitoring of damage growth. Fracture of art can be caused by centuries of stress fatigue, pollution, seismic activity, and dynamic impact due to theft or bad custodial care. Modern designers should consider structural design and composite material in the selection process to ensure longevity of the art.

KEYWORDS: composite, mural, sculpture, fracture in art

A culture expresses itself through art. This art describes the culture and often remains long after the artists have died. When investigating fracture of art, it is useful to consider three modes of art: sculpture, mosaics, and murals. Sculpture is an art form created through fracture caused by impact. Material for this sculpture has long been specially selected wood and stone. Modern sculpture now includes combinations of these materials. The artist must plan the work to minimize stress, which can cause unintended fracture. The intrinsic value of the work is determined by the artist's gift in the use of fracture to create art such as sculpture.

A good example is the fine cedar wood sculptures of animals painted using pigments from berries by natives of Alaska, British Columbia, and Washington State. This style of art, including selection of wood, the angles the artist draws, and the method of carving and painting, was standardized by regional tribes. The tradition is practiced today so that each individual mask and animal is distinguishable. Figure 1 is an example of this art. Recently, chiefs have changed the style with the addition of the color yellow and the use of acrylic paints.

Mosaic art uses pieces of fractured materials to create art, while mural art consists of a painting on a wall, ceiling, or other large surface. A mosaic uses stone and a cement to compose a picture on a wall, while a mural uses paint and a bonding technique to create a picture. There are fine mosaics in Istanbul, Turkey that date back many centuries, some of which used wax as a bonding agent. Other fine mosaics can be found in Italy. Modern mosaics use a cement to bond stones and ceramic tiles into place. A strong bonding material can help provide longevity.

---

[1] Research scientist, Lyndell, PA 19354.

FIG. 1—*Carved wood totem pole in Quebec City, Canada.*

Masonry Murals

Masonry murals are an important art form made from composite materials. Three examples include: the 1000-year-old Cacaxtla murals of Mexico, the Renaissance Era Sistine Chapel ceiling in the Vatican, and the 1000- to 1500-year-old Buddhist cave murals of China, of which there are 653. Each of these mural collections used a unique matrix of material and paint that provides for both artistic beauty and longevity.

The Buddhist caves along the ancient silk road at the Kumtura, Xinjianjs region of China were created from the fifth to the eighth century by applying a straw and mud mixture to cave ceilings and walls. These walls were then painted in colors of green, red, and blue using pigments including azurite, lapis, and lazul [1]. The straw provided a reinforcing structure for mud or clay and the mural.

The recently discovered Cacaxtla murals are made from Nopal cactus juice and paint applied to adobe walls. They were buried carefully in sand, possibly by the Maya Indian elders, for protection and preservation [2]. Diana Magaloni did chemical analyses of the paint and determined that they were made from precise mixtures of yellow ocher, hematite, carbon, and Nopal cactus juice. Lime was also used. Paints bonded with the stucco of the walls. Examination reveals both cracks and some failure in bonding to the wall.

The mural art of the Renaissance had been created in the fresco craft style by applying a thin layer of plaster, intonaco, onto a base plaster wall, arriccio; paint was then applied to the wet plaster. The artist's initial sketch was made by transferring from paper a sketch using charcoal. It was sometimes changed during the process by the master. The pigment formed a chemical bond with the calcium carbonate of the plaster. This process became the standard mural technique during the High Renaissance period.

Some murals such as those in Florence, Italy, which were restored from flood damage in the last decade, were made from two intonaco layers. The first layer contained the artist's drawing, and the second contained the final painting, including changes made after reflection on the entire group of mural sketches. This was an important discovery of an old technique.

Art can also combine various modes. Figure 2 is a photograph of a bronze wall sculpture that uses composite materials in a form that benefits from the styles of sculpture, mosaic, and mural. This unique art is located in Toronto, Canada, near the CN tower. It incorporates the use of structural balance for a unique composite structure at a reasonably high elevation.

Fracture by Impact

A review of the investigative work by Reza [1] of the Buddhist caves in China provides examples of impact fracture on murals created by apparent theft and removal of the blue material lazuli, or of a painted figure. There is also evidence of cracks caused by seismic events and failure in bonding of the mural to the cave walls along fracture lines, which would have been created by a seismic event. The murals were created out of a composite of mud and straw. This composite was bonded to the walls and was then painted. The composite material facilitates theft by cutting. Most of these caves have sections of art removed.

FIG. 2—*Stone and metal composite sculpture in Toronto, Canada.*

Fracture by impact was carefully prevented through the 1000-year burial of the Cacaxtla Murals in Mexico, which were discovered only recently. These murals also show evidence of fracture due to seismic events and some failure of bonding along the line of fracture. The red temple stair mural shows significant linear cracks, probably caused by ground shifting.

Fracture through impact is the mode by which sculpture in wood and stone has been created. The ability of the artist to create using this skill determines much of the intrinsic value of sculpture. The type and size of the material as well as the style provide both limitations and freedoms.

Thermal Fatigue and Fracture

Figure 1 shows a photograph of a totem pole carved 25 years ago by natives of British Columbia and presented as a gift to the Province of Quebec, Canada. Located in Quebec City, outside the province parliament building, the art has suffered large cracks along grains in the cedar wood from the cold winters and hot summers. The cracks are caused by temperature cycles that are more severe than those found on the coasts of the Pacific Ocean, where temperature is kept more consistent by the ocean. This sculpture should have been placed within the parliament building, which is designed to contain sculpture and other art forms.

Multiaxial Loading Failures

Mechanical fatigue and fracture of ceiling plaster can be found in the Sistine Chapel of the Vatican City [3].

The Sistine Chapel was built in the late fifteenth century. Michelangelo painted the ceiling in the beginning of the sixteenth century. By the middle of the sixteenth century the settling of uneven soil under the chapel caused fatigue-induced fracture, and part of Jeremiah and the left hand of God fell to the floor. Bronze T-shaped clamps were used to anchor plaster during an eighteenth-century restoration. Polyvinyl was inserted to fill voids of separated plaster in the most recent restoration. The hand and face of Jeremiah have been repainted and restored.

A life-size sculpture outside a police station in Toronto shows a young workman cast in heavy bronze on a stone slab. Several pieces of slab have shifted and cracked from the weight of the bronze. It is a modern example of failure from unbalanced static forces. Uneven ground settling is the probable cause of this failure.

Sculpture Composites

Marble sculptures are of interest particularly in life size. A very large piece of marble, greater than life size, was sculpted by Michelangelo for his "David" during the Renaissance period.

Verdite is a green stone found in Zimbabwe, Africa, in the industrial ruby mines. It is considered to be an interesting and valuable medium for sculpture. It is a semi-precious stone estimated to be over three and a half million years old; it was used by both ancient tribesmen and modern African artists for small sculpture pieces and jewelry. The largest pieces can weigh 3000 kg (~6600 lb).

Recent stone sculptures made by the Eskimo and Inuit natives of North America are made from small pieces of onyx and soap stone. These smaller pieces of sculpture are now considered to be important art for collection and museum display. The art represents life with the animals and people of the Arctic North.

Monitoring of Damage Growth

Monitoring of damage growth was performed and recorded in Italy at the Sistine Chapel. Since the middle of the sixteenth century, when painted plaster fell from the Sistine Chapel, the mural had often been incorrectly monitored and incorrectly repaired. Today a history is kept of the building and murals, and maintenance is performed.

Art preservation programs are an important practice. New technology is a significant improvement. For example, computer mapping and program analysis is used. Advanced cleaning methods are also used. In addition, security has become very important to protect art from damage, especially after the bombing of the Uffizzi Gallery in Florence, Italy.

Portrait and Icon Composites

The ancient Egyptian portraits of important people were made from the composite materials of pigment, gold, wax, and linen onto wood. From this art developed the icon painting. Eggshells and seashells were used for white and purple pigment. Gold, linen, and wood were again also used. During the fifth century the practice of icon painting became standardized in Egypt and Greece. This standardization process was similar to that of the native tribes in North America. It included style, color, and consistency of the icon. Some groups, such as the Damascene Studio in Vorep, France, and the French Benedictine Monasteries, have insisted on the use of specific materials such as linen, dry wood of birch, alder, oak, cypress, or pine. These groups have also exaggerated the importance and the need for materials such as garlic, beer, beef gall, and egg yolk, which were not part of the original process. Modern iconography, in fact, uses acrylic and oil paints. The technique of applyng a fiber-reinforced icon, which has been painted in the artist's studio, onto a wall is a currently accepted practice.

Modern Mural Composite Materials

Bonding of mural paints to walls has become of interest. Much that had been forgotten has been rediscovered. The pigment bonding to plaster that was practiced during the Renaissance has recently become understood. Modern paints are currently applied to dry surfaces such as concrete or sheet rock. Bonding failure can occur in humid and in storm conditions. A strong bond provides for longevity and preservation.

There are many new paints available for murals, including acrylic and polyvinyl-based paints [4]. The polyvinyl coating contains high-strength microfibers, polyvinyl acetate, freeze-thaw stability additive, surfactants, coleasing agents, and an antimildew agent called barium metaborate. Mildew was a significant problem during the Renaissance. These paints provide a strong bond to the masonry wall surface.

For masonry and concrete walls, coatings containing hydraulic cement with tobermorite gel provide superior bonding, above 100 000 Pa, and tensile strength tested at 65 000 Pa. The catalytic component is hydraulic cement, which is a finely interground mixture of calcium silicates, calcium aluminates, and calcium ferrites. The product final is calcium silicate hydrate, known as tobermorite gel. It has considerable strength and durability. This new technology provides for waterproofing upon which an acrylic paint can be applied [5]. Curing occurs up to one month, and in stages, giving the artist an opportunity to paint on a material during the curing process.

Another technique includes applying an asphalt primer containing microfibers to reinforce and strengthen the mural base.

Conclusion

Large sculptures from natural materials are unique art objects. These art objects need to be protected from fracture. Modern large sculpture includes art made from composite materials, including wood, stone, and metal. These must be carefully designed to avoid unbalanced stress forces that could cause fracture. Small sculpture made from stone is equally important and requires great skill in the use of fracture to create an artwork of interest.

The recent discovery of ancient murals in China and Mexico reveals the use of composite materials and pigments bonded together. The murals show cracks from seismic activity. The murals in Mexico were protected from further damage by being buried in fine sand. The Buddhist murals were damaged by the dynamic impact of theft of a pigment or painting.

After flooding in Florence, Italy, many murals were removed from outdoor walls for restoration, and a new understanding of this art was learned. The Sistine Chapel murals have been carefully monitored throughout the centuries, but have small cracks either from poor maintenance or century-long stresses. The Renaissance fresco craft of mural art was rediscovered recently and revealed an interesting use of composite materials.

Iconography is an old art that makes use of new materials such as acrylic paints. Modern icons can be found in many buildings of North America. The large group of icons in the new Greek Orthodox Church of Fort Lauderdale, Florida, for example, uses the traditional style with new composite materials.

Artists have created many fine examples of modern sculpture from wood and stone in Alaska and Canada that are of interest today. The recent sculptures using the verdite stone from Africa is of interest and of unusual value.

Modern technology can assist the artist with structural analysis and new composite materials, which provide a new opportunity for longevity and resistance to failure by fracture.

References

[1] Reza, "Pilgrimage to China's Buddhist Caves," *National Geographic Magazine,* Vol. 189, No. 4, April 1996, pp. 52–63.

[2] Stuart, G. E. and Ferorcelli, E., "Murals of Ancient Cacaxtla," *National Geographic Magazine,* Vol. 182, No. 3, September 1992, pp. 120–136.

[3] Jeffrey, D., Woolfitt, A., and Boswell Jr., V. R., "The Sistine Restoration: A Renaissance for Michelangelo," *National Geographic Magazine,* Vol. 176, No. 6, December 1989, pp. 688–713.

[4] Southwestern Petroleum Corporation, "Masonry Coating," SWEPCO Product Data Information, pp. 1–4.

[5] Southwestern Petroleum Corporation, "Hydrocoat," SWEPCO Product Data Information, pp. 1–4.

Author Index

Subject Index